Sense and Goodness without God

A Defense of Metaphysical Naturalism

By

Richard Carrier

authorHOUSE™

1663 LIBERTY DRIVE, SUITE 200
BLOOMINGTON, INDIANA 47403
(800) 839-8640
WWW.AUTHORHOUSE.COM

© 2005 by Richard Carrier.
All Rights Reserved.

No part of this book may be reproduced, stored in a retrieval system, or transmitted by any means without the written permission of the author.

First published by AuthorHouse 02/03/05

ISBN: 1-4208-0293-3 (sc)

Library of Congress Control Number: 2004097444

Printed in the United States of America
Bloomington, Indiana

This book is printed on acid-free paper.

Cover Photo: *One and Only Oil* by Richard C. Carrier

For Jen...

        My buxom brunette
        My wellspring of joy
        My north star of sanity

Table of Contents

I. Introduction 1
  1. What This Is 3
  2. How I Got Here 9

II. How We Know 21
  1. Philosophy: What It Is and Why You Should Care 23
  2. Understanding the Meaning in What We Think and Say 27
    2.1 The Meaning of Words 29
      2.1.1 Reducible and Irreducible Sensation 30
      2.1.2 Meaning, Reality and Illusion 31
      2.1.3 Experience is the Font of Knowledge 33
      2.1.4 Getting at the Real Meaning of Words 33
    2.2 The Meaning of Statements 35
      2.2.1 Stipulations and References 35
      2.2.2 Descriptions 37
      2.2.3 Opinions 37
      2.2.4 Moral Imperatives 39
      2.2.5 Wishes and Commands 40
      2.2.6 Facts and Hypotheses 40
      2.2.7 The Nature of a Contradiction 42
      2.2.8 Naturally Warranted Belief 43
  3. Method 49
    3.1 Finding the Good Method 51
    3.2 The Method of Reason 53
    3.3 The Method of Science 54
    3.4 The Method of Experience 55
    3.5 The Method of History 57
    3.6 The Method of Expert Testimony 58
    3.7 The Method of Plausible Inference 59
    3.8 The Method of Pure Faith 60
    3.9 Final Remarks on Method 60

III. What There Is 63

1. The Idea of a "Worldview" 65

2. A General Outline of Metaphysical Naturalism 67

3. The Nature and Origin of the Universe 71
- 3.1 Plausibility and the God Hypothesis 71
- 3.2 God and the Big Bang 74
- 3.3 Modern Multiverse Theory 75
- 3.4 The Multiverse as Ultimate Being 81
- 3.5 Answering the Big Questions 83
  - 3.5.1 The First Cause 84
  - 3.5.2 The Origin of Order 86
- 3.6 Time and the Multiverse 88

4. The Fixed Universe and Freedom of the Will 97
- 4.1 Why Determinism? 98
- 4.2 The Alternative: Libertarian Free Will 100
- 4.3 Why Libertarian Free Will Eliminates Responsibility 102
- 4.4 Compatibilism: The Only Sensible Notion of Free Will 103
  - 4.4.1 The Ability Condition 104
  - 4.4.2 The Control Condition 105
  - 4.4.3 The Rationality Condition 106
  - 4.4.4 The Cause Condition 108
- 4.5 What Free Will Really Is 109
- 4.6 The Fatalist Fallacy vs. Improving Self and Society 115

5. What Everything is Made of 119
- 5.1 Space-Time 119
- 5.2 Matter-Energy 120
- 5.3 Physical Laws 122
- 5.4 Abstract Objects 124
  - 5.4.1 Numbers, Logic, and Mathematics 125
  - 5.4.2 Colors and Processes 127
  - 5.4.3 Modal Properties 128
- 5.5 Reductionism 130

6. The Nature of Mind 135
6.1 The Mind as Brain in Action 135
6.2 The Mind as Virtual Reality 136
6.3 The Chinese Room 139
6.4 The Mind as Machine 144
6.4.1 Thoughts 144
6.4.2 Abilities, Memories, and Traits 145
6.4.3 What Machines Can't Yet Do 145
6.4.4 Qualia 146
6.5 The Nature of Knowledge 148
6.6 The Evidence for Mind-Body Physicalism 150
6.6.1 General Brain Function Correlation 150
6.6.2 Specific Brain Function Correlation 151
6.6.3 Positive Evidence Mapping the Mind to the Brain 151
6.6.4 Negative Evidence Mapping the Mind to the Brain 152
6.6.5 Brain Chemistry and Mental Function 152
6.6.6 Comparative Anatomy and Explicability 153
6.7 Evidence Against Mind-Body Physicalism? 154
6.8 Immortality and Life After Death 157

7. The Meaning of Life 161

8. How Did We Get Here? 165
8.1 Biogenesis 166
8.2 Evolution by Natural Selection 168
8.3 The Evolution of Mind 173
8.4 Memetic Evolution 175

9. The Nature of Reason 177
9.1 Reason vs. Intuition 178
9.2 Why Trust the Machine of Reason? 181
9.3 Contradiction Revisited 188
9.4 Alternative Accounts Are Not Credible 191

10. The Nature of Emotion 193
10.1 Emotion as Appraisal 194
10.2 Reason as the Servant of Desire 196

  10.3 The Nature of Love 197
  10.4 The Nature of Spirituality 202

IV. What There Isn't 209
1. Not Much Place for the Paranormal 211
  1.1 Science and the Supernatural 213
    1.1.1 The "Scientific Method" 214
    1.1.2 The Advantage of Doubt 216
    1.1.3 The Science of Faith 217
    1.1.4 The Power of Artifice 218
    1.1.5 Distinguishing Fact from Theory 219
    1.1.6 The Marriage of Creativity with Truth 220
    1.1.7 The Lessons of History and the Burden of Proof 221
    1.1.8 The Balance of Proof and Proof of the Extraordinary 222
    1.1.9 Simplicity and Occam's Razor 224
  1.2 Miracles and Historical Method 227
    1.2.1 The Rain Miracle of Marcus Aurelius 228
    1.2.2 Understanding the Ancient Milieu 231
    1.2.3 Historical Method Saves the Day 236
    1.2.4 The Argument to the Best Explanation 238
    1.2.5 The Argument from Evidence 242
    1.2.6 The Criteria of the Good Historian 246
    1.2.7 Prophecy and History 247

2. Atheism: Seven Reasons to be Godless 253
  2.1 Metaphysical Naturalism is True 256
  2.2 The Religious Landscape is Confused and Mundane 257
    2.2.1 Religion Didn't Win by Playing Fair 258
    2.2.2 To the Victor Goes the Spoil 261
    2.2.3 Dissent is Checked at the Door 268
    2.2.4 Religion as Medicine 270
  2.3 The Universe is a Moron 273
  2.4 The Idea of God Doesn't Make Any Sense 275
  2.5 Too Much Needless Cruelty and Misery 277
  2.6 Not Enough Good from God 280
  2.7 Anything Defended with Such Absurdities Must be False 282
    2.7.1 The Argument from Mystery 283

    2.7.2 The "Free Will Defense" Deployment Number One 284
    2.7.3 The "Free Will Defense" Deployment Number Two 285
    2.7.4 The "Arrogance Defense" Deployment Number One 286
    2.7.5 The "Arrogance Defense" Deployment Number Two 287
    2.7.6 The Great Deceiver Defense 287
    2.7.7 Facing the Absurd and Calling it Bunk 288

V. Natural Morality 291

1. Secular Humanism vs. Christian Theism 293

  1.1 Do Secular Humanists Have No Reason to be Moral? 293
    1.1.1 Love as Reason to be Moral 294
    1.1.2 Debt as Reason to be Moral 296
    1.1.3 Goodness as Reason to be Moral 297
    1.1.4 Self Interest as Reason to be Moral 297
    1.1.5 Trust as Reason to be Moral 299
    1.1.6 Self-Image as Reason to be Moral 300
    1.1.7 Worldly Self-Interest as Reason to be Moral 302
  1.2 What's Wrong with Secular Humanism? 302
    1.2.1 Do We Live in a Sick Society? 303
    1.2.2 Does Believing in Evolution Make Us Immoral? 308
    1.2.3 Selfish Genes and Selfish Memes 309

2. Morality in Metaphysical Naturalism 313

  2.1 Outlining a Moral Theory 313
    2.1.1 The Goal Theory of Moral Value 315
    2.1.2 Happiness and the Moral Life 316
    2.1.3 Self Worth and the Need for a Moral Life 320
    2.1.4 The Futility of Secret Violations 323
  2.2 How Naturalism Accounts for Value 324
    2.2.1 Evolution of Moral Values 326
    2.2.2 Human Nature 328
    2.2.3 Personhood 329
    2.2.4 Speciesism 330
    2.2.5 The Meaning of Normative Propositions 331
    2.2.6 Moral Relativism and Moral Controversy 336
    2.2.7 Defining Good and Evil 337
    2.2.8 Moral Reason and Moral Intuition 339
  2.3 Eliminating Some Metaethical Defeaters 341
    2.3.1 What About Moral Suicide? 341

  2.3.2 What About Weird Aliens and Psychopathic Robots? 342
 3. Moral Conclusions: Tying it All Together 345

VI. Natural Beauty 349
 1. Beauty as Emotional Appraisal 351
 2. Eight Rules of Beauty 353
  2.1 The Peak Shift Effect 353
  2.2 The Correlation Effect 354
  2.3 The Stand-Out Effect 355
  2.4 The Contrast Effect 356
  2.5 The Symmetry Effect 357
  2.6 The Counter-Symmetry Effect 358
  2.7 The Analogy Effect 358
  2.8 The Anticipation Effect 360
 3. Beauty in Human Life 361
  3.1 Is Beauty Bunk? 361
  3.2 The Subjective Nature of Beauty 362
  3.3 The Higher Virtues of Art 363
   3.3.1 Art as Communication 364
   3.3.2 Art as Education 364
   3.3.3 Art as Skill 365

VII. Natural Politics 367
 1. Morality vs. Politics 369
 2. The Rationality of the Moderate 371
 3. Basic Political Theory 375
 4. The Politics of Metaphysical Naturalism 381
  4.1 Political Method 381
  4.2 The Best Polity 383
  4.3 Choosing Our Leaders 384
 5. My Politics 389

    5.1 A Commitment to Freedom 389
    5.2 A Commitment to Social Reform 391
    5.3 A Commitment to Executive Reform 394
    5.4 A Commitment to Education 396
    5.5 A Commitment to Defense 400
    5.6 A Commitment to Secularism 402
6. The Secular Humanist's Heaven 405

VIII. Conclusion 409

I. Introduction

1. What This Is

Philosophy is not a word game or hairsplitting contest, nor a grand scheme to rationalize this or that. Philosophy is what we believe, about ourselves, about the universe and our place in it. Philosophy is the Answer to every Big Question, and the ground we stand on when finding answers to every small one. Our values, our morals, our goals, our identities, who we are, where we are, and above all how we know any of these things, it all comes from our philosophy of life—whether we know it or not.

Since this makes philosophy fundamental to everything in our lives, it is odd that people give it so little attention. Philosophers are largely to blame. They have reduced their craft to the very thing it should not be: a jargonized verbal dance around largely useless minutiae. Philosophy is supposed to be the science of explaining to everyone the meaning and implications of what we say and think, aiding us all in understanding ourselves and the world. Yet philosophers have all but abandoned this calling, abandoning their only useful role in society. They have retreated behind ivory walls, talking over the heads of the uninitiated, and doing nothing useful for the everyman. So it is no surprise the general population has lost interest. And when pundits lament a spiritual aimlessness in modern culture, what they see is not the loss of faith in any particular religion, but the divorce of human beings from a devoted exploration of philosophy—philosophy as it should be. That divorce was a serious mistake.

Many people call their philosophy a "Religion." But that does not excuse them from their responsibility as philosophers. You either have a coherent, sensible, complete philosophy that is well-supported by all the evidence that humans have yet mustered, or you do not. Yet most people cannot even tell you which of those two camps their religion, their philosophy, is in. Hardly anyone has spent a single serious moment exploring their philosophy of life. Far fewer have made any significant effort to get it right. Instead, "Religion" has become a factory-made commodity, sold off the shelf to the masses, who assume it must be good if it is really old and lots of smarter and better educated people say it's a good buy ("8 out of 10 experts recommend Christian Brand Salvation!"). People think they can just plug such a goodie into their lives, maybe with a few unskilled adjustments of their own, and never have to think about whether it is well-constructed, well-thought-out, or even true. Some people, more creative but no wiser, take a shallow glance around and tear pieces from existing products, or grab whatever pops into their heads, and throw together something of their own, with little in the way of careful investigation or analysis. It would require more than the Luck of the Irish for either approach to succeed. It is the rare bird (and the humble one, who never claims to know more than they do) who can hit upon wisdom without taking more serious care.

I have taken a different approach, and wish to recommend it to everyone. My religion is Philosophy Itself. Every hour that devout believers spend praying, reading scripture, attending sermons and masses, I spend reading, thinking, honing my skill at getting at the truth and rooting out error. I imagine by most standards I have been far more devout than your average churchgoer. For I have spent over an hour every day of my life, since I began my teen years, on this serious task of inquiry and reflection.

I am no guru. But I have gotten pretty far. Now, nearing middle-age, I have found myself with that coherent, sensible, complete, evidentially well-supported philosophy of life that I had been looking for. Though I know there is a lot I still don't know, and many mistakes yet to be corrected, I am always learning. I have spent a long time pulling pieces together, correcting my errors, backing up from dead ends and starting over, making sense of it all. Now I can actually say I have something to say. I might not be right. I might be only partly on target. But at least I gave it as good a try as anyone could.

This book surveys my philosophy of life, my "worldview," and explains why I believe it is true. The formal category it falls into is "Metaphysical Naturalism," a daunting bit of jargon, not of my choosing, whose meaning will become clear as you read on through this tome. It is essentially an explanation of everything without recourse to anything supernatural, a view that takes reason and science seriously, and expects nothing from you that you cannot judge for yourself.

Though I am an atheist, in the basic sense that I do not believe there are any gods, you will find after completing this book that whether God exists or not really doesn't matter all that much. Every component of my philosophy can be arrived at independently, and stands on evidence and reasoning that would not change tomorrow if a god announced himself to the world today. Rather than being a starting assumption, my atheism is but an incidental conclusion from applying my worldview to the current state of evidence. If I should ever become convinced a god does exist after all, most likely very little adjustment of the philosophy I defend here would be needed.

Even so, since most people assume God is the answer to every deep question, the same concerns always come up. If God does not exist, then what does? Is there good and evil? Should we even care? How do we know what's true anyway? And can we make any sense of this universe? Can we even make sense of our own lives? Answering questions like these is essentially what this book is about. I build and defend a complete worldview by covering every fundamental subject—from knowledge to art, from metaphysics to morality, from theology to politics. Along the way I discuss free will, the nature of the universe, the meaning of life, and much more. At every step of the way I use sound reason and scientific evidence to argue that there is probably only a physical, natural world without gods or spirits, but that we can still live a life of love, meaning, and joy. So the conclusion of this book is not negative. Nature is all there is. But life is still good. This book aims to show how Metaphysical Naturalism satisfies all our concerns—about existence, meaning, right and wrong—without need of any gods or mystical secrets.

I have attempted to write this work with the average college-level reader in mind, and not just for specialists. I avoided using any language you won't find in a half-decent dictionary, unless I explain a word's meaning myself. My vocabulary and mode of expression is as colloquial as the subject permits, though my style is a bit old-school. I do not assume the reader knows anything about philosophy, or any more about anything that a quality high school education wouldn't teach an attentive student. I use no footnotes or endnotes, but I often back up my claims by referring readers to the relevant literature that presents the supporting evidence, in bibliographies placed at the end of their respective chapters or sections.

For all readers, I ask that my work be approached with the same intellectual charity you would expect from anyone else. First and foremost, this book describes and defends only one kind of Metaphysical Naturalism—there are many other varieties, and mine should not be confused with them. But more generally, ordinary language is necessarily ambiguous and open to many different interpretations. If what I say anywhere in this book appears to contradict, directly or indirectly, something else I say here, the principle of interpretive charity should be applied: assume you are misreading the meaning of what I said in each or either case. Whatever interpretation would eliminate the contradiction and produce agreement is probably correct. So you are encouraged in every problem that may trouble you to find that interpretation. If all attempts at this fail, and you cannot but see a contradiction remaining, you should write to me about this at once, for the manner of my expression may need expansion or correction in a future edition to remove the difficulty, or I might really have goofed up and need to correct a mistake. I am most easily found by email (*naturalism@secular.org*) but any regular mail sent to my publisher will eventually find my door.

On the other hand, if what I say appears to contradict, even indirectly, something someone *else* says, whose work I otherwise cite or recommend, you should take my meaning as the one I intend, and not add to it what others have added to the same or similar ideas. In short, do not attribute to me beliefs I do not declare, especially those that are not compatible with what I *do* assert. Yet if such an outside declaration can be interpreted in a way that *is* compatible with all I say, you may take that as conforming to my belief system, as I would probably endorse it.

In contrast, if you find any case where something I write is factually *false*, or in need of qualification, as established by repeated, confirmed, empirical investigation by relevant experts, something that has enough evidential support to persuade the vast majority of a profession that it is true, then please inform me of that, too. For my philosophy, or at least my presentation of it here, will then stand in need of revision, and I hold nothing so dear as the desire to correct my mistakes and get things right—in short, to grow and improve myself and my beliefs.

Though the writing of this work was a private passion, it would not have been possible without a great number of people who affected my life in important ways. My mother and father—Monica and Hal—and my wife Jennifer, were most important of all. The Internet Infidels were also instrumental in helping to complete the latest phase of my intellectual development, especially Jeff Lowder and many affiliated colleagues: Evan Fales, Victor Stenger, Keith Augustine, Dan Barker, just to name a few, who also gave advice about improving this work specifically. My education, especially my skill as a writer, researcher and critical thinker, would not have been what it is without the tutelage of Drs. David Keightley and William Harris, to name only my most important teachers, in terms of time spent and effort on my abilities. I must also thank Bob Scott, who should go down in history as the best boss anyone could have.

I would be remiss if I did not also thank the now-dead men whose philosophical genius has most impressed me. Though their names are too numerous to list, the Honor Roll would run from Lao Tzu to Hsün Tzu, and from Epicurus and Seneca, on to David Hume, A. J. Ayer, and Bertrand Russell. I hope their best ideas live on in me.

2. How I Got Here

No philosophy is created in a vacuum. Any reader should know, at least in some general sense, where I've been and how I got here, to gain perspective on why I believe what I do, and what my possible biases may be. Certainly, my intellectual journey is relevant to where I've ended up. What influences have I been under? Have I really shopped around? What experiences have I had? What philosophies have I explored? Did I really think things through? Above all, many readers will wonder how I can be an atheist, especially in an overwhelmingly Christian society. So this short chapter will tell my story.

My experiences with religion as a child were all good. My mother was a church secretary at a First Methodist Church only a block from our home, and I attended Sunday School fairly regularly, though my parents rarely insisted that I attend any sermons. The religion sold at this local business was a very liberal brand of Christianity. It was more like a preschool and social club, and that made it an excellent asset to the community, and a place of fond memories for me. Amidst arts and crafts, lunches, running and climbing about, and basic learning, the alphabet and numbers and whatnot, Sunday School had its story time. Bible stories were always on the menu, intermingled with other popular fables and parables, and it was never even suggested there was any difference.

The Good Book was always treated as a collection of handy tales used as springboards for teaching moral lessons, not as a history book. Indeed, I was never once told that unbelievers go to hell or that I had to "believe on Christ" to be saved or anything like that. All good people went to heaven, so you'd better be good. That was it. Jesus in this version of Christianity was little more than a moral teacher. Being the Son of God made him an authority on the subject but had no other importance. Perhaps it was no accident that everyone who attended this church was very kind and jovial and all around just good folk.

During my first few grades, whenever I had free time in school (and wasn't running and climbing about) I read for myself the New Testament (red letter edition, of course—I think any child loves books with different colors in them). But the moment I got home my nose was in much bigger and better books: all manner of encyclopedias, my favorite reading material. The Bible was boring and not very informative, and hardly intelligible to a child, but it was the only book anyone ever gave me that would fit in my pocket.

Yet I never had the feeling that I was doing anything religious, or what I was reading was special in any way, apart from the fact that everyone seemed happy or impressed to see me reading it, which I never understood since these same people thought I was weird for reading encyclopedias, which I knew, even at that age, were more educational. As I grew older, my social life expanded, and my spare time at school was spent completing homework, leaving no time for idle reading, while my appetite for knowledge grew to deeper levels of sophistication.

The New Testament had given me no useful information about the meaning of life or the nature of the universe. Later I learned that people extracted from it such things, but they only did so by importing ideas and concepts that aren't in the book itself, and so just reading it alone I found it to be shallow and unsatisfying. Its message was obsessed with strange moral rules that no one around me ever followed. Instead of turning the other cheek, people called for more cops and longer prison terms. Far from giving thieves their cloaks, people kept baseball bats by their beds and hung signs that said Beware of Dog. While the very Son of God Himself defended a whore from moral condemnation, whores were routinely morally condemned, most ardently by the Devout.

Then there was all this talk about the worm that never dies and morbid metaphors about washing with blood, and so forth, that weren't very relevant to the world I saw and wanted to understand. Littered everywhere was exultation about the Good News, but *God forbid* should any passage ever clearly explain just which news that was supposed to be. At one moment it seemed to be the moral message, which I just observed was nonsensical, at another it was about a horrible End Times that hardly sounded good. No one around me thought a Nuclear War was good news, yet it sounded like much the same thing. At yet another moment it had something to do with Jesus dying for something called sin, even though it was never explained how he could die for it when I was always taught that I had to seek my own forgiveness from any person I'd wronged. At yet another time it was the fact that there was an afterlife "so don't despair," which even as a child I found to be rather childish. And so on.

The Bible was confused, illogical, often unintelligible, but always irrelevant to the social and political reality in which I lived. Where was any explanation and defense of democratic values? Where was gender equality? What was wisdom? What was virtue? How come all my encyclopedias were full of the beautiful, wonderful things of the universe, yet not a single peep about them from the Son of God Himself? One would think he of all people would have had a kick ass science education, having the most powerful and knowledgeable father in the universe and all. I wanted to know what the fundamental nature of the universe was, what the fundamentals of a moral life really were, how to achieve happiness in this life. The Bible didn't help. Better moral wisdom came from mortal word of mouth around me, and far more knowledge from other books, and from school, where I majored in science and took and mastered every science course offered. So, with the other childish things I put away as I approached my teen years, the Good Book was among them.

And so I became a seeker. Rather stereotypically, I entered my teen years hungry for truth, for something that made sense of it all, for direction. The universe just didn't seem right. Hypocrisy was everywhere, problems abounded, along with contradictory opinions about how to solve them, and the most basic facts about the world were, or so I thought, unexplained by scientists, who were clearly those best able to get the answers. And yet the one book everyone said had all the answers was shallow, frequently confused or uninformative, unnecessarily verbose and obscure, and contradicted the society I found myself in. Worse, it read like a preachy fable: no logical arguments, no demonstrations of evidence, just assertions, and vague ones at that. It had nothing to say about democracy or science or technology, the three things that most defined my world. How useless. So I lived a life of the mind, and thought and studied, always anchored by a stable home life and a good circle of friends. Logic alone led me to what I would later discover was an ambiguous form of agnostic deism.

Then a miracle happened. At least, it was what believers would call a miracle. In a bookstore hunting down a dictionary for school, I had a feeling that told me to turn. I did, and the first thing I saw was a Jane English translation of the *Tao Te Ching*. I took it up, and, like a modern-day Augustine, turned to a page at random and read. What it said was so simple, so true, so wise, so elegantly and concisely put, I knew this was the answer. I bought the book and read it all through, and from that day I declared my faith in Taoism, my first real religion.

Christianity was never a religion for me—it was simply a fixture in my cultural atmosphere, and I never affirmed any faith in its principles. But I had faith in Taoism. I was a True Believer. And I am glad that, unlike most people, I made an informed choice, at an age when I had the capacity to choose sensibly. Religion was never imposed on me and no one in my family ever assumed I had to be Christian, and consequently I can say my one chosen religion was born neither of peer pressure nor indoctrination.

I studied Taoism avidly. At one point I had eight different English translations of the *Tao Te Ching* and a few of the *Chuang Tzu*, and my Taoism became full and sophisticated: I was a Philosophical Taoist, a Chinese tradition that adhered to the texts and the wisdom alone, scorning the surrounding superstitions and religious cult that grew around it, as being against the very message of Taoism. In time I also discovered how Taoism was a response to Confucianism, and the relationship the two religious philosophies had, and in the course of things I acquired some acquaintance with Buddhism as well.

My life was transformed. I acquired a sense of discipline and focus I never had before, an attraction to quiet, simple living, and a strong yet humble moral conception of things. All finally made sense, and I was happier than I ever imagined possible. In my holy text I had a toolbox for dealing easily and sensibly with every problem, from sexual angst to metaphysical doubt, from political debate to material danger. There was a verse in the *Tao Te Ching* for everything, and it was written beautifully and simply, often appealing to the only truly universal Bible for evidence of its truths: the world itself, as well as the undeniable evidence within the reader's own soul. It had a train of thought, an implied logical argument. In time I created my own version of the *Tao Te Ching*, selecting my favorite translations of every line from among the many I knew, and carried this with me as the one devotional item we were allowed in boot camp. I read it nightly.

The proof that this was the one true religion was manyfold, and seemingly irrefutable. Apart from the "clearly" supernatural miracle of my discovering the faith, and the "self-evident" perfection of its sacred text, following its tenets I was led to peace of mind and a balanced life, to friendships and goodness. With it, all harm was defeated or of no consequence, and every benefit came easily and naturally. I learned to have fewer expectations, to care more about others and to worry less about what I didn't yet know. Things were of little importance next to contentment itself, and the good life was a life of friends and the mind, not of luxury or power.

Above all, the Tao told me the simple truth: that my humanity was a good and natural thing. From sex to humor, all had an accepted place, without being forced into unnatural modes of thought or behavior. Sin was the artificial deviation from the harmony of nature, and if you would simply stop meddling with things you would be free of sin. Taoism explained everything, even the existence and nature of the universe, in a way that made perfect and beautiful sense. And it cultivated a tolerant mind like I had never seen Christianity do.

The Chinese had known this for over two thousand years. I still cherish the memory of seeing a picture of three holy men traveling a road together, all laughing with each other. One was a Buddhist, another a Taoist, and the third a Confucian. This image is a regular motif in China. There, the three religions, despite being so doctrinally and intellectually at odds, get along peacefully, even happily, a friendship that is celebrated in such artwork everywhere. What better proof is there of the goodness and truth of a creed that it inspires such jovial tolerance? Instead of holy wars, condemnations and combative debates, these religions interact in dialogues, and each accepts the other as possibly different facets of the same coin. They live comfortably with doubt and uncertainty, even thriving on it. They condemn no one to an eternal hell, and require no belief: they simply tell it like it is, take it or leave it.

I was a happy Taoist for many years. Burned out on schooling I chose to live a simple life, contented at gardening or ditch-digging for a living, doing everything from installing electrical fixtures to waiting tables. But eventually I signed up for a life in the Coast Guard, studying electronics and sonar and living at sea, until I yearned again for an education and thus embarked on a long career as a student of ancient science and history.

During all this, in cultivating the mental life that Taoism taught, I had powerful mystical visions, which only confirmed further that I was on the right track. These ranged from the simple to the fantastic. The simplest and most common was that clarity of an almost drug-like wonder, perceiving everything striking the senses as one unified whole. It is hard to describe this. Normally, your attention is focused, on something you are looking at or listening to, or in a semi-dream-state of reverie, but with a meditative sense of attention this focus and dreaminess vanishes and you are immersed in a total, holistic sense of the real. It is both magnificent and calming. It humbles you, and brings you to the realization of how beautiful simply living is, and how trivial all your worries and difficulties are. Profound insights about the world would strike me whenever in such a state, leading far more readily and powerfully to an understanding of myself and the world than studying or reasoning ever did.

The most fantastic experience I had was like that times ten. It happened at sea, well past midnight on the flight deck of a cutter, in international waters two hundred miles from the nearest land. Sleep deprivation affected my consciousness like a New Age shaman. I had not slept for over 36 hours, thanks to a common misfortune of overlapping duty schedules and emergency rescue operations. For hours we had been practicing helicopter landing and refueling drills and at long last the chopper was away and everything was calm. The ship was rocking slowly in a gentle, black sea, and I was alone beneath the starriest of skies that most people have never seen. I fell so deeply into the clear, total immersion in the real that I left my body, and my soul expanded to the size of the universe, so that I could at one thought perceive, almost 'feel', everything that existed in perfect and total clarity. It was like a Vulcan Mind Meld with God.

Naturally, words cannot do justice to something like this. It cannot really be described, only experienced, or hinted at. What did I see? A beautiful, vast, harmonious and wonderful universe all at peace with the Tao. There was plenty of life scattered like tiny seeds everywhere, but no supernatural beings, no gods or demons or souls floating about, no heaven or hell. Just a perfect, complete universe, with no need for anything more. The experience was absolutely real to me. There was nothing about it that would suggest it was a dream or a mere flight of imagination. And it was magnificent.

But I had never stopped my private readings in the sciences, and it did not take long for me to realize that everything I had experienced through Taoism had a natural explanation. At the same time, the more I studied my religious text the more I came to disagree with certain parts of it. Since the One True Religion could not be faulty even in part, this brought me to realize that Taoism was not sacred or divine, but just an outpouring of very admirable and ingenious, but ultimately fallible human wisdom. That did not diminish its merit, but it did lead me to think outside the box.

More and more I found I agreed with Confucians against the Taoists, but still sided with the Taoists against the Confucians on other issues, and in the dance of thesis and antithesis I came to my own synthesis, which can now be described as a science-based Secular Humanism rooted in Metaphysical Naturalism, which this book shall describe and defend. More and more I found brilliant wisdom in Western philosophers like Epicurus or Seneca, or Ayer or Hume, and so my worldview became more eclectic and for that reason more complete: by drawing the best from many points of view, I was purging myself of the faults of relying on only one, all the while seeking carefully for a coherent and complete philosophy of life.

Inevitably, I had to confront the Christian question. There was a point in sonar school when I was regularly pestered by a Christian who was bothered by my Taoism, even more than my agnosticism (it didn't matter to a Taoist whether a god existed—an answer to his question "Do you believe in God?" that frustrated the hell out of him). Eventually he argued that you have to read the whole Bible before you can make an informed decision about it. He recommended the *NIV Student Bible*, which I purchased, and still have. I set down to read it all through, every word, front to back, Old Testament and New (I have since read the entire New Testament in the original Greek).

I figured now, with my greater understanding and maturity, I might receive more from it than I did as a child. Instead, now that I could understand it, I was able to see far worse things in it than I ever did before. I saw a terrible, sinful God by the standards of the simple, kind wisdom of Taoism—a jealous, violent, short-tempered, vengeful being whose behavior is nonsensical and overly meddlesome and unenlightening. Later I was to find that the vast majority of Christians never actually read the Bible, and have no idea what is really in there, and the hypocrisy of them telling me I had to read the whole thing before I could make an informed choice is still palpable.

In all I can say that the Old Testament disgusted me, while the New Testament disappointed me. In general, no divinely inspired text would be so long and rambling and hard to understand. Wise men speak clearly, brilliantly, their ability at communication is measured by their success at making themselves readily understood. The Bible spans over a thousand pages of tiny, multi-columned text, and yet says nowhere near as much, certainly nothing as well, as the *Tao Te Ching* does in a mere eighty-one stanzas. The Bible is full of the superfluous—extensive genealogies of no relevance to the meaning of life or the nature of the universe, long digressions on barbaric rituals of bloodletting and taboo that have nothing to do with being a good person or advancing society toward greater happiness, lengthy diatribes against long-dead nations and constant harping on a coming doom and gloom. I asked myself: would any wise, compassionate being even allow this book to be attributed to him, much less be its author? Certainly not. How could Lao Tzu, a mere mortal, who never claimed any superior powers or status, write better, more thoroughly, more concisely, about so much more, than the Inspired Prophets of God?

It was not only this that struck me. What was most pungent was the immorality of the Bible. Though called a wise father, there is not a single example in the Old Testament of God sitting down and kindly teaching anyone, and when asked by Job, the best of men, to explain why He went out of His way to hurt a good man by every possible means, including killing Job's loved ones, this "wise father" spews arrogant rhetorical questions, ultimately implying nothing more than "might makes right" as his only excuse. I looked in horror at the demonic monster being portrayed here. He was worthy of universal condemnation, not worship. He who thinks he can do whatever he wants because he *can* is as loathsome and untrustworthy as any psychopath.

It was bad enough that this god's idea of the "best" in man is a willingness to murder one's own child on demand. It is inconceivable that any kind being would ever test Abraham's loyalty that way. To the contrary, from any compassionate point of view, Abraham failed this test: he was willing to kill for faith, setting morality aside for a god. A decent being would reward instead the man who responded to such a request with "Go to hell! Only a demon would ask such a thing, and no compassionate man would do it!" But the Bible's message is exactly the opposite. How frightening. It was no surprise, then, to find that this same cruel god orders people to be stoned to death for picking up sticks on Saturday (Numbers 15:32-36), and commands that those who follow other religions be slaughtered (Deuteronomy 13:6-16). Indeed, genocide (Deuteronomy 2:31-34, 7:1-2, 20:10-15, and Joshua, e.g. 10:33) and fascism (Deuteronomy 22:23-24, Leviticus 20:13, 24:13-16, Numbers 15:32-6) were the very law and standard practice of God, right next to the Ten Commandments. Instead of condemning slavery, God condones it (Leviticus 25:44, cf. Deuteronomy 5:13-14, 21:10-13). And so on. All fairly repugnant.

I could go on at length about the many horrible passages that praise the immoral, the cruel, as the height of righteous goodness. It does no good to try in desperation to make excuses for it. A good and wise man's message would not need such excuses. It follows that the Bible was written neither by the wise nor the good. And the New Testament was only marginally better, though it too had its inexcusable features, from commands to hate (Luke 14:26) to arrogantly sexist teachings about women (1 Timothy 2:12), from Jesus saying he "came not to bring peace, but the sword," setting even families against each other (Matthew 10:34-36) and approving the murder of disobedient children (Mark 7:6-13), to making blasphemy the worst possible crime (Matthew 12:31-32), even worse than murder or molesting a child. It, too, supported slavery rather than condemning it (Luke 12:47, 1 Timothy 6:1-2).

Worse, its entire message is not "be good and go to heaven," itself a naive and childish concern (the good are good because they care, not because they want a reward), but "believe or be damned" (Mark 16:16, Matthew 10:33, Luke 12:9, John 3:18), a fundamentally wicked doctrine.

The good judge others by their character, not their beliefs, and punish deeds, not thoughts, and punish only to teach, not to torture. But none of this moral truth was in the Bible, and the New Testament had none of the humanistic wisdom of the *Tao Te Ching* which spoke to all ages, but instead drones on about subjection to kings and acceptance of slavery, while having no knowledge of the needs of a democratic society, the benefits of science, or the proper uses of technology. It even promotes superstition over science, with all its talk about demonic possession and faith healing and speaking in tongues, and assertions that believers will be immune to poison (Mark 16:17-18).

The Bible is plagued with a general obscurity and ambiguity, and illogicality, which I had already noted as a child, and though I did understand more and saw it as less confused than I once had, the improvement was minimal and not encouraging. It still taught a morality that is unlivable, and above all contained hardly a hint of humor or any mature acceptance of sexuality or anything distinctly and naturally *human*.

When I finished the last page, though alone in my room, I declared aloud: "Yep, I'm an atheist." It was the question I had sought to answer by reading this book revered by 85% of the American public as the paragon of religious truth. I had never before been so acquainted with how hundreds of millions of people could be so embarrassingly wrong. This revelation led me on a quest to find out more about this matter. It seemed inconceivable that I was the only one who noticed what a total pile of baloney the Bible was, the only one who could see that all the evidence, and the simple process of well-thought logic, led to the conclusion that there was no god, or certainly none around here.

But my search in bookstores for anything about atheism came up with nothing. No one I knew had even given the matter any real thought. As far as I could tell, I was alone. That was annoying, but as the lone Taoist in a sea of nominal apathetic Christians it was nothing new. Eventually I stumbled across two old books in a used book store, Bertrand Russell's *Why I Am Not a Christian* and Corliss Lamont's *The Philosophy of Humanism*, and each gave me an excellent introduction to the thoughts of like-minded men. In time, a booth at a street-fair introduced me with much excitement to American Atheists, which later, disappointed with their attitude, I traded for the more human and sensible Freedom From Religion Foundation.

And though I had been "on the internet" since the mid-eighties, with the rise of national online communities through services like Prodigy and Compuserve I found several atheists to share notes with, and encountered for the first time ardent and avid Christian missionaries and arguers. This was largely new to me—apart, of course, for the perpetual seasonal barrage of Jehova's Witnesses and Mormons who had been knocking at our doors no doubt since my conception. But they were rarely willing to debate, excusing themselves faster than if I were a leper the moment I raised an intellectual question. They were especially confused when hearing I was a devout Taoist, and so already had a religion and didn't need another, thankyouverymuch.

In time two things happened. On the one hand, my studies led me to a more Western humanist philosophy. Though I never abandoned the best of my Eastern intellectual heritage, I fell in love with knowledge and science and logic and the quest and fight for truth. Yet, though I no longer call myself a Taoist, I have not lost any of the joy, wonder, and happiness of life. I retain the lessons that always brought peace, tranquility and simplicity, and my life remains as spiritual as it had been. I live joyfully in a free society with a loving wife and good friends, with no real problems to speak of. And in this lucky position, having struggled my way from poverty to a doctoral fellowship at an Ivy League university, I joined a movement. With compassion for the welfare and enlightenment of the human race, many people like me devote much of their free time to defeating lies, correcting errors, and informing the unknowing. For which we are condemned regularly. Perhaps some day such behavior will instead be an object of emulation and praise, though I don't see Christianity doing anything to make that so.

On the other hand, I became ever more acquainted with the horrible history of Christianity and the sorts of things Christians have done and are still doing around even this country in places less liberal than my First Methodist neighborhood, from trying to pass blasphemy laws to murdering doctors, from throwing eggs at atheists to killing their cats, from trying to dumb-down science education to acting holier-than-thou in pushing their skewed moral agenda upon government and industry alike.

For the first time, rather than being merely constantly pestered, I was being called names, and having hellfire wished upon me. It was a rude awakening. I knew of the eccentricities of Christian Fundamentalism from my high school days, but it was more humorous then than anything: from Jack Chick tracts informing the world—with melodramatically absurd story lines—that role-playing games were a form of ritual Satan-worship, to my friend putting his I-Love-Jesus girlfriend in tears because she was certain he was going to hell for believing there might be life on other planets. But I was generally spared the nasty effects of such nonsense, which was always a fringe minority in my town.

Not so elsewhere. When I heard the horror stories, saw the machinations on Capitol Hill, read the news, I found it was not so funny as I thought it was. So great is the threat of this superstition against individuals, against society, against knowledge, against general human happiness, that it would be immoral not to fight it. It did no good that most nominal Christians disavow all this behavior, for I discovered all too quickly that hardly any of them had the moral fiber to stand up to it, an ominous echo of a phenomenon of apathy and spinelessness we find quite amplified in the Islamic world. Few make much effort to defend in public their apparently kinder, gentler message of tolerance and love against the Righteous Hoard, and fewer still would call me ally. Why would they? Jesus himself tells everyone I am damned, and if the most informed, wise and compassionate being in the universe condemns me utterly, deeming me worthy of unquenchable fire and immortal worms, far be it for any mortal to have a kinder opinion of me.

Worse, the liberal Christians have no text. In any Bible debate, the liberal interpreter always loses, for he must admit he is putting human interpretation, indeed bold-faced speculation, before the Divine Word of God. Appeals to "direct inspiration by the Holy Spirit" win no one over, for the rest of us call that opinionized guessing. And without a believable Revelation or the Bible to stand on, a Christian can be condemned as an unbeliever in disguise. Since being thought an atheist is worse than being thought a whore, not many believers raise their head against Fundamentalism.

It was then that I realized, because of this threat, and because of my own experience in not being able to find like-minded people to share thoughts with, I had to state my case and publish as much as I could. And so I wrote to help others like me, and to defeat the nonsense and lies that I saw being spread everywhere, and to answer the constant barrage of redundant questions I had faced ever since I allowed the Christian public to know I'm an atheist.

This crusade eventually landed me for a time as Editor-in-Chief of The Secular Web (www.infidels.org). And now, this book represents the culmination of research and thought that spans all the way back even to my pre-Taoist years. Though compiled with the wisdom of maturity and hindsight, this book helps to explain the intellectual journey just described, and my arrival at a humanistic atheism.

II. How We Know

1. Philosophy: What It Is and Why You Should Care

Philosophy means "the love of wisdom." The true philosopher is anyone inspired by a passion for pursuing wisdom and truth. Many a non-expert is a true philosopher. But after two thousand and five hundred years of trial, error, inquiry and debate, we now know there is a certain sequence this pursuit should follow. We must always begin our self-examination by looking at our 'theory of knowledge' or in philosopher's jargon, our 'epistemology'. Why? Because anything you intend to investigate, or assert, first requires that you have some criteria on hand to distinguish the true from the false—or in the most basic sense, what can reasonably be asserted and believed, and what cannot. In other words, if you ever assert something ("My wife is a brunette" or "Truth is good"), are you being reasonable? Do you have enough reasons to trust you are right about that?

Constructing an epistemology that answers these questions in a reasonable and thoughtful way—believing or asserting anything trustworthy at all—requires at least three steps. First, you must have some sound and clear idea of what you are investigating or asserting ("What is a 'wife' or a 'brunette'? What is 'truth'? What does 'good' mean?"). Second, you must have some sound and clear idea of how you would go about discovering whether it can be asserted or not ("How do I prove my wife is a brunette, or that truth is good?"). And third, you must actually follow through on that procedure, at least a little, before asserting anything. In practice, we all learn a bit about all three steps and intuitively apply such a procedure, relying on an epistemology we did not examine or construct, one that we simply borrowed from our parents and teachers and peers, or picked up intuitively from our own life experience or professional training—sometimes carelessly, sometimes not.

But the outcome of this process will only be a sound and trustworthy knowledge of ourselves and the universe (and then, eventually, of proper conduct), if our procedures, at all three stages, are sound and trustworthy. But if you've never even thought about it, much less carefully examined and tested what you know about "how" you know stuff, then how likely is it that what you are using, your unexamined assumptions, will just "by chance" be sound and trustworthy? Everything you do, everything you believe, everything you aim at, depends entirely on your knowledge being correct. Certainly, it is better to make sure your methods are the best they can be, that they are even consistent and effective at all. Philosophy is thus your most important business.

So we need a 'theory of knowledge'. But how do we know ours is correct? How do we know its results will actually get close to what really is the case, and continually get closer and closer still, the more we apply it? The real test will be its results in practice. But prior to this we must begin with some first principles that make sense on their own. Only then can we embark upon putting those principles to the test, in order to refine them by studying their results. How we arrive at those first principles is not important (indeed, we may, as time goes by, start all over again with new ones), so long as they first make sense to us, and then are vindicated by their results in practice. This is a little known secret of thinking like a genius: it doesn't matter where your ideas come from, or how many turn out to be harebrained, so long as you only trust the ones that are soundly proved.

Of course, it should be obvious that if you don't have a sound or clear idea of what it is you or others are *saying*, then you will be unable to know if it is really true, for you won't even really know what it *is* that's being said. Thus, every belief you have that is based on such a vague, unexamined notion will be confused from the start. It is therefore of greatest importance to understand your own ideas and beliefs. This is mastered by a serious study of language and logic, though any good schooling in speaking and thinking will serve you better than none. For if you are not clear on these matters, your reasoning cannot be clear, your understanding will be muddled, and you will be less able to pin down what you really mean when you say this or that. Consequently, anyone who simply ignores the study of grammar and logic shall make only slow and faulty progress in understanding anything, and shall inevitably be unknowingly committed to many falsehoods, with little skill in rooting them out. Though many of us might get lucky and do fairly well on natural talent—if we are clever and well-educated—even this will be nothing in comparison with a skill well-honed and well-trained.

It should also be obvious that if you don't have a sound or clear idea of how to go about sorting the true from the false, then you will only fumble about ineptly in every investigation, and again end up committing yourself to many false beliefs. For without a clear conception of method, you will never complete any investigation well or thoroughly, and your findings will often be inconsistent or misleading. By being poorly tested and ill-thought-out, even your methods shall inevitably contain many faults. So anyone who ignores the study of method will have less success in rooting out false beliefs and arriving at genuinely true ones. Such a person has no good reason to be confident in their beliefs. Think of why scientists and expert criminal investigators, or the master of any craft, from mechanics to typists, do so well at their profession, and so easily and consistently, even when meeting unprecedented challenges. Think of how we wouldn't trust anyone else at their jobs. Why not have such a skill in life and thought itself? The advantages are clear.

It should be equally obvious that if you don't employ a sound method routinely and vigorously, then your entire belief system will be unsteady and imperfect. For you cannot be successful in anything of importance if you have a poor or even incorrect grasp of yourself, of what truly exists, and of what you ought to do about it. It does no good to have understanding and skill and then use them only occasionally. You have to use them as often as you can, certainly on every matter of great importance in your life, so that the facts and beliefs you most rely on can be assured of being as reliable and accurate as you are able to make them. Philosophy is not just about thought. Philosophy must be lived.

This is why it is folly to ignore philosophy. For this, the "pursuit of wisdom," is the very activity of studying language, logic, and method, and of employing these tools to construct a comprehensive and intelligible—and ultimately useful—view of yourself and the world. If you love wisdom, this should be your path. And you need this well-found wisdom, so you may fulfill all your desires consistently, successfully, and in the proper order, and thereby achieve a happiness of tranquil contentment, wherein the anguish of injury, disappointment and want is minimized or vanquished.

From long experience, I can vouch for the fact that the study of philosophy has steadily improved my ability to identify and correct my own errors, and to identify (and thus avoid) the errors of others. It has thereby improved my ability to think well and clearly, and has made my continuing education in all other things easier and more fruitful. Philosophy has also led to an understanding of myself and the world that has made my general happiness easier, more constant, and more profound, while rendering every sort of misery ever easier to avoid or endure. Above all, I have a clear sense of always improving myself and my worldview, a sign that I am indeed approaching the truth, and am with every step closer to it.

Philosophy is therefore no idle pastime, but a serious business, fundamental to our lives. It should be our first if not our only religion: a religion wherein worship is replaced with curiosity, devotion with diligence, holiness with sincerity, ritual with study, and scripture with the whole world and the whole of human learning. The philosopher regards it as tantamount to a religious duty to question all things, and to ground her faith in what is well-investigated and well-proved, rather than what is merely well-asserted or well-liked. Instead of keeping her nose ever in one book, she reads widely and constantly. Instead of aligning herself with this or that view and keeping only like-minded company, she mingles and discusses all views with everyone. And above all, she commits herself to the constant study and application of language, logic, and method, and seeks always to perfect, by testing and correcting, her total view of all things.

> There is a lot more I could say about this, and I recommend reading two other essays on the subject of philosophy's value: Bertrand Russell's "The Value of Philosophy" appears in *The Problems of Philosophy* (1912), pp. 153-61; and Charles Sanders Peirce's "The Fixation of Belief" appears in *Popular Science Monthly* 15 (1877), pp. 1-15. Both can be found in James Gould's *Classical Philosophical Questions*, 8th ed. (1994), pp. 39-56.

2. Understanding the Meaning in What We Think and Say

First principles. I shall begin by proposing that all assertions, all statements of fact, convey some "proposition" about what exists, which can only have meaning to the extent that it entails predictions—that is, predictions about what will and what won't happen, if we go here or there or do this or that. These predictions could never all be fulfilled if the proposition were false, but could all be fulfilled if the proposition were true. So the more predictions entailed by a proposition that are fulfilled, the more reasonable it is for us to believe it. And vice versa: the more predictions a claim entails that actually fail to transpire when investigated, the less reasonable it is for us to believe it.

If I believe I have a pet cat, the truth of this belief, in fact the very meaning of the statement "I have a cat," will hinge on whether we have the expected experiences of a cat hanging around me: with all its sights and sounds and furriness and behaviors, its mass and shape, and so on. If not, I don't have a cat. Having a cat can cause all manner of things to exist that wouldn't if there were no cat: vet bills, cat toys strewn about, cat hair on my clothes. None of these things would be inevitable, nor decisive, but if found they would add to the overall proof that I had a cat. As would seeing and petting a cat on my lap. These are all things that "I have a cat" predicts, at least in conjunction with other propositions. For instance, the existence of vet bills is only predicted by "I have a cat" if "I took my cat to the vet" were also true, among other things.

This requires us to know fairly well what predictions a proposition entails. But, to complicate matters, it also requires us to abandon tunnel vision: for the same predictions can be made by wildly different claims, and we have the tough task, through all our lives and in everything we do, of trying to figure out which of several equally plausible explanations of a particular thing is right. Moreover, many true claims won't have successful predictions to their credit only because we never checked, not because their predictions failed. So we are faced with an impossible task: we can't check every possible prediction made by every possible claim. These problems can only be overcome by developing a practical method, which is something we will explore in the next chapter. But this heavy paragraph first needs some unpacking and explaining.

When I say "predictions" in this context, I refer to those events in our direct experience that would at least partly fulfill a proposition or its negation. In other words, in the case of all assertions to fact, "prediction" refers to all experiences, of whatever kind, that can be said to verify or falsify, confirm or disconfirm, a proposition (in whole or in part). This includes the experience of 'equivalence' or 'tautology', which is the experience that "this" is identical to "that," and things of such sort, and it includes historical statements, since an explanation of something that has already happened is just a prediction projected back in time.

When considering statements that do not assert anything, such as those that merely request, demand, ask, etc., such as stating some wish or desire, "prediction" refers to all experiences, of whatever kind, that can be said to fulfill or satisfy the wish, demand, request, question, etc. Though these are not the sort of statements that are ever true or false, they still have meaning, and since any first principle of meaning, if correct, ought to encompass the meaning of all propositions, it follows that my first principle's success in this regard is one indication of its correctness. Another indication is the fact that despite many years of diligent searching, I have yet to discover anything that disconfirms this principle, while everything I have discovered confirms it.

Fellow philosophers should note that I stop short of proposing here that a proposition (which constitutes the meaning of a statement) *only* means what it predicts, though that may well be true. I am proposing rather that, first, a proposition must predict something to mean anything (and therefore a proposition that predicts nothing is meaningless), and, second, that a proposition's truth can only be determined by going out and seeing whether those predictions are or are not fulfilled. Therefore, if we know what a statement means, what it proposes, then we will know how to investigate whether or not it is true. And vice versa, for we cannot claim to know what a statement means if we don't even know what would confirm or refute it. Since understanding what words and sentences mean requires a mastery of language (including the principles of grammar), and since understanding the predictions entailed by any given sentence requires a mastery of logic, these two studies must be fundamental to our education.

Naturally, applying this first principle to itself, it follows that if we can find any proposition that has meaning but does not make any predictions, or that makes predictions but does not have any meaning, or that can be confirmed as true or false without any reference to what it predicts, then this principle would have to be revised, and my entire philosophy reconstructed from the ground up (unless the revision had no other consequence than to expand or qualify what was already established). So it is important to see if I've got it right here, and equally important that I help you grasp what I am talking about. In the process, you will get a taste of different aspects of my whole philosophy, on which I expand in later chapters.

I must first apologize for how didactic and dry this chapter will feel, for here I am only building the scaffolding for the rest of the following project, and scaffolds are never pretty.

2.1 The Meaning of Words

We must begin with words. What do words mean? Words are code signals that human societies made up, because it was useful in thought and communication to categorize everything in some consistent way. Words are the names of things that we experience or imagine, and by sharing the same codebook we can use these codes to communicate our imaginings and experiences to others, and we can organize and study our own thoughts more effectively this way, too—in effect, communicating with our selves: our inner voice, our train of thought, the manifestation of our reason.

As long as our codebook is the same, and as long as we know what experiences or imaginings the code words refer to, we will be successful in our communications. Of course, this is only an ideal dream. In reality, no two codebooks are entirely the same, and most code words are ambiguous, having uncertain meanings, and most have many different meanings, which can only occasionally be determined from context or inflection. Moreover, although humans share a great many of the same experiences, from which they can imagine a great deal more, it must happen that some people will have experiences that no one else has ever had, experiences others cannot even imagine. When they look up the codes for these things, these people will not know what they refer to, only that they refer to something unknown—until, that is, they seek out and acquire the missing experience for themselves, or something sufficiently similar that advances their understanding.

2.1.1 Reducible and Irreducible Sensation

How does this work? Experiences come in many different kinds: thoughts, feelings, sights, sounds, and many others. All experiences, however, are of one or another basic type: those that are irreducible, and those that can be reduced to other components. Several irreducible sensations can combine to form a reducible pattern of sensation.

For example, again, take a cat: when we experience a 'cat' what we experience is a pattern of many component sensations, such as colors, textures, sounds, smells, even thoughts and feelings, or in a less general sense, specific colors, patches of colors, physical relationships and proportions, behaviors, specific sounds and patterns of sounds, and so on. Each of these components can be experienced on its own, even rearranged into new patterns. Once we have seen black cats and white cats, we can imagine black and white cats. Once we have heard a cat meow and a dog bark, we can imagine a dog meowing and a cat barking. We can even imagine a creature, half-dog and half-cat. We can reassemble the components of each sensation, comprising every experience we have had, to imagine something new.

This talent becomes exponentially vast in power, for the number of things we can imagine equals the number of ways we can arrange into patterns all the sensations we have experienced, and that list is itself vast beyond imagining. And if someone has experienced, in different patterns, all the sensations comprising a cat, and if we can refer to all these component sensations and describe the pattern into which they fall, it will be possible to communicate the idea of a cat even to someone who has never seen one.

Of course, as one will quickly discover, the whole idea of a cat is so complex, including huge numbers of component sensations and abstractions about sensations, that this communication of ideas is always partial and incomplete—it never suits as well as drawing a cat, or pointing to one, or living with one. But the ability to communicate even a little bit of what a cat 'is' gives us a great deal of intellectual and practical power. It is what makes us humans so darned smart, and education so darned useful.

As an example of how abstraction complicates things, cats are not simply black or white, but can be both, or any other combination of many kinds of colors, thus they only 'have a color'. But do only certain colors and patterns count? Is a green cat not a cat? This is not always easy to answer, since our codebook was cobbled together over time by no one person and a lot of loose ends remain, especially when it comes to unfamiliar, unusual, or unanticipated circumstances. For example, what if it is a cat in every respect but one: it has green scales? Thus, not only complexity, but ambiguity plagues the process of communication. But even more important, if someone has never seen colors, it will not even be possible to communicate this idea. Such a person's idea of a cat will never be quite the same as anyone else's, though it will otherwise share many of the same elements.

On the other hand, we cannot reassemble the color green. This sensation is irreducible. We cannot describe it by referring to component sensations, which a person can then assemble in their mind and thus imagine 'green' even without having experienced it. To know what green is, a person must have experienced it. Lest we focus too much on vision, we should know that these features, of reducibility or irreducibility, of complexity and ambiguity, are as true for all other kinds of experiences—love, democracy, the number two. All words refer to experiences, either irreducible sensations (like green, or the feeling of attraction, or the number one), or reducible ones (like checkerboards, or love, or the number two), even while many words are ambiguous, or refer to patterns so complex that even though reducible in theory, they are hardly so in practice.

Numbers and mathematical and logical terms are examples of exceptions to this, being neither ambiguous nor complex in themselves. Though in combination the total complexity increases, the absence of ambiguity and the presence of component simplicity is a defining characteristic of mathematics as a language, a point we shall revisit in a later chapter.

2.1.2 Meaning, Reality and Illusion

Now an early mistake can be made, a wrong turn taken. If this idea of meaning is not realized, it is easy to forget what things really mean, to get confused about what words refer to. I think Plato got confused. By not realizing that words are names that people made up to describe their sensations, he got around to thinking that ideas, even numbers, had a supernatural substance, that they were themselves distinct 'things', more than mere labels or codes. There are people who still think this even today, and various things like this, all because they took a wrong turn— forgetting, or never seeing, that words are nothing more than names for human experiences. This is something we will return to in a later chapter (III.5.4, "Abstract Objects").

Worse, one can slide from this into the mistaken belief that our whole world is an illusion, that reality is somewhere else. Of course, the word 'cat' can refer to different kinds of experiences, for instance a picture of a cat, a hallucination of a cat, or an actual cat. What is the difference? The difference will always lie in some aspect of experience—a picture will not meow, and a hallucination will not leave physical traces of having existed at all. When you examine the conditions under which the cat is observed, like 'while watching TV' or 'with a schizophrenic brain chemistry', you will have other clues to what it actually is that you are seeing. However, since something that can never be experienced in any way can never affect you in any way (because any noticeable 'effect' would be an 'experience'), it is possible for there to be no relevant difference between an actual cat and a hallucinated cat.

To understand this, consider the movie *The Matrix*: if life in the 'matrix' is in every way the same as a real life, there will be no relevant difference. It will be a real life. Pain, pleasure, knowledge, life, death, good noodles, are all real in this world. The rules are exactly the same. But the movie described certain things that make this world different: glitches, superhuman 'agents', and groups who can 'wake you up' to the higher reality, where you can go even further and see the machinery and circuitry that produces it all. As long as things like this exist, not only does it remain possible to discover the true nature of reality, but this matrix would *not* be the same as the real world—for the very reason that it had these differences.

The importance of this digression is that it only makes sense to talk about the world as an illusion (or 'computer simulation', etc.) by reference to other possible experiences that would justify the label. Even if the experiences (of the 'true reality') are not in practice possible, they must at least be possible in theory, or else the term 'illusion' would not be applicable. If there is no way, even in theory, to tell that this world is not what it seems, then it is meaningless to claim that this world is not what it seems. For if no experience of any kind can in any way be had, even in theory, *except* an experience in this world, then, by definition, no other experiences exist—there is no other world. It follows that all our ideas of a 'physical' reality (as opposed to, say, a 'matrix') are rooted in the observation that, as of yet, what we have experienced makes the most sense by appealing to a natural organism called a universe, rather than by appealing to computer simulations—or any other kind of Cartesian Demon, a monster we will discuss in a later chapter (and for more on the dream-reality distinction, see III.6.2, "The Mind as Virtual Reality").

2.1.3 Experience is the Font of Knowledge

We must now start with the realization that in the simplest sense, knowledge is the possession of experiences, both reducible and irreducible, which can be recalled and assembled or reassembled by the imagination. Without experience, there is no knowledge. Without knowledge, there is no experience. By 'experience' I mean all experiences, mental and sensory, including emotions and thoughts. And by 'knowledge' I am speaking of 'cognitive knowledge', a particular kind of thing which does not include, for example, reflexes and intuitive skills. The infant's 'knowledge' of how to suckle or to breathe, or our learned ability to ride a bike without thinking about it, is not cognitive knowledge, but noncognitive knowledge.

With noncognitive knowledge, our nervous system is automatically trained to recognize or react to certain data "without thinking," a fact that cannot be communicated. Only our experiences of the application of this knowledge can be communicated, and only if the experiences have been shared and named. For humans cannot actually transmit experiences, but only propositions *about* experiences. Since a proposition is not the same thing as an experience, a proposition can only be understood if it refers to noncognitive knowledge that the audience *already has*. In other words, propositions cannot directly cause another's nervous system to rearrange itself as needed—for instance, in order to truly learn how to ride a bike or recognize a sensation. Thus, knowing every true proposition about bicycling will never train your nervous system to ride a bike, just as knowing every true proposition about a heart will never pump blood. Noncognitive knowledge is central to understanding "intuition," which we will discuss in III.9.1 ("Reason vs. Intuition"). But most of philosophy is concerned with cognitive knowledge, and with that we proceed.

2.1.4 Getting at the Real Meaning of Words

The typical dictionary is a lexicon, a rulebook for how words are to be spelled, spoken, and employed, based upon an empirical investigation of how people most generally speak and write. But this usually amounts to nothing more than describing a word's various meanings with a synonymous word or phrase, which is hardly less ambiguous than the original word itself. The philosopher cannot settle for this. In her business she seeks precision in her own definitions of words. But when she tries to enter ordinary human discourse she often becomes a fish out of water, since she can no longer create the definitions herself but must resort to discovering how social convention has defined things. But there is nothing comparable here to what she normally does in defining words, when she specifies a definition and then analyses it.

People, much less the collective force of blindly produced convention, do not think like philosophers. They rarely analyze their words at all, and often hold all manner of superstitions about what those words mean or how they use them. I call these added beliefs "superstitions" because in practice people do not employ these meanings but different ones altogether, of which they are usually not consciously aware. And though linguistic intuition has a tremendous utility, it is very hard to interrogate.

What meaning does our intuition employ when we use words in the English language? The answer is rarely much explained by what is found in your typical dictionary. Thus, philosophers must master and analyze language in a different way, so they can better communicate with the public and themselves and better study human beliefs and conventions. Philosophers would also be doing us all a service if they would use this skill to aid the general public in acquiring a better understanding of how we really use words, so we can begin using them more accurately and without improper assumptions about what those words entail or suggest.

The method to be employed is this. The proper definition of a word, as opposed to its lexical definition found in dictionaries and on the tip of everyone's tongue, is fixed by how that word is actually used in practice. What people claim the word means is not much good if their use of the word does not in fact align with this claim, or if the claim is too vague to pin down. Thus, to identify what words properly mean, it is necessary to observe the word being used. If a word in one specific connotation is X, then whenever people say X exists *here*, but does *not* exist *there*, we must analyze what properties or circumstances were present in the one case and absent in the other. After examining countless cases, we should find a consistency of behavior: every time Y exists people say X exists, and every time Y does not exist people say X does not exist. At that point we have identified Y as the proper meaning of the word. Whatever else people say that word means (within the particular connotation being investigated) is what I call a linguistic superstition.

These "additional" or superstitious meanings must either be accepted as a separate concept, not necessarily true of every instance of X (and thus not part of the word's meaning) or else they must be abandoned as spurious, or relegated to a *new* connotation of the word, one restricted to rare, speculative contexts not relevant in normal discourse. In some cases these superfluous "meanings" are actually meaningless, since people cannot even imagine a single observation that would ever in principle confirm or

Sense and Goodness Without God

refute that the thing they think the word describes exists 'here' but not over there. But in most cases, these criteria *can* be imagined, but have never yet been met, or might never be. Whatever the case, to understand what words really mean, we must learn how to engage in this new method of inquiry, rather than leaning on "one part dictionary" and "two parts assumption."

For example, take the concept of a "life-force." We observe a certain behavior of chemical systems that is complex, active, self-perpetuating and self-maintaining, and that exhibits a kind of healthy harmony, and we call it a "life-force," the "power" of a certain pattern of matter and energy to be active and alive. But we often go beyond this and attribute some mystical property to it, or imagine that it is a kind of 'force field' that can extend outside the physical body. But there is no need to make such assumptions—they are not essential to what the word "life-force" really means in practice, what it actually refers to—and in fact these are false attributions, a superstition about living systems produced by our ignorance of, or inability to comprehend, what is really going on.

But everything else we believe about a "life-force," such as what it looks like, what it can do, and so on, remains true, and we can still point to where there is one and where there isn't. Indeed, the fact that people still use the word to refer to actual, observable distinctions in the world is why the word has a useful meaning worth getting at and preserving in the first place. Though a life-force's cause, its abilities, its subliminal nature, is different than our superstitions would have it, it still exists, and is still basically what people observe it to be. So a right thinker must never confuse the superstitions we attach to an idea with its essential meaning. The means to disentangle them is the procedure of enquiry just described: go and see what is *really* observed, what people *actually* point to in practice when they say something is over here but not over there.

2.2 The Meaning of Statements

So much for words. Now we move up a notch. For words are most useful when arranged into sentences, which express those "propositions" I was talking about earlier. What follows represents almost every kind of statement, or at least such a variety that any statements not included can easily be understood in reference to the following categories.

2.2.1 Stipulations and References

"Cats are felines" is the most basic. It is a simple statement of definition: if we take the "are" to mean "equivalent to" then it communicates that one code word represents another. It reports that any time the name "cat" is used, the name "feline" (in American English at any rate) can be used instead, with exactly the same meaning. But this statement can actually have two distinct meanings, and it is important to understand this, since it may not always be clear which meaning this and similar statements have, which is a major cause of misunderstanding between people.

The first meaning is formal, a meaning we call 'stipulation', like a command: a cat is a feline because I say so, and for no other reason ("so every time I use the word, you will take it to mean this"). The second meaning is lexical, a meaning we call 'reference': a cat is a feline because the dictionaries, the codebooks, possessed in the brains and libraries of a certain group of people—in this case, all speakers of Contemporary American English—say so, and for no other reason. Of course, there are many lexicons: every profession has its own way of using words, as does every dialect, and fashion and effort are always changing every lexicon over time. But context usually tells us which lexicon is relevant, and a good communicator will make sure of it.

This distinction is important. It can at times be useful to stipulate new definitions, but it is often counter-productive if there is no good reason to do so. Indeed, it is self-defeating if you do not explain, or know, that this is what you are doing—unless, of course, your object is in fact to deceive. But changing what names refer to will never change the things themselves, and since our desires and plans depend upon actual things, not on what they are called, stipulation is of limited utility, except when naming things that have not already been given names, or when clarifying which elements or connotations of a lexical definition you are referring to. On the other hand, it is essential to know the lexical meaning of words in order to communicate ideas successfully to others, since in every case where stipulation is not declared, most people will assume the current and relevant lexical meaning is the only proper one (even your own intuition will tend to do this)—and in all cases where no definition is stated, the lexical meaning (like the meaning found in a dictionary) will be the only one your audience knows.

Likewise, since words must refer to experiences to have any meaning at all, and since desires, worries, plans, and predictions are all inexorably connected with experiences, it is important to remember that it is always these experiences that really matter, not what they are called. Those who forget this will often get tangled up in arguments about what things 'should' be called, or what certain words 'should' mean. But words should only mean what they do in fact mean—that is, lexically. The codebook already exists, in billions of copies, inside everyone's head and on their bookshelves. It is a foolhardy quest to try and rewrite a codebook that is already in print. It is better, and far easier, to work with what words already mean, than to try and change their meaning altogether. As has been said, it is easier to wear shoes than to pave the earth with leather.

2.2.2 Descriptions

What about other categories? They each report something, and to know the meaning of a statement is to know what it communicates. For example, "cats are mammals" uses the word "are" in a different sense. It reports that all patterns of sensation referred to by the word 'cat' belong to another, larger or more abstract pattern of sensation referred to by the word 'mammal'. A mammal is not always a cat—the pattern of sensations corresponding to "mammal" contains that of a cat, but also that of dogs and other things. But a cat is always a mammal, because the pattern 'mammal' includes all patterns that are called 'cat'.

This kind of statement can also come as either a stipulation or a lexical reference. But there is another meaning this and other statements can have that is distinct from stipulations and references: a reductive definition, which we call a 'description'. For instance, a dictionary will say more than that a cat is a mammal. It will list other key features, like 'carnivorous'. The correctness of these definitions will be tied not only to the lexical traditions, but to actual experience. If a dictionary listed 'cat' as 'a feline herbivore' it would be wrong, because when we go and look at cats, we will not observe herbivores, but carnivores.

The problem here lies in the fact that a reductive definition attempts to take a known pattern of sensation ('cat') and to describe the component sensations as well as the pattern they are arranged in ('furry quadruped'). In doing this, it is possible to agree on the pattern being described ('a cat is a popular uncaged pet with pointed ears' which everyone can name the moment they see it) but to err in describing the properties or features of that thing ('a cat is herbivorous'). This, therefore, gives us three kinds of statements so far: formal stipulations, lexical references, and reductive definitions.

2.2.3 Opinions

We enter a new area here. "Cats are cute" reports that all patterns of sensation referred to by the word 'cat' belong to another, larger or more abstract pattern of sensation referred to by the word 'cute'. Unlike 'mammal', however, 'cute' is a conditional term. Since 'mammal' always refers to the same set of sensations, its meaning is not conditional on any other fact. Once known, its meaning never changes from one circumstance to another, so long as the codebook remains unchanged. However, 'cute' refers to the pattern of sensation that we describe as 'what someone thinks is charmingly attractive'.

In other words, the meaning of 'cute' will always be conditional on another, ever-variable fact: namely, what a particular *individual* refers to with the word 'cute'. We can thus know when something is cute to us, but we cannot automatically know whether it will be cute to someone else. We shall call this an 'evaluative' term. So why are 'cute' and other evaluative terms not unconditionally defined in a common lexicon? Because these words refer to values (see parts V and VI, "Natural Morality" and "Natural Beauty"), and different values are often possessed by different individuals. To call something 'cute' is to say something about what *you* think or feel about the thing being so called—it communicates something not just about the cat, but about *you*.

Couldn't there be some 'objective' sense in which cats are cute, in the same way they are mammals? In the sense of being "charmingly attractive to someone" perhaps, though that still describes a connection between cats and humans, and not just cats—and it is not complete. To answer the question, we must inquire into what experiences people actually refer to with the word 'cute'. It is certainly possible to define 'cute' in any way we want, by means of stipulation, such as 'cute is being furry', and we can thus create an unconditional definition like we have for 'mammal'. But as I've said (in II.2.2.1, "Stipulations and References"), this is not very useful, because what we want to know are *the experiences* people refer to with the word. And if we redefine cute, we will not have changed those experiences. So people will continue to refer to those now-unnamed experiences by coining a new term, and consequently we will have learned nothing by redefining the original one.

When we examine the actual experiences that 'cute' refers to, we always find that it has something to do with the speaker's attitude toward the thing being called 'cute' (or any other evaluative term). "Cute means charmingly attractive," for instance, leads us to this: "a cat is cute because it charms me and is pleasing to my eye." We might say, instead, "a cat is cute because it charms and is pleasing to the eye," but we would risk reporting a falsehood, for the truth of this statement then hinges on whether the cat actually charms, and actually pleases the eye. A single counter-instance will falsify this claim. If we find anyone whom a cat doesn't charm or please, the statement "cats are cute" in such a sense would be false.

We usually allow for unstated exceptions when making sweeping statements, so that "cats are cute" often means "cats are cute to most people." But if we are going to do this, we had better remember what it is we are doing. No matter what, 'cute' must always refer to subjectively defined qualities—the cuteness of a cat is contingent on its ability to charm or please, and since only individuals are charmed or pleased, one can only know if a cat is in any way cute by finding out if any individuals are charmed and pleased by it. There is no way to find this out by examining just the cat by itself.

Couldn't there be a sense in which a person 'should be' or 'ought to be' charmed and pleased by a cat? To answer this requires figuring out what 'should' and 'ought' mean, which we will discuss below, and in parts V and VI ("Natural Morality" and "Natural Beauty"). Consider, for now, that the nature of "artistic criticism" is not to say what we *should* think, but to point out the features of something that an individual finds attractive or ugly, so that others who perhaps missed them can notice those details as well and make a more informed decision about whether it is attractive or ugly to them.

Often there is a consensus of values, at least within certain groups large enough that one can 'join' or 'listen-in on' the most like-minded group. In this fashion, the communication of aesthetic opinions becomes easier among those who share common aesthetic values. But it is not possible to change someone's values by merely pointing out features of the thing to be evaluated. Through artistic criticism it is possible to call attention to features that went unnoticed before, and which relate to someone's aesthetic values as they already exist. In this fashion, for example, it is possible to change someone's mind about whether cats are cute. But this is not a changing of their values, but rather an appeal to values they already have, by noting facts they had previously overlooked.

2.2.4 Moral Imperatives

"I shouldn't torture my cat" is another kind of statement with evaluative features, called a "moral imperative" because it is believed to apply to everyone with 'moral force', something much more powerful than a mere demand on opinion. Since I devote a substantial part of this book to the subject (Part V, "Natural Morality"), I won't digress too much on it here. It will be enough to summarize what I will later argue to be the *meaning* of imperative statements.

A statement like "you ought not to torture cats" is partly evaluative (like "cats are cute") and partly factual (like "cats are mammals"). The evaluative part is the implied claim that certain values (like "you don't want to cause pain") are possessed by *everyone* to whom the statement is addressed (so, "you ought not" because, if you think about it, you really won't want to). The factual part is the implied claim that certain actions will probably have effects (like "causing pain") that fulfill or contradict those values. How all this works out as far as what is really right and wrong and why you should care does not matter for now, since the *meaning* of moral statements ends up the same no matter what your worldview. For example, "You must adopt our values or go to hell" entails the same twofold meaning: that we do not want to go to hell (the evaluative claim), and that there not only is a hell, but we will actually avoid that hell by adopting the values in question (the factual claim).

2.2.5 Wishes and Commands

Another kind of statement that we will cover only briefly is the wish or command, since these are so simple in their meaning. Statements like "get my cat" are commands, meaning something like "I want you to get my cat for me." Such things express a desire for someone to perform some action. Just as with instructions for the fulfillment of some goal ("glue the edges, then assemble the model"), the desire implied is for the reader to achieve a certain result (a properly-assembled model). The meaning is thus connected to certain fulfillment conditions: the fulfillment conditions of "get my cat" are the getting of my cat. There is a future outcome that is imagined, and the meaning of the command is that this "outcome" is to be sought.

The same applies to wishes, such as "if only I had a cat," which simply express the desire for an event without specifying that the desire be fulfilled by anyone in particular. For example, "I wish I had a cat" differs from "get me a cat" in that the desire in the second case is two-fold: to have a cat, and for someone (perhaps someone in particular) to get that cat for us. Note that all stipulations are, in essence, wishes or commands.

2.2.6 Facts and Hypotheses

Now for the most crucial of all categories. "I have a cat" is a factual claim. Such "hypotheses" put forward a proposition about human experience. In particular, "I have a cat" means that if someone were to look at all my belongings, among all those things they would find a cat.

All factual statements have this meaning: they predict that under certain conditions, certain experiences will be had, and the conditions and experiences in question are inherent in the meaning of the words and the manner of their arrangement. "I have" means that one of the experiences being predicted is that a cat is connected to me in some way that would constitute "having" it (like being in my house, or being fed by me on a regular basis, or being described in a bill of sale to me), and one of the conditions for this experience is that someone must examine the things that are connected to me in that way in order to experience the "cat" I am referring to. Likewise, "a cat" means that one of the experiences being predicted is of something that could be called a "cat" (like a small furry quadruped that purrs), and one of the conditions for this experience is that a person have the required senses, and go to where the cat is, to experience the "cat." And all this is *inherent*, it is understood, in the very *meaning* of the words and the way they are put together.

This brings us to the question of truth. What do words like "true" and "truth" mean? Truth most commonly means "correspondence with fact" such that a statement is true if the experiences it predicts will actually be experienced under the implied conditions. Anything is in some respect a fact if it can actually be experienced as described. In a simple sense, a hypothesis is false when the conditions are met but the predicted experience is not experienced as described, and true when the conditions are met and the predicted experience *is* experienced as described. The words "true" and "false" thus refer to either potential: given the absence of errors, interfering circumstances, and so on, and accounting for modifiers when appropriate (like "sometimes" or "often" or "probably"), "this is false" means that in every case, when we satisfy the implied conditions, we will find our prediction fails, and "this is true" means that in every such case we will find our prediction succeeds.

Now for the punch line. All the other types of proposition can actually be described as types of hypothesis, as different kinds of claims to fact. For they all make predictions. References (lexical definitions) predict that if any copy of the implied lexicon is consulted (whether in print or in the mind of someone who speaks that language), the stated meaning will be experienced. So "cats are felines" is true if the code word 'cat' is actually found to be equivalent to the code word 'feline' in some actual lexicon. Stipulations, on the other hand, predict something far more obvious: that whenever we comply with the stipulated meaning of a term, then that will be the meaning of that term for those who so comply. So "cat shall mean feline" is 'true' for someone if they decide to let 'cat' mean 'feline', fulfilling the request.

Descriptions, however, predict that if the thing being described is experienced, the features stated in the description will also be experienced.

Sometimes descriptions are more like stipulations in their simplicity: that "the thing described is the thing experienced" is sometimes obvious if, for example, the description is something like "a red rose is red." In this case, the truth of the prediction is inherent in the statement itself: even if there were no red roses, it would still be true that a red rose would be red, because if there were any red roses, then they would be red by definition. This is a claim to fact (about the meaning of words) that can be automatically true. At other times, descriptions are more like references and thus not inherently true: "all roses are red" can be false if there are any things called "roses" (whether lexically or by stipulation, depending on the context and thus the meaning of the original assertion) that are not red.

In like fashion, opinions and imperatives are also claims to fact. In the case of opinions, they claim that someone actually has the stated opinion. So "cats are cute" is true for me if I find that I actually think cats are cute, and is true in general (though not for everyone) if we discover that many people think cats are cute. And imperatives claim that a certain action will have some kind of consequence that will be desired or not desired by everyone, whether they know it or not. So "you ought not to torture cats" is true if there are actually any consequences to torturing cats that you actually do not desire—or would not, if you knew all the facts.

Wishes and commands predict that if the wish or command is fulfilled, the described events will take place. Thus, "get my cat" is 'true' in an analogous sense if, when someone obeys, someone has actually gotten my cat, and "if only I had a cat" is 'true' if I actually want to have a cat. As might be apparent, whereas the truth of wishes depends on there actually being a desire for the thing wished for, the 'truth' of commands, like stipulations, is inherent in the statement itself. This shows that stipulations and commands do not actually make claims to fact of the sort that we bother calling "true" or "false" because there is no way they can ever be false in our usual understanding. Rather, they can only be fulfilled or unfulfilled—or more fundamentally, they can be meaningful or meaningless: "a flissle is a prundle" is a meaningless stipulation, and "flissle my prundle" is a meaningless command.

2.2.7 The Nature of a Contradiction

This brings us to the final feature of meaning: the contradiction. If we say "cats are carnivores and cats are herbivores," we know right away at least one of these claims must be false. Why? Because, stating them together, we would be asserting a fact as true *and* false, which means *something* we are saying must be false. Of course, such statements, taken together, do not describe anything that can ever be experienced in the implied conditions. We will never experience a cat that is a carnivore and a herbivore. We could experience a cat that is an omnivore, but an omnivore is neither just a carnivore nor just a herbivore. However, each statement taken alone could describe something that can actually be experienced: "cats are herbivores" may be false, but it is not meaningless. So we might not know right away where the problem lies. This can only be discovered by checking each statement for factual truth.

From the rules of language we know "cats are carnivores and herbivores" means "cats are carnivores and not carnivores" and, given the lexical meaning of these words, this statement is meaningless: it describes nothing that can be experienced by the human mind. Since it refers to nothing, except something that is *not* the case, it cannot be called true. In that sense, we can be certain it is false without even investigating the matter, because from the start, merely by analyzing the meaning of this sentence, we can see there is nothing to investigate: there is no conceivable experience that would confirm it, and our immediate experience (of the content of the assertions) refutes it.

Still, every term and their arrangement is still meaningful. We know what each claim refers to. So we can also say the statement is false because what it asserts entails predictions that will always fail to pan out. That is, "cats are carnivores and not carnivores" asserts (predicts) that there is no distinction between being a carnivore and not being one. But this is false: the very words themselves mean there is a distinction. In this sense, a contradiction asserts something about the meaning of its terms that is false (either by stipulation or lexical fact). Either way, contradictions, once identified, immediately betray themselves as untrue, even before we bother to engage any method of investigating their claims beyond perceiving their content in our thoughts (I say more in section III.9.3, "Contradiction Revisited").

> For essential reading on the basis for my analysis of meaning, see the relevant part of the bibliography to II.3.2, "The Method of Reason.."

2.2.8 Naturally Warranted Belief

A well-known Christian philosopher, Alvin Plantinga, has attacked the very foundations of the kind of "theory of knowledge" I have just laid out, and something must be said about his case before we move on. Much of this book renders many of his arguments moot. But I will address two of his claims here: that we can have warranted beliefs without evidence; and that naturalists cannot have warranted beliefs.

First, evidence. Plantinga claims that it "is entirely right, rational, reasonable, and proper to believe in God," indeed, all the basic elements of the Christian worldview, which if true would entail from the start that naturalism is false, "without any evidence or argument at all." In other words, no one needs evidence to believe Christianity, contrary to what I have been saying (that one cannot properly believe *anything* without evidence). However, Plantinga thinks "belief in the past, in the existence of other persons, and in the existence of material objects" are also things we believe without evidence, but that is clearly not true. Ask anyone why they believe in these things and they can start rattling off a long list of experiences they base these beliefs on. A clever person could even say what experiences would cause him to *no longer* believe in them.

Belief in the past is based on the evidence of our memories, as well as of our selves and surroundings, which show the effects of past events, thus confirming our memories. Hence when people need to identify false memories, they turn to such physical evidence, and scientists and historians have generally confirmed the range of trust we can place in memories alone as evidence, by studying the usual correlation of memory to the observed facts of the effects of remembered events. The evidentially-based hypothesis that there is a past explains all this evidence better than any alternative. We always need evidence to trust our beliefs about a past. The same is true of other persons. We have lots of experiences confirming that belief, and can imagine experiences that would refute it, convincing us that other people were merely illusions, or robots, or computer generated images. And the same goes for material objects: the hypothesis that they exist explains, better than any alternative, the experiences we do have (of the phenomena of mass and solidity, the findings of particle physics, etc.). Thus, again, our belief is based on evidence.

Plantinga uses at least three arguments altogether. The first is that we have evidence that all people believe things without evidence, but as we just saw there is no such evidence. There is nothing we can find that is believed in without *some* evidence, except when belief is not warranted, as when we believe for no reason at all, but on a mere whim, or when we believe for reasons that have nothing to do with the truth, such as a mere desire that something be true. So his evidential case fails.

His second argument is hypothetical: if a superbeing arranged the world so that we would automatically know certain things, then we would be warranted in believing them. But a superbeing would do this by making these things *undeniable* (just as raw, uninterpreted experiences are undeniable) or *obvious* (like the existence of the moon), which hasn't happened, therefore we know no superbeing has done this. In our reality, if we can doubt a belief, if it can be false, then we do not "automatically know it," for we must have a reason overcoming the doubt, ensuring that our belief isn't false.

Indeed, a superbeing could arrange the world so that we *think* we automatically know things that are in fact *false* (see my discussion later of the Cartesian Demon). There would be no way for us to tell the difference. Indeed, there is no real way for us to tell the difference between a world containing either an honest or a dishonest superbeing, or a world containing nothing of either kind. So we cannot appeal to the existence of such a being, since we see that this does nothing to make belief in it undoubtable, and we do not even know if such a being exists, nor if one did would we be able to know if it was aiding us or tricking us. Either way we are back where we started: having no clue which if any of our beliefs are warranted, *unless* we have evidence on which to establish that warrant.

Plantinga's third argument is that some propositions must be believed without evidence, for otherwise we would have to believe an infinity of propositions, which is impossible. But propositions are not evidence. Propositions are *claims about* evidence. In that respect, they can stand in as place-holders for evidence in any process of reasoning, but their truth ultimately rests on the evidence itself, not on other propositions *per se*. The buck stops with the evidence: which means experience, for there is no other kind of evidence. The buck does not stop with any propositions, since they are mere formulations about the evidence, which manifests to us non-propositionally, prior to thinking or making any statement about those experiences. The moment we start talking *about* them we are unavoidably engaged in using evidence as a basis for further belief.

In his vocabulary, "basic" propositions are what constitute the logical and conceptual foundations on which all other knowledge is built. Such propositions must in turn be based on the direct and manifest evidence of the senses (which includes internal sensation, like thought and emotion). But these "basic" propositions state nothing but that those experiences exist, not what those experiences mean or entail, or what caused them. To propose things like that we must formulate *other* propositions, which are *about* the experiences and go beyond them, making claims about what exists *apart* from the experiences themselves. And such propositions cannot be "basic" in the present sense, because they must be built on other propositions. After all, the only thing we can trust without evidence is what cannot be denied, and the only thing we cannot deny is that certain experiences exist. Everything else is fair game. So all other beliefs must be based *on* something in order to stand as anything other than random or arbitrary.

Key to the distinction between evidence and propositions is that it is not possible to be mistaken that experiences exist (at the moment they exist). But it *is* possible to be mistaken when you claim they exist, when the belief is formulated in propositional form (as a statement), since you could have misformulated the statement and thus the proposition. For example, that there is an experience of me typing now is a properly basic belief for me. But the proposition "there is an experience of me typing now" could still be false, if for instance I am really writing instead, and included the wrong concept, by using the wrong word ("typing"). All other propositions (such as that I really am typing now, and not merely dreaming it, or that keyboards really exist and are made of plastic) are built on this first proposition (in conjunction with others), which rests on properly basic *evidence*.

Ultimately, it is not reasonable for anything to be a "basic" proposition if it could be false, since only a proposition that can never be false can reasonably be believed on its own, without evidence to support it. And yet all propositions can be false in principle (since, at the very least, they can be mistakenly formulated). Therefore, we can only have basic *beliefs*, which precede any proposition about them, and since propositions merely propose what those beliefs are, there is no need for an infinite series of propositions. Of course, any belief that can be false needs some assurance it is true. For otherwise it could as well be false as true. But a belief that *cannot* be false needs no assurance.

But those are beliefs. We can state countless *propositions* about any given object, yet only a very small proportion of those will be true, which means the probability in any given case, absent all evidence of any kind, is that a proposition is false. Even in the best conditions the odds are no better than 50/50 that an unsupported proposition will be true rather than false. Therefore, *only* evidence (ultimately in the form of basic *experiences*, which support the probability or explanatory merits of a claim) can ground any belief whatsoever, for there is nothing else. These demonstrations need not be absolute, or even detailed. It is enough that I have any evidence at all. My belief can then be proportioned to the amount and strength of evidence I have, and amended if and when it turns out to be mistaken.

So much for the claim that belief can be warranted without evidence. What about the claim that naturalism cannot warrant belief in anything? Plantinga's claim here is that given our worldview, there is no reason to expect that our truth-finding abilities are at all reliable, because they arose without any intelligent engineer, but essentially by accident, with the natural, blind aim not of finding truth, but mere survival. I refute this argument in III.9.2, "Why Trust the Machine of Reason?" There I show that no substantial difference remains between the development of a truth-finding mechanism and the advancement of survival, and that, in fact, cognitive powers can usually only be selected for insofar as they do not mislead, meaning evolution will inevitably produce better and better truth-seeking abilities, within obvious practical limits. Thus, on our worldview, we *can* expect our truth-finding abilities to be reliable—not perfect, but reliable.

So what do I believe establishes a belief as warranted? If it is produced by any causal process that is substantially truth-selective (for example, any process that generates a feeling of confidence that is in proportion to probably genuine realities), then a belief has warrant. Metaphysical Naturalism as described in this book entails that we very probably possess and employ such a process. And when we observe the facts, we will find that Metaphysical Naturalism is indeed probably true, and that such processes do indeed exist in us. A radical skeptic could still charge that we are somehow fantastically deceived in all this, but that is an issue I will tackle more thoroughly in the next section.

> Quotes and paraphrases from Alvin Plantinga from "Reason and Belief in God," Nicholas Wolterstorff and Alvin Plantinga, eds., *Faith and Rationality: Reason and Belief in God* (1983), p. 17. The argument is developed fully in Plantinga's books *Warrant and Proper Function* (1993), pp. 194-237, and *Warranted Christian Belief* (2000), pp. 227-240.
>
> For more discussion, see my critical reviews of similar arguments by Michael Rea and Victor Reppert (www.infidels.org/library/modern/ richard_carrier/rea.shtml and www.infidels.org/library/modern/richard_ carrier/reppert.shtml). Note that I reject outright Plantinga's gratuitous claim that "proper" function is necessary for warrant. To the contrary, any truth-finding function that is *functioning* will suffice. The word "proper" has no business in any formulation of the criteria of warrant.
>
> For a recent, powerful defense of my view that beliefs must be based on evidence, see Jonathan Adler, *Belief's Own Ethics* (2002), while the philosophical foundations of "evidence," and what does and does not count as such, is addressed in technical detail by Peter Achinstein in *The Book of Evidence* (2001) and Susan Haack, *Evidence and Inquiry: Towards Reconstruction in Epistemology* (1995).

3. Method

Once we understand that knowledge of the true and the false depends upon our knowledge of the fulfillment of what a proposition predicts, we will soon learn that a fairly sophisticated method is needed to approach the truth and avoid mistaking the false for the true. For many propositions have sets of predictions that overlap, so when we experience a prediction we cannot be sure which hypothesis it confirms. Even from the start, that we have experiences *at all* can be explained in more than one way. In fact, many explanations can be correct, different ones on different occasions. Thus, for example, having a particular experience of an object in our visual field can confirm that we are seeing an object, or that we are hallucinating an object. The experiences by themselves are a simple, irrefutable fact, at least for ourselves when we have them, but what they mean for us besides an uninterpreted experience requires some detective work. This is where method comes in. And there is more than one.

The end result of any method will never be absolute certainty, for we can always be in error. Even what we think we can be most certain of, like mathematical equations, is subject to doubt—for what we believe to be a correct equation can always turn out incorrect after all. A mathematician can arrive at a proof that is confirmed by thousands of colleagues, yet he can never be absolutely certain he has made no errors in that proof. Even equations so simple that they can be comprehended in their entirety in a single experience (e.g., $1 = 1$) can be founded on a faulty sense of what is being experienced or of what it means.

This is not a misfortune so much as an inevitability. For it could not be any other way. Even a god could never have certain knowledge, since he could be in error about his being infallible. Consider what philosophers call a "Cartesian Demon." Suppose some demon were actually solely responsible for sending you all your experiences, whatever they were, and this demon made sure you never knew their real cause. You could even be fooled into thinking you were an all-powerful, all-knowing Creator God, and you would never be the wiser. It follows that if there is a god, he could well be a victim of such a Cartesian Demon. He could never be certain he wasn't. It is therefore irrational to demand certainty for any of our knowledge. Not even a god could have that! What is rational is to assign degrees of conviction to degrees of certainty established by a tried-and-tested method. What is rational is *reasonable* certainty, not absolute certainty.

The methods of logic and mathematics are well-developed and provide the greatest certainty we have yet been able to find regarding anything, other than a present, uninterpreted experience. The next greatest certainty has been found in the application of scientific methods to empirical problems. In third place is our own daily experience, when interpreted with a logical or scientific mindset. Fourth is the application of critical-historical methods to claims about past events. Fifth is the application of the criteria of trust to the claims of experts. Sixth is the untested but logical application of inferential generalizations from incomplete facts—that is, plausible deductions. Such is the scale of methods that we have historically been able to discover and confirm as effective.

Experience shows that our degree of certainty will generally be weaker with regard to facts at each stage down this six-rung ladder, though within each category lies its own continuum of certainty and uncertainty, and the ladder itself is a continuum of precision and access to information: the more data we have to ground our conclusions, the farther up the ladder we find ourselves. Thus, mathematics is just perfected science; science, perfected experience; experience, perfected history; and history, perfected attention to experts; while plausible inference is what we are left with when we have none of those things.

Lacking any of the above approaches to the truth, we are faced with untrustworthy hearsay and pure speculation, where only the feeblest of certainty can ever be justified, if at all. In the absence of *all* certainty—certainty of *any* degree that a proposition is true or false—there is an absence of knowledge (of whether that proposition is true or false), and in such a state one can only assert, at best, that the uninterpreted experiences or claims exist, not that they refer to anything true.

This is a state of agnosticism, which entails an absence of belief in what is unknown. Unless we still assert belief on pure faith alone. However, though belief on faith alone may be comforting, it is wholly arbitrary and thus does nothing to ensure that you are more correct than anyone else.

So it cannot properly be described as knowledge, but rather as a mere wish, a desire that something be true or false, or else it is a naive trust in guesswork or hearsay. In contrast, the more *justifiably* certain you are that a proposition is true, the more likely it is that you genuinely know it is true. So we can say that real knowledge statistically exists among our beliefs in proportion to how accurately those beliefs were formed—to how well, thoroughly, and routinely our method was applied, producing conclusions approximate to what really is the case (see also III.6.5, "The Nature of Knowledge").

3.1 Finding the Good Method

It is reasonable to predict that an accurate method, a method that leads significantly more often than not to the discovery of genuinely true and false propositions, will exhibit two particular features, which an inaccurate method will not exhibit: predictive success and convergent accumulation of consistent results. We can even expect that a more accurate method will exhibit these features more often than a less accurate one. And this is how we can test out different methods and choose the best from among them, and throw away the ones we don't need.

First is predictive success. If we use an inaccurate method we should expect our desires and expectations to be routinely frustrated, as what our trusted propositions predict fails to transpire. This failure, in fact, is what it would *mean* for those propositions to be false (at least before we start making too many assumptions about the underlying reality), so this conclusion follows necessarily from the very meaning of truth itself. Therefore, if our method is *correct*, then we can expect to routinely produce propositions whose predicted experiences do in fact take place (since that is what it means for them to be true, prior to our adopting any metaphysics).

This is *especially* true for those experiences that would otherwise be a complete *surprise*. Why should we expect this? Because this sort of result would not likely occur if our trusted propositions were false, but could easily occur if they are true. Either way, a bad method will lead us to conclusions that fail to anticipate the future. In short, its results will fail every real test. A good method, because it succeeds in getting at the truth, must necessarily produce assertions that do successfully anticipate the future, to a degree and with a frequency not at all possible by chance. Thus we can identify a good method when we see one.

Of course, we can always explain such success as the machinations of a Cartesian Demon, but this eventually becomes quite implausible, for two reasons. On the one hand, there is no reason to believe there is such a demon. Even the view that *we* are that demon, constructing the world subconsciously, has no evidence to give it any credit. Indeed, we can distinguish accurate and inaccurate mental constructs, so even though our brains do generate a virtual reality (which will be explored in III.6.2, "The Mind as Virtual Reality"), there still remains an observable distinction between true and false constructions, and thus there remains no reason to believe it is *all* "just a construct."

On the other hand, if the demon were really this consistent in giving us results, through which we satisfy our every goal and desire, there would hardly be any intelligible difference between what we call "reality" and the world the demon is inventing for us. As noted in II.2.1.2 ("Meaning, Reality, and Illusion"), such a construct would *be* reality, in every sense of the word we normally use. And since we observe some methods to work better than others, and indeed some work best of all, a Cartesian Demon would have to be arranging it this way, constructing reality for us solely in accord with a fixed plan it has chosen. In that case we have just as much reason to pursue the relevant methods for discovering *that plan*, and to abandon the bad ones, so we can gain the reward of a successful life experience from this mischievous demon. In other words, there is no reason to trust that any Cartesian Demon theory is true, and even if it is, nothing significant changes for us regarding method.

The second criterion of good method is convergent accumulation of consistent results. If we use an inaccurate method we should expect that when we investigate a proposition from several angles we will get inconsistent results, and the more propositions we accumulate the more contradictions we would encounter and the more complex our belief-system would have to become to accommodate them. But if our method is correct, we should expect that when we investigate a proposition from several angles we will get the same results, which would be an improbable coincidence if there were no stable truth being hit upon. At the same time, the more propositions we accumulated, the more consistent our system of propositions would become, with researches in various areas all confirming and supporting each other and permitting cumulative advances in practical knowledge. In contrast to a bad method, with a good method we would find ourselves eliminating rather than accumulating contradictions, and our belief-system would become less convoluted.

All this would be a very good sign that our method is sound and successful and gaining us access to what is really true. For this result could never happen by chance. And even for a Cartesian Demon to pull it off, it would have to have from the very beginning a set, preconceived idea how everything will turn out, or else make everything, including our memories, conform as if that were so. For only in such a way could everything accumulate and coincide so well over a long period of time, or appear to have done so. But if the demon has such a complete game plan already in place, or is so adept at inventing one at any given moment, then there would again be no practical difference between this "truth" and a truth that was just "there." We could even keep talking of 'Metaphysical Naturalism' as the best description of the demon-made world and its contents.

On the other hand, a method that meets both criteria would also stand as evidence of a truth that was not of a Cartesian Demon's manufacture. Not only would there be no reason to believe a demon was at work (for merely being possible does not make something credible), and not only would there be no advantage gained by believing so (for insofar as our happiness is procured anyway by mastering the rules of the universe established by a fixed reality or by a Cartesian Demon, our propositions always remain true in an operational sense), but it would be altogether improbable. For it would be hard to imagine what the motive of such a demon would be, or why it exists, or how it acquired or employs its powers in the first place.

So what methods have turned out successful?

3.2 The Method of Reason

The reason why the logical-mathematical method is supremely successful is that it has, in respect to the two features of an accurate method, produced the broadest, most complete, and most consistent success. Moreover, when a proposition of logic or mathematics is challenged and seriously debated, the most widespread and solid agreement is achieved in comparison with any other method or subject. This is because the predictions entailed by such propositions are comparatively few, simple, and precisely defined, as well as thoroughly interrelated, and therefore these propositions are very easy to test. For example, to test that a proof is valid one need only validate by direct experience each step of the proof, including its axioms and the steps by which each step leads to the next. In effect, by being the least ambiguous or laborious of all the sciences, it has made the most progress the quickest.

On the one hand, we ourselves can in most cases duplicate the investigations and thus directly confirm logical-mathematical propositions. On the other hand, propositions of logic and mathematics only make claims about the meaning of concepts. So the only empirical inquiry they require is conceptual, and therefore inexpensive and immediate. For they can all be tested in the laboratory of the mind, where concepts exist.

See chapter II.2, as well as III.5.4, "Abstract Objects," III.6, "The Nature of Mind," and III.9, "The Nature of Reason." See also the bibliographies concluding section III.5.5, "Reductionism," as well as II.3.3 and II.3.4.

To learn the methods of logic and mathematics, any college-level textbook will do. A good one to start with, which touches on everything, is Ronald Staszkow and Robert Bradshaw, *The Mathematical Palette* (1991); another excellent introduction, relating math to reason, is Edward Burger & Michael Starbird, *The Heart of Mathematics: An Invitation to Effective Thinking* (2000). On the underlying philosophy of logic (what logic is and why): Susan Haack, *Philosophy of Logics* (1978).

On the nature and importance of meaning, and the connection between logic and language, see: A. J. Ayer, *Language, Truth, and Logic*, 2nd ed. (1946); Wesley Salmon, *Logic*, 3rd ed. (1984); Paul Horwich, *Truth*, 2nd ed. (1999); Patrick Suppes, *Introduction to Logic* (1999); Ernest Lepore, *Meaning and Argument: An Introduction to Logic through Language* (2000).

For applying logic and critical thought to real-life experiences and problems, see J. Anthony Blair and Ralph Henry Johnson, *Logical Self-Defense* (1994); Gregory Colomb & Joseph Williams, *The Craft of Argument* (2000); Merrilee Salmon, *Introduction to Logic and Critical Thinking*, 4th ed. (2001); Anne Thomson, *Critical Reasoning: A Practical Introduction* (2002); and Deborah Bennett, *Logic Made Easy: How to Know When Language Deceives You* (2004).

3.3 The Method of Science

The reason why the scientific method, which is in fact a whole complex of empirical methods, is penultimate in success is the same. It falls short of logical-mathematical certainty because the predictions entailed by scientific propositions are vast, complicated and often difficult to pin down precisely, they are less thoroughly interrelated, and require much more expensive and active investigations that must range far beyond the laboratory of the mind.

Therefore, these propositions are very hard to test, requiring long effort, special care, and extensive duplication. But when these standards are met, well and properly, our conclusions will be the most certain we can achieve about facts outside the human mind, correcting even our own errors in direct experience. We have proven again and again that the results of thorough scientific investigation are more reliable than the results of our own casual observation, producing far more extensive agreement and far more surprising successes, with the most impressive examples of convergent knowledge in history. It should already be obvious that without the refinement of effort and method, and the comparison of many different perspectives, our observations are always inferior to what they are *with* such refinement.

> On the methods of science, I say a lot more in chapter IV.1.1, "Science and the Supernatural." For a practical guide: Stephen Carey, *A Beginner's Guide to Scientific Method* (1997). But for the heavy detail, see Ernest Nagel and Morris Cohen, *An Introduction to Logic and Scientific Method* (1934), which is greatly expanded upon by more recent works like Richard Boyd, Philip Gasper, and J. D. Trout, *The Philosophy of Science* (1991), and Mario Bunge, *Philosophy of Science I: From Problem to Theory* and *Philosophy of Science II: From Explanation to Justification* (1998).
>
> See also the very important work: Susan Haack, *Defending Science—Within Reason: Between Scientism and Cynicism* (2003), which explains the place of science on my scale of methods. Also important for the underlying logic of science (and everyday reasoning): Brian Skyrms, *Choice and Chance: An Introduction to Inductive Logic*, 4th ed. (1999); Ronald Giere, *Understanding Scientific Reasoning*, 4th ed. (1996); and Paul Horwich, *Probability and Evidence* (1982).

3.4 The Method of Experience

The reason why our own daily experience, even when unanalyzed by a stricter logic or scientific procedure, gets third place, is that we know from experience that it is more reliable than anything else besides. Indeed, *everything* we know ultimately comes from our own life experience, and on most things we are all in complete agreement about what we encounter there, which would be unlikely for so many very different people unless we were largely right about a lot of those agreed facts.

But if someone comes up with a scientifically or logically well-proven claim that contradicts our direct experience, then we have good reason to believe our experience is in error, because a single unexamined experience cannot possibly be more trustworthy than a hundred well analyzed and tested ones. This is all the more the case if our faulty experience can itself be explained by the proven facts of science or logic.

In contrast, if someone comes up with a well-proven historical claim, or the credible assertion of an expert, or a valid but unproven inference, which contradicts our direct experience, then we have good reason to believe our experience over this claim. For there can be no doubt for us that a direct experience is genuine, even if we have not examined it scientifically or logically and were thus misled by it. But we know from experience that historical claims can easily turn up the result of lies or mistakes, while even experts get things wrong (deliberately or not), and anything that is unproven can claim no authority over what has an inherently greater proof by being directly in our perception. This is all the more the case if we can prove there are faults or defects in these contrary claims, or even a plausible suspicion thereof.

Still, it is always better than not to examine your own experiences with reason and scientific acumen, even when you cannot grant to them the same authority as a rigorous proof or scientifically established fact. For us, if we want greater certainty rather than less, the method of personal experience ought to be the simple practice of living a life of reason, applying scientific and logical principles whenever and wherever possible. This will ensure your life experience produces more reliable knowledge, and is more flexible (by being more open-minded and skeptical), and thus less challenged by the findings of science and logic.

> See last part of the bibliographies for II.3.2 and II.3.3 above. For a very basic introduction: Theodore Schick & Lewis Vaughn, *How to Think About Weird Things: Critical Thinking for a New Age*, 3rd ed. (2001).
>
> But on the perfection of the method of personal experience, one would do best to understand the sorts of errors we are prone to, so you can avoid or compensate for them. See Massimo Piattelli-Palmarini, *Inevitable Illusions: How Mistakes of Reason Rule Our Minds* (1994), and Thomas Gilovich, *How We Know What Isn't So: The Fallibility of Human Reason in Everyday Life* (1993), just for starters. Perhaps also: Thomas Gilovich, Dale Griffin, and Daniel Kahneman, eds., *Heuristics and Biases: The Psychology of Intuitive Judgment* (2002) and Dietrich Dörner, *The Logic of Failure: Why Things Go Wrong and What We Can Do to Make Them Right* (1996). Also relevant to this issue is section III.10.4, "The Nature of Spirituality."
>
> It is also handy to study research on eye-witness testimony, since it contains clues to how much or little you can trust your own perception and memory and how you can improve them. See: Daniel Schacter, *The Seven Sins of Memory: How the Mind Forgets and Remembers* (2002); Elizabeth Loftus & James Doyle, *Eyewitness Testimony: Civil and Criminal*, 3rd ed. (1997); Daniel Schacter & Joseph Coyle, eds., *Memory Distortion: How Minds, Brains, and Societies Reconstruct the Past* (1995); and Gary Wells & Elizabeth Loftus, eds., *Eyewitness Testimony: Psychological Perspectives* (1984). But the real secret is learning on your own how to adapt ideas borrowed from logic and science to everyday life (see previous section bibliographies).

3.5 The Method of History

The reason why the critical-historical method takes fourth place is that, lacking the ability to observe its object directly, its results are as indirect as its evidence, and by being less direct, is less certain. But given a report of something we did not observe and are unable to observe (since it happened in the past), if we apply a tried-and-tested method of critical, historical analysis we will be able to sort reports that are more believable from those that are less, and the most believable from those that are the least.

And we know our present historical methods are good at this, and have improved remarkably, because they now, in respect to the two features of an accurate method, produce the broadest, most complete, and most consistent success with regard to historical claims, and the widest and most uniform agreement than any other method. But even the most certain of historical propositions will not be more certain than well-established scientific facts, since the variety of ways historical propositions can be in error is much greater, and the means by which to confirm them far less secure.

> On the methods of history, which are my professional specialty, I say a good deal more in Chapter IV.1.2, "Miracles and the Historical Method." But for the real detail, see: Walter Prevenier and Martha Howell, *From Reliable Sources: An Introduction to Historical Methodology* (2001); J. Tosh, *The Pursuit of History: Aims, Methods and Directions in the Study of Modern History*, 3rd ed. (1999); Joyce Appleby, Lynn Hunt and Margaret Jacob, *Telling the Truth About History* (1995); C. Behan McCullagh, *The Truth of History* (1998) and *Justifying Historical Descriptions* (1984); Robert Shafer, *A Guide to Historical Method*, 3rd ed. (1980).
>
> Works of lesser merit include: John Gaddis, *The Landscape of History: How Historians Map the Past* (2002); David Cannadine, ed., *What Is History Now?* (2002); Richard Evans, *In Defense of History* (2000); Paul Veyne, *Writing History: Essay on Epistemology* (1984); John Lukacs, *Historical Consciousness: The Remembered Past* (1968); William Dray, ed., *Philosophical Analysis and History* (1966); Marc Bloch, *The Historian's Craft* (1964); Edward Carr, *What Is History?* (1961); Homer Hockett, *The Critical Method in Historical Research and Writing* (1955; cf. Part I); Louis Gottschalk, *Understanding History: A Primer of Historical Method* (1950); and Gilbert Garraghan, *A Guide to Historical Method* (1946).

3.6 The Method of Expert Testimony

The last two methods, expert testimony and plausible inference, are the least reliable because, on the one hand, they are the least involved in direct analysis of the relevant evidence and, on the other hand, they are essentially derivative of the more accurate methods discussed above. Expert testimony is fifth in rank, because experts have experience and knowledge from which to make far more accurate and trustworthy inferences in their field than non-experts do.

It is reasonable to trust the claims of experts, whom we have good evidential reasons to believe have applied one of the more accurate methods to a problem and reached a well-founded conclusion from them, if those experts meet certain tests of reliability. These tests include, but are not limited to, possessing genuine qualifications suited to investigating the proposition at issue, corroboration by many other experts, and proof that the expert's biases are regularly controlled by strict adherence to one or more of the other methods delineated above. The more such criteria an expert source meets in each particular case, the more trustworthy are her claims on relevant matters. Yet even the most trusted expert testimony is not as certain as the more widely confirmed and more evidentially-supported results of logic, math or science, or our own direct investigation of the relevant facts, or the direct results of historical research. For there are countless ways an expert can be in error, which can only be checked by other, more direct means.

It is also important to distinguish just what someone has expertise in. For example, a theologian may be an expert on theology, but that only means he has genuine expertise in the concepts of theology, not that he is an expert on factual questions like whether a god exists or whether Catholicism is the One True Religion. No one can really be an expert on these questions because no one has any real evidence for them, at least evidence properly produced by one or more of the superior methods above. A theologian can hardly claim any more experience with an actual god than we can.

Moreover, logic and science precede expertise, and insofar as any expert claims something is true contrary to logic or science, his expertise alone counts for nothing—in fact, he throws into suspicion anything else he might claim on his authority. The same holds when an expert attempts to assert something that contradicts our own personal experience or sound historical investigation. Worse, in fields like theology we find very little agreement among qualified experts, and a vast influence of ideological bias that is rarely placed under any objective control.

Thus, when we examine expert claims, we must account for whether they really are expert in the matter at hand, and whether their claims meet the criteria of trustworthy expert testimony.

> Relevant material on how to assess an expert's merit can be gleaned from legal guides to the issue: for example, Steven Lubet, *Expert Testimony: A Guide for Expert Witnesses and the Lawyers Who Examine Them* (1999) and James Richardson, *Modern Scientific Evidence, Civil and Criminal: Weight and Sufficiency, Admissibility, Objectives of Law and Science, Scientific Tests and Experiments, Specific Methods of Proof*, 2nd ed. (1981). See also: Benjamin Radford, *Media Mythmakers: How Journalists, Activists, and Advertisers Mislead Us* (2003).

3.7 The Method of Plausible Inference

Likewise, it is reasonable to trust the untested but logical outcomes of valid inferential generalizations from incomplete facts. At least, so long as their inductive force is compelling and we don't grant them greater certainty than the results of more accurate methods. What on earth does all that mean?

A "generalization" is a claim that infers from a few instances that something common to those cases will in fact be true of everything similar, using what is called "inductive logic" (as opposed to "deductive logic" whose results are conceptual and thus effectively certain, rather than empirical and thus relatively less certain). A generalization has a "compelling inductive force" when there is no trustworthy evidence that places it in doubt, and an overwhelming body of evidence from disparate fields or sources that implies it is true.

For instance, many of the beliefs of metaphysical naturalists on the nature of the universe are as yet untested by any method but plausible inference. But as the facts are compellingly in its favor, and its explanatory scope and power is enormous, achieving consistency and convergence with the results of all five other methods, while alternative worldviews come nowhere near it in these respects, it is reasonable to believe it with appreciable certainty. Still, this could all change, if the results of any of the superior methods turn up a contrary fact.

3.8 The Method of Pure Faith

The method of pure faith refers to basing beliefs solely on tradition, hearsay, desire or mere speculation. That is, faith in this sense is trusting what we are told, or just 'guess' or want to be true, without requiring any proof. In other words, believing an ungrounded assertion. Naturalists reject this method, for two important reasons.

First, we know as a matter of experience that ungrounded assertions like these are usually false. We know they are caused by processes that are generally *not* truth-finding. Tradition, hearsay or desire easily transmit beliefs irregardless of their correspondence with anything real. They convey false beliefs just as easily as true ones. In contrast, the other methods we have discussed *are* generally truth-finding. They are certainly more truth-selective than random chance.

Whenever we have put claims to the test, in fact, we have found that coming solely from tradition or desire, or any other source with no support, ideas are more often false than true. We can see this from the great number of traditional myths, legends, and beliefs that have been exploded or overturned throughout history. So we cannot trust these things by themselves. We need something more. And therefore we need something more than faith (see IV.1.1.3, "The Science of Faith").

Second, blind faith is inherently self-defeating. The number of false beliefs always vastly outnumbers the true. It follows that any arbitrary method of selection will be maximally successful at selecting *false* beliefs. So the probability is always very high that a belief based on mere faith will be false. History is rife with examples of the sad consequences of misplaced faith. So, again, we need something more. And that 'something' is what I have described above: a belief system based on applying proven truth-finding methods to basic, direct, undeniable experiences.

3.9 Final Remarks on Method

As already noted, the methods defended above are to be regarded as superior to pure hearsay and speculation, which from long experience we know we can't trust. We know the latter are rarely arrived at by anything having to do with the truth, but often by unknown and chance factors unrelated to truth.

In contrast, among the valid methods one thing is held in common: a thorough reliance on evidence and reason. Reason, because we must think carefully and not erroneously, and evidence, because by no other means do we have any access to the truth. Since our access to truth is therefore generally in direct proportion to the abundance and quality of evidence, we align our beliefs to this, and nothing else.

There is no sense in replacing reason and evidence with anything else, like tradition or faith. As we have seen, such a mistake will fill our minds with untrustworthy and largely false beliefs. For there is nothing in the methods of tradition or faith, or any other procedure, that suggests they are reliable, certainly not in any way that can compete, logically or in practice, with the six methods described above (see also II.2.9, "Naturally Warranted Belief" and III.10.4, "The Nature of Spirituality").

Finally, before going on I must briefly clarify two connotations of "proof" or "proving" in the context of method. Scientists, judges, and the average Joe mean by "proof" any body of evidence sufficient to justify belief, i.e. "proof" that something is very probably true. Not enough proof means not enough reason to believe. Only in logic and mathematics does "proof" mean a decisive demonstration that something is *certainly* true. Of course, even that is not so certain as it seems (we can always be mistaken), but it is a much different kind of thing than "proof" in empirical practice. Since the facts, the truth about what exists, can only be known probably, never certainly, more often than not I will be talking about empirical rather than logical proof throughout the rest of this book.

> For more on the inadequacy of faith against reason, and on the proper use of logic and science, see the "Faith and Reason" section of the Secular Web (www.infidels.org/library/modern/reason). Also interesting is a short but apt discussion by Herbert Feigl in "Naturalism and Humanism," *American Quarterly* 2 (1949), pp. 135-48; the relevant part is reproduced on pp. 59-67 of James Gould's *Classical Philosophical Questions*, 8th ed. (1994). See also the bibliography concluding II.2.2.8, "Naturally Warranted Belief."

III. What There Is

1. The Idea of a "Worldview"

Your "worldview" is your complete philosophy of life: what exists, who we are, why we are, how we should behave—everything. Everyone has a worldview, whether they are conscious of it or not. If you are unaware of this, if you cannot articulate your worldview, then you probably have a poorly-reasoned one, mostly borrowed from your surroundings, your peers, and random experiences, rather than from careful thought and observation.

We should abandon such carelessness. We should pay attention to the evidence, and construct a worldview that makes the most sense of it, using all we have learned. As we follow the best methods, consistently and correctly, a particular range of justified beliefs about ourselves and the universe will emerge. This is what scholars and scientists have done, more and more, for several centuries now. You will find that their collection of well-demonstrated knowledge altogether supports Metaphysical Naturalism, as the general worldview that most readily and easily makes the most sense of everything we can honestly say we know with any certainty. Many people dispute that. But they often do so out of ignorance or open denial of the findings of science and history, especially those of these past fifty years and more.

> For more on the idea of 'worldview', see David Naugle, *Worldview: The History of a Concept* (2002) and *A History and Theory of the Concept of 'Weltanschauung' (Worldview)* (1998); William Cobern, *World View Theory and Science Education Research* (1991); Leon McKenzie, *Adult Education and Worldview Construction* (1991). Using Christianity as an example: Michael Palmer and Stanley Horton, eds., *Elements of a Christian Worldview* (1998) and Ronald Nash, *Worldviews in Conflict: Choosing Christianity in a World of Ideas* (1992).

2. A General Outline of Metaphysical Naturalism

Imagine the naturalist's worldview as a ship. The first plank, what we might call the very keel of our beliefs about the universe, is the whole body of conclusions that we call science, which is itself built from mathematics and logic applied to careful observation. Upon this foundation we fit everything else. First, our own personal experiences, which form the hull. Then, the body of conclusions most agreed upon by those expertly employing critical-historical methods, which is like the deck. The trusted claims of experts form the superstructure. And the sea in which the whole edifice floats is what we conclude through the application of reason and evidence to everything altogether—our metaphysics.

So if you want to know what we believe on almost any subject, you need merely read authoritative works on science and history—which means, first, college-level textbooks of good quality and, second, all the other literature on which their contents are based. The vast bulk of what you find there we believe in. The evidence and reason for those beliefs is presented in such works and need not be repeated here. But as I discuss our metaphysics, I will also survey some of the most controversial aspects of this treasury of sound knowledge, especially as they relate to distinguishing Metaphysical Naturalism from other worldviews.

But it is that sea on which the ship floats that distinguishes us the most. For we do not fill it with anything not justified by the rest. And the only thing we see justified is that the natural world exists. Thus, all metaphysical naturalists believe that if anything exists in our universe, it is a part of nature, and has a natural cause or origin, and there is no need of any other explanation. This belief is not asserted or assumed as a first principle, but is arrived at from a careful and open-minded investigation of all evidence and reason, using the methods surveyed in the previous chapter.

As we see it, the progress of science and other critical methods has consistently found natural causes and origins for everything we have been able to investigate thoroughly—for so long, so widely, on so many subjects, both disparate and related. Indeed, it has never once failed in this regard whenever a problem or question could be properly investigated. So it is a thoroughly reasonable inference that this shall continue unabated. We have every reason to believe that the results of future investigations will most probably be the same for every subject once we have access to sufficient evidence to decide the matter.

So wherever we have a vast body of evidence, we find nothing else but a very strong basis for belief in naturalism, and since this is never observed to be the reverse, naturalism is the most sensible conclusion. Should any change in this pattern occur in the future, we may be justified in changing our worldview. But until then, this is the most reasonable view to take. Why? Because with a complete system of Metaphysical Naturalism it is possible to offer a plausible hypothetical answer to every question science can't get at yet, which means it is a very robust and useful worldview. It means we are on to something.

Now, by "nature" we mean a non-sentient universe, with all its properties and behaviors. Basically, we mean nothing more than space, time, material, and physical law. There may be other dimensions besides space and time, but these would still be nothing more than mindless extensions of the same physical being, much as time may be a mere extension of space, and all three dimensions of space mere extensions of one. There may also be materials other than matter and energy, and many metaphysical naturalists posit such things, though I do not, for there is neither evidence nor need for anything else (a view that is generally called "physicalism"). Nevertheless, if there are other materials, these would again be nothing more than mindless things populating the cosmos in just the same way as particles of matter and energy or the extension of being.

Likewise, there are certainly other physical "laws" besides those we know—which may even permit things beyond our imagining, things we would otherwise call miraculous, just as a tribal shaman would call a jumbo jet's flight—but these would be no different than the laws we already know: brute properties of the universe that describe how its dimensions and materials manifest and behave. And the cause or origin of all these things we believe to be natural in turn: a simple, non-sentient fact. There is no evidence of anything more.

This is as far as all metaphysical naturalists agree. What follows in the rest of this chapter, and beyond, is now my own worldview, and should not be regarded as defining Metaphysical Naturalism as such, but merely one complete kind of it. Nor is the following survey of "Big Questions" entirely even or thorough. Some topics will be briefly treated, others given more extensive attention. Space limits this book to a summary account, and the whole project will certainly benefit from being expanded in the future. But here is a needed start.

For further reading, see the "Naturalism" and "Materialism" sections of the Secular Web (in www.infidels.org/library/modern/nontheism). For a survey of the total scientific view of humanity and the universe that all naturalists share, see for example Greg Reinking, *Cosmic Legacy: Space, Time, and the Human Mind* (2003).

Books defending versions of Metaphysical Naturalism are legion: Andrew Melnyk, *A Physicalist Manifesto: Thoroughly Modern Materialism* (2003); Daniel Dennett, *Freedom Evolves* (2003); Taner Edis, *The Ghost in the Universe: God in the Light of Modern Science* (2002); Simon Altmann, *Is Nature Supernatural? A Philosophical Exploration of Science and Nature* (2002); Joseph Rouse, *How Scientific Practices Matter: Reclaiming Philosophical Naturalism*, (2002); John Shook, *Pragmatic Naturalism and Realism* (2002); Matt Young, *No Sense of Obligation: Science and Religion in an Impersonal Universe* (2001); Robert Nozick, *Invariances: the Structure of the Objective World* (2001); Lewis Edwin Hahn, *A Contextualistic Worldview: Essays* (2001); Kai Nielson, *Naturalism and Religion* (2001).

That's just from the 21st century. From the 20th century: J. T. Fraser, *Time, Conflict, and Human Values* (1999); Roy Bhaskar, *The Possibility of Naturalism: A Philosophical Critique of the Contemporary Human Science*s, 3rd ed. (New York: Routledge, 1999); Willem Drees, *Religion, Science and Naturalism* (1996); Sidney Hook, *The Metaphysics of Pragmatism* (1996); Kai Nielson, *Naturalism without Foundations: Prometheus Lectures* (1996); Richard C. Vitzthum, *Materialism: An Affirmative History and Definition* (1995); Paul K. Moser and J. D. Trout, *Contemporary Materialism: A Reader* (1995); Jeffrey Poland, *Physicalism: The Philosophical Foundation* (1994); Peter French, Theodore Uehling, and Howard Wettstein, eds., *Philosophical Naturalism* (1994); John Ryder, *American Philosophic Naturalism in the Twentieth Century* (1994); David Papineau, *Philosophical Naturalism* (1993); Jeffrey Walther, *Religious Naturalism, the First Secular Western Culture Religion in History: An Introduction to a Significantly Modern, New and Different Religion* (1991); Paul Kurtz, *Philosophical Essays in Pragmatic Naturalism* (1990); Paul Kurtz, *In Defense of Secular Humanism* (1983); Wilfrid Sellars, *Naturalism and Ontology* (1979); Sterling Lamprecht,

The Metaphysics of Naturalism (1967); Yervant Krikorian, *Naturalism and the Human Spirit* (1944).

Also relevant are Paul Draper's chapter "God, Science, and Naturalism," in Bill Wainwright, ed., *Oxford Handbook of Philosophy of Religion* (2004) and Alex Rosenberg's article "A Field Guide to Recent Species of Naturalism," *British Journal for the Philosophy of Science* 47:2 (1996), pp. 1-29.

Several books criticize naturalism. None respond or relate to the form of naturalism I am defending, but rather an incoherent caricature of naturalism generally. I have also composed this book in a way that exposes the ineffectiveness of their criticisms. See: Victor Reppert, *C. S. Lewis's Dangerous Idea: In Defense of the Argument from Reason* (2003) and Michael Rea, *World without Design: The Ontological Consequences of Naturalism* (2002), on which see my critical reviews at the Secular Web (www.infidels.org/library/modern/richard_carrier/rea.shtml and reppert. shtml); James Beilby, ed., *Naturalism Defeated? Essays on Plantinga's Evolutionary Argument Against Naturalism* (2002); Frederick Olafson, *Naturalism and the Human Condition: Against Scientism* (2001); William Lane Craig & J. P. Moreland, eds., *Naturalism: A Critical Analysis* (2000); Mark Steiner, *The Applicability of Mathematics as a Philosophical Problem* (1998), on which see my critical review on the Secular Web; Steven Wagner & Richard Warner, eds., *Naturalism: A Critical Appraisal* (1994); Ronald Nash, *Worldviews in Conflict: Choosing Christianity in a World of Ideas* (1992); William Shea, *The Naturalists and the Supernatural* (1984).

There are also many bigoted critiques of naturalism, too, both ignorant and inept, lacking in honesty or sophistication. For example: Phillip Johnson's diatribes in *Reason in the Balance: The Case Against Naturalism in Science, Law & Education* (1998) and *The Wedge of Truth: Splitting the Foundations of Naturalism* (2000); or David Noebel's *Understanding the Times: The Religious Worldviews of Our Day and the Search for Truth* (1994) and (with Tim LaHaye) *Mind Siege: The Battle for Truth in the New Millennium* (2000) or (with J. F. Baldwin & Kevin Bywater) *Clergy in the Classroom: The Religion of Secular Humanism*, 2nd ed. (2002). What they say has little to do with what I defend here.

3. The Nature and Origin of the Universe

Where did the universe come from? Why does it exist at all? Why is it the particular way it is? These are some of the Big Questions, which scientists are now attempting to answer. But no one really knows the answers yet. We have too little information, and too many possibilities. Any explanation that anyone can offer, including "God did it," would only be hypothetical at this point. But our worldview must suggest some answers, it must make predictions, or else it won't be sound. No one can claim scientific certainty yet. But our answers can still be plausible, a good metaphysical inference from all the facts at hand—maybe even the most plausible, the best inference yet.

3.1 Plausibility and the God Hypothesis

To be plausible, a hypothesis should make sense, it should not conflict with any evidence, it should be a logical inference from the evidence we do have, and it should contain as few unproven elements as possible. In other words, it shouldn't require us to "make up" too much out of whole cloth—what we call *"ad hoc"* assumptions. Furthermore, if several theories are plausible, any one of them could turn out to be true, but until we get more evidence our best bet would be the *most* plausible theory. So, after meeting the criteria of plausibility, the *most* plausible explanation will be the one that has the greatest explanatory *scope* and *power*. A hypothesis with "explanatory scope" explains many facts, not just one or two, and thus would explain a great deal about why *this* universe exists rather than some other, why the universe has the properties it does rather than others. And a hypothesis with "explanatory power" makes the facts it explains highly probable. In other words, given that explanation as a fact, we would very likely, if not almost certainly *expect* this universe to exist and to be the way it is.

Historically, most people have resorted to some God or Creator as the explanation for the universe. But as explanations go, this isn't a very good one. The problem arises in the details, where people start making up way too much to make the god hypothesis agreeable (see IV.2, "Atheism: Seven Reasons to be Godless"). Indeed, the theory either contradicts a lot of evidence or, to avoid that, entails a huge complex of *ad hoc* assumptions to explain those contradictions away. Worse, the idea that there was a god around before there was a universe—in other words the idea that something existed when there was no place for it to exist, that something acted when there was no time in which it could act—does not make much sense, either. The idea of a god thinking and acting before time existed, even though no "before" can exist without time, and no "thought" or "action" can exist without a time in which to occur, is pretty much unintelligible. Some theologians thus invent a second layer of uncreated time or space for God to work in, but this is yet another *ad hoc* assumption for which we have no evidence and that only creates a new problem: what caused *that* layer of space or time to exist?

And even if we can come up with an intelligible theory of creation, it still isn't the best logical inference to make. Can we infer from what we see as a completely natural universe that a sentient Creator is behind it? Not really. Given the lack of any clear evidence for a god, and the fact that everything we have seen happen, which was not caused by humans, has been caused by immutable natural elements and forces, we should sooner infer the opposite: that immutable natural elements and forces are behind it all. We will explore this point more later (see IV.2.3, "The Universe is a Moron"). For now, observe that the only things we have ever proven to exist are matter, energy, space, and time (on which see III.5, "What Everything Is Made of"). Since we can explain everything by appealing to only those things and their properties, then (all else being equal) such an explanation is the most plausible one around—leaving no need and no sound reason to go beyond them and invent all manner of unproven entities, like gods and spirits and miraculous powers.

Hence even if the 'god hypothesis' were plausible, it would not by any means be the most plausible. For one thing, it does not have much explanatory power. It does not follow from "there is a God" to "that God will create this universe, just as we see it," since any description of God that people would agree with would sooner entail a quite different universe (a problem we will again explore later). Second, this theory does not have much explanatory scope. Almost none of the features of the universe are explained by saying "God did it!" We will look at this problem in the next section, where we shall find other explanations with greater explanatory scope and power, and without appealing to any supernatural beings. But we can already ask perplexing questions of the God hypothesis. For example, if heaven is the best possible place to be and God already knows everything he needs to know and can do anything he wants to do, why does this universe need to exist at all, much less take the peculiar form that it does?

Another problem is what philosophers call "infinite regress." If everything must have an explanation, then you do not really get anywhere by explaining the universe by proposing a god. For then that god needs an explanation. Why does a god exist at all? Why that particular god and not some other? And where did this god come from? Even if he was 'always' around, why does he have the attributes he has instead of different ones? Why does he have those attributes at all? And even supposing you can answer all these questions with an explanation for God (which already adds a load of *ad hoc* assumptions onto the theory, just to make sense of it), you will then need an explanation for that explanation—and so on, forever. So either there is an eternal string of endless explanations, in which case there is no "ultimate" explanation because the explanations never end (and so the universe remains ultimately unexplained), or else there is something that has no explanation, something that just "is," what we would call a "brute fact." There are no other possibilities.

It follows that no matter how you look at it, there can be no "ultimate" explanation in the sense people want. The very concept is a logical impossibility. Even the idea of an infinity of explanations would itself be in need of some inexplicable "brute fact." Why should such an infinite series of explanations exist for something as relatively simple as a single universe? Why that series instead of some other? And so on. Thus, the question for us really is: Where do we stop? What is the one, ultimate "brute fact" that needs no explanation? Certainly, most people say this is God, that God is self-explanatory, having no origin, that God exists necessarily as the one brute fact. But that requires resting on a huge number of assumptions. Why not just stop with what we actually *know*— the natural world? Certainly this is just as viable. After all, if a god needs no explanation, then why does nature need one? (see sections IV.2.1 and IV.2.3, and the following).

> For more on the implausibility of the god hypothesis, or the equal or greater plausibility of alternatives, see the sections at the Secular Web on the

"Atheistic Cosmological Argument" (www.infidels.org/library/modern/nontheism/atheism/cosmological.html), the "Cosmological Argument" for God (www.infidels.org/library/modern/theism/cosmological.html), and "Physics and Religion" (www.infidels.org/library/modern/science/physics/). See also Victor Stenger's books on the subject: *Has Science Found God? The Latest Results in the Search for Purpose in the Universe* (2003), *Timeless Reality: Symmetry, Simplicity, and Multiple Universes* (2000), and *Not by Design: The Origin of the Universe* (1988).

3.2 God and the Big Bang

Scientists have amassed a wealth of evidence confirming that the universe *we can see* began about fourteen billion years ago as an incredibly hot, dense kernel of energy, which inflated under pressure, expanding and cooling to produce the known world. This event is called the Big Bang, from which scientists can explain almost everything that has happened since: how the immutable particles and forces of nature led all by themselves to a vast expanse of billions of complex galaxies and sheets of galaxy clusters stretching farther than the eye can see; how within each galaxy there arose billions of stars, which altogether sport billions of planets; and how, on at least one of those, life could arise by a natural accident, evolving into us. This is the story as scientists have so far been able to suss it. The facts support it fairly widely and well.

The important questions that remain are why this Big Bang happened at all, and why it produced this particular universe, rather than some other. And this is where "God did it" is now often proposed. But it is an awkward fit. What does God need a Big Bang for? That's a terribly slow, messy, complicated way to create a universe, much less people. Why the long, complex process of condensation from energy to matter to stars to galaxies? Why the vast expanse of the end result? You would think a god would simply create the whole universe at once, or much more quickly at least, and only make it as large as would suit us. There would be no need of long drawn-out processes, nor of other planets or galaxies, much less all the hundreds of subatomic particles we know of.

"God did it" doesn't predict any of these things, nor does it explain them very well. God has no need of quarks, for example, or neutrinos, or galaxies, or billions of years of slow, mechanical processes. Nor can we make any predictions about any of these things from the "God did it" hypothesis. Can we deduce from "God did it" how many types of quark there are? Or that there should even be quarks? Or how long it would take that god, from the initial moment of creation, to make a human being? Or that there would be such things as galaxies? Or such thongs as neutrinos?

Sure, you might invent a vast and clever array of detailed assumptions about a god or his plans that could predict or explain all this, or you can resort to something vacuous like "God's ways are a mystery," but either way you would only be making the god hypothesis *less* plausible than any naturalist theory that *already* predicts and explains all these strange things. And there are several theories that do that. Scientists are testing them even as we speak. The god hypothesis, by contrast, makes few if any testable predictions.

When we have a number of theories, and all of them account for all the evidence we have, then we cannot say which of those theories is correct. That means even the best god theory imaginable cannot yet be confirmed. We cannot claim to know for sure whether *any* god theory is true until the many leading alternatives are actually *refuted*, to the point when it is scientifically reasonable to reject them. And that has yet to happen. Nor is it reasonable to accept "God did it" as an explanation until someone formulates that theory in such a way that it *actually explains things*, like why there are quarks, and those particular varieties of quark, and why there are galaxies, and why God used all these things instead of something else, and why God took billions of years to make us—instead of, say, a few years or days. And so on.

Several naturalist theories already provide us with these explanations, and right now that makes them more plausible than the god theory. They meet all the other criteria of plausibility, too: they make sense, fit all the evidence, are logical inferences from what we know, and require the fewest *ad hoc* assumptions. They also have more explanatory power and scope, since they explain almost every strange feature of this universe, and predict those features to be highly probable—while the god hypothesis does neither.

> On the Big Bang theory, see: Joseph Silk, *The Big Bang*, 3rd ed. (2000); Alan Guth, *The Inflationary Universe: The Quest for a New Theory of Cosmic Origins* (1998); and Barry Parker, *The Vindication of the Big Bang: Breakthroughs and Barriers* (1993).

3.3 Modern Multiverse Theory

Currently the most credible explanations of the nature and origin of the universe belong to "multiverse theory," the idea that our universe is just one of many. The two most reasonable theories that scientists are exploring at present I shall call the Smolin Selection theory and the Chaotic Inflation theory. I think the former is the most interesting and has the most evidential support. But most scientists prefer the latter, even though it does not have as much supporting evidence, because it relies on fewer *ad hoc* assumptions. Still, both theories fit all the criteria of plausibility, and do so better than any god theory I know. Surely, until we can rule them out, we can't say whether any god is responsible.

Chaotic Inflation theory is a reasonable inference from contemporary scientific observations and understanding, and predicts everything we observe. It holds that those properties of the universe that can be different than they are, like the mass of quarks, "froze" into place when the universe cooled, and due to chaotic or quantum indeterminism, different parts of the universe randomly ended up with different features—some with no quarks, some with quarks of a different mass, and so on. Yet the universe inflated so quickly, that once these properties froze in place in each tiny spot, that area grew to a size thousands of times larger than we could ever see. Thus, the universe we observe appears everywhere the same—but if we could see far enough, we would see different parts of the universe with completely different properties. It follows from the same theory that many regions of this multi-faceted universe will collapse and start the whole process over again, causing more multi-faceted universes to emerge from the original one. And so on. There is nothing we know that could stop this process, so it must go on forever—and may already have. So if inflation did occur, and it was chaotic, then nearly every possible universe would exist, including ours.

The above theory is not mere speculation. Every element builds entirely on known science. Inflation itself, chaotic or random behavior at small scales, "freezing" at larger scales, collapsing regions jump-starting inflation again, inflated regions being much larger than any distance we could see, etc. All these things actually follow from known scientific facts and established theories, based on empirical observations. Most scientists are in agreement about this. And that makes for a pretty strong argument, although we have yet to find direct evidence for this kind of mechanism at the origin of our universe, and that is what scientists are now looking for.

This theory also predicts the range of possible things that can arise. In particular, according to Chaotic Inflation and present scientific knowledge, only dimensions and particles ever come to exist from this process, which would follow a pattern now called symmetry-breaking. Thus, the fact that only dimensions and particles have ever been proven to exist, explaining all phenomena (including the fundamental forces and all physical laws, as explained in III.5.3, "Physical Laws"), is strong support for Chaotic Inflation theory. Scientists also have a decent understanding of the ways in which symmetry could be broken, and from this they can predict all the different kinds of 'universes' that might arise—and ours happens to be one of those explicable and predictable outcomes. Since there is no known force that could stop inflation from continuing to happen again and again all over the cosmos, there would be something like an infinite number of 'universes', and from this it follows that the probability our universe would exist, and have exactly the features it does, is extremely high. You can think of it like a lottery: to win a lottery is incredible, but if you buy all the tickets there are, then winning is not surprising at all, but certain.

While Chaotic Inflation theory was primarily developed by Andrei Linde in the 1980's, and it has gained wide support today, another theory was developed by Lee Smolin in the 1990's, which most scientists agree has some merit, too. Unlike Chaotic Inflation, Smolin's 'selection theory' requires one completely *ad hoc* assumption: a new physical law. This is still fewer than the many *ad hoc* assumptions required for a god theory, involving numerous new laws governing his many amazing powers and properties, and countless maneuverings to explain, or explain away, all the strange evidence we actually have. So even if Smolin Selection is less plausible than Chaotic Inflation, it is still more plausible than a god. However, unlike Chaotic Inflation again (and very much unlike any god theory), Smolin's theory actually has some evidence that supports it over all contenders. And for that reason I think it is, at present, the strongest candidate, the best prediction yet.

Smolin's theory follows from two peculiar observations: a strong connection between black holes and the Big Bang, and the amazing suitability of this universe for black holes. First of all, the Big Bang looks just like a black hole. When we crash big atoms into each other in particle accelerators, they break into a whole crazy mess of particles, and it turns out that the ratio among those different particles, every time we do this, is exactly the same ratio of particles that apparently came out of the Big Bang. So the Big Bang looks exactly like what always happens when you squeeze a big enough atom until it explodes. It is strange that we can produce the same ratio of particles as came out of the Big Bang every single time we explode atoms in a laboratory—unless that's simply what happens. And that would mean the Big Bang is something that inevitably happens, whenever enough material is crushed under enough pressure. Yet this happens all the time in our universe, for that's exactly what black holes are—stars that have been crushed by gravity so much that even light can't escape them. Our universe is chock full of these things. There are trillions and trillions of them.

Now, that's pretty odd. It is almost as if the very purpose of the universe was to create black holes. And that's the second thing. Simply look at the facts: first, there are a lot more black holes than life-bearing planets—a *lot* more; second, a lot more material in this universe is devoted to creating black holes than to creating life—again, a *lot* more; third, this universe is almost entirely a vacuum—far, far more of that exists than any warm, breathable atmosphere—yet black holes thrive in a vacuum, while life is killed by it; fourth, even in this very rare, habitable pond called earth, life has a really difficult time surviving—we barely struggle along on this tiny little planet, in brutal competition for scarce resources, on a microscopic island that will be melted by the sun in a relatively short time, if it isn't wiped out by meteors or interstellar radiation before then; and that leads to fact number five: space is chock full of that deadly radiation and debris, which happen to be food for black holes, but death for us; and sixth, life can be wiped out easily, and has a very hard time even getting started, yet black holes are *inevitable* products of this universe, and then it is almost impossible to get rid of them—they will stick around a heck of a lot longer than any life ever will (unless we do something about that before it's too late, as I discuss in VII.6, "The Secular Humanist's Heaven").

Clearly, we are not made for this universe. But black holes are right at home. "God did it" does not explain this. It doesn't even make sense of it. God could have made the simple geocentric universe of ancient imagination, with only four or five fundamental particles, and a cosmos no bigger than the solar system, filled with breathable air. Or any of an infinite array of universes much more hospitable to life, with far less needless baggage. Even if God had some strange reason to make a universe that is almost entirely deadly to life, makes the arrival of life very rare and difficult, and its survival even harder, in a universe far larger than it needs to be, providing only a narrow window of time in which life has a chance, and so on—even then there is no particular reason why a god would also make that universe tailor-made for black holes.

The bottom line? Most of this universe—by far—serves the function of producing and sustaining countless numbers of black holes. We are, by comparison, like a lone flea stuck in a fleeting bubble of air at the bottom of the ocean, lucky even to be alive, soon to expire (unless we take matters into our own hands), gazing in awe at the millions of Big Fish who swim the vast and deadly sea with natural ease. Which would make more sense? That Neptune created that sea for the flea? Or for the Big Fish? After all, a flea may have adapted himself to the dog, yet the fact remains: the master only intends to feed the dog.

This was the peculiar fact that led to Smolin's idea. A good explanation of why the universe exists and takes the form it does must predict a universe perfectly suited to black holes, and only barely suited for life. So

Smolin put those two observations together: if a Big Bang looks exactly like a really big mass crushed to an extreme point, and black holes *are* really big masses crushed to an extreme point, then it may well be that inside every black hole is a new Big Bang. Every time a star collapses, a new universe explodes, in another direction, outside our universe. Like Chaotic Inflation, this already follows from known physics. In fact, Chaotic Inflation incorporates exactly the same prediction: that black holes (either all of them or certain kinds) spawn new inflation events—new Big Bangs. This is a logical inference from what we already know, and there is no known law of nature that would prevent it.

It is at this point, however, that Smolin introduces a new physical law, in order to explain why our universe is so well-made for black holes. Chaotic Inflation predicts this, too, but only as an accident. Indeed, there could be regions of the universe that are even better suited to life than to black holes. And if there are, that puts us essentially in the cosmic ghetto, so to speak. That would explain a lot—and there is, so far, no evidence against it. But Smolin's theory actually predicts that a universe *exactly* like ours is *inherently* probable, by drawing again on known scientific phenomena. And that gives it some special appeal.

Smolin's new 'law' holds that when a star collapses into a black hole and triggers a new Big Bang, one hidden out of sight from us, then, as a result of the same quantum indeterminism scientists have experimentally confirmed (and from which Chaotic Inflation was deduced), the fundamental properties of that new universe will vary slightly from those of its mother universe. No new law is really needed to propose that they will vary—that most if not all broken symmetries are eventually erased within a black hole and that new symmetries might randomly break in different ways in the resulting expansion, can already be inferred from known science, and is already an element of Chaotic Inflation theory. Where Smolin's theory differs is this: he proposes that some of the most fundamental properties are 'remembered' and retained by the new universe. The random variations they undergo may be small deviations around the original value, so the result is not totally random.

This is an *ad hoc* feature. But it has two pieces of evidence in support of it: it predicts exactly what we observe (a universe tailor-made for black holes), and does so by appealing to the only natural process that we know *for a fact* can produce such specific complexity—evolution by natural selection. For Smolin's one single assumption produces all three ingredients (as explained in III.8.2, "Evolution by Natural Selection"): reproduction— as every universe producing black holes spawns new universes, and many of those spawn countless more, and so on; mutation—as each universe is randomly just a little different than the next; and selection—as only those universes that are rich producers of black holes will multiply.

From this single assumption it follows that universes exactly like ours are *inevitable*. Just as for the evolution of life from very simple beginnings, the natural, unguided tendency of this process, even if beginning with a very simple, basic, tiny universe, is to end up with very large, very old, very complex, star-rich universes, in countless abundance. In fact, almost all such worlds are certain to arise eventually—since there is no known force that could stop this process, it would go on forever, producing an infinite array of universes. And it is inevitable that some of those universes would also produce life, since all the properties that are necessary for that (large size, great age, and complex chemistry), also belong to the set of properties conducive to black hole formation. And given infinite tries, the laws of probability entail that a universe exactly like ours will come about eventually.

And this is where Smolin's theory has its greatest merit. We have never found a single example of a proven case of a god causing anything, much less any fine-tuning of the properties of our universe, or anything else. But we *have* found overwhelming evidence for a process that produces very amazing fine-tuning without any intelligence behind it: evolution by natural selection. This is a known precedent—unlike bodiless minds or divine causation, for example. And a theory based on a known precedent is always less *ad hoc* than a theory based on completely novel and unobserved mechanisms. We also *know* black holes produce conditions resembling a Big Bang. And we *know* chaotic and quantum effects can produce random variation. And we *know* that symmetry can theoretically break in various ways, explaining the formation of all the dimensions and fundamental particles that explain all physical laws (see III.5.3, "Physical Laws"). And we *know* a new envelope of space-time can grow within a black hole where we can't see it. And so on.

So, like Chaotic Inflation, the Smolin Selection theory is a logical inference from known science, there is no evidence against it, and it predicts exactly the kind of universe we have, including its most unusual properties—not only its amazing suitability for black holes, but all the specific details that make for that suitability, such as many of the physical constants (which, if they varied even slightly, would produce fewer black holes, or none at all). And *both* theories are more than credible. Both explain why everything seems made of nothing but dimensions and particles. Both theories explain why there are quarks and neutrinos, and why they have the properties they do. Both theories explain why the universe is so big and so old, and why most of it is so inhospitable to life. Both theories also explain why our universe is so complex, and why it is organized in such a way as to make life possible in the first place. And both theories predict not only that the universe would be brought about by a Big Bang (instead of, say, instantaneous creation or intelligent assembly), but they predict why that *particular* Big Bang, with its very peculiar result. So they both do a much better job of explaining things than "God did it" ever has...

> For Smolin's Selection theory, see: Lee Smolin, *The Life of the Cosmos* (1997), with Lee Smolin, "Did the Universe Evolve?" *Classical and Quantum Gravity* 9 (1992), pp. 173-192; and Damien Easson & Robert Brandenberger, "Universe Generation from Black Hole Interiors," *Journal of High Energy Physics* 6.24 (2001). For a discussion of all the plausible 'multiverse' concepts currently accepted as viable scientific theories, see: Paul Davies, "Multiverse Cosmological Models," *Modern Physics Letters A*, 19.10 (2004), pp. 727-743; and Martin Rees, *Before the Beginning: Our Universe and Others* (1998). For Chaotic Inflation theory, see the highly technical work by Andrei Linde, *Particle Physics and Inflationary Cosmology* (1990).

3.4 The Multiverse as Ultimate Being

In the realm of cosmology, the debate between theism and atheism is really only a quibble over details. Both sides agree there must be some ultimate entity, which is the eternal first cause and ground of all being, the end point of all explanations. They only disagree over what properties this "ultimate being" has. Theists think it has a whole plethora of amazing powers and attributes, including the most complex mind imaginable. But as atheists point out, there is no evidence for any of those tacked-on assumptions (see IV.2, "Atheism: Seven Reasons to be Godless"). There are only *two* properties we can be sure the ultimate being has: its nature is to exist, and it had a reasonable chance of producing our universe exactly as we see it. We can't say anything more than that without sufficient evidence. And there is no actual evidence for any of the traditional divine attributes.

We've already examined how the belief that there is an infinite array of different 'universes' is more plausible than a god, and we've seen there is at least as much evidence for that as there is for a god, if not substantially more. For this reason (as well as others to be examined in subsequent chapters), metaphysical naturalists conclude that the "ultimate being" is probably this ensemble of universes. We have no evidence the ultimate being consists of anything more than that, and this being is sufficient to explain everything else we observe, better than any alternative so far.

Such an ensemble of universes is formally called a "multiverse." That can be misleading, since in neither of the most plausible theories is there really any more than one 'universe' *per se*. In Chaotic Inflation, different 'universes' are really just different regions of the same universe, even when in some cases these other regions are hidden within black holes. This is also true in Smolin Selection, the only difference being that in that theory *every* distinct region of the universe is hidden within a black hole. Nevertheless, it has become the convention to employ "universe" to refer to a single distinct region of what is in turn called the "multiverse." And that is how I will use these terms from now on.

The multiverse explains everything that exists, and so even from the start it is just as good as "God did it." It is even better than that, since the multiverse fits and follows from known scientific facts, and it makes the exact features of this universe highly probable—whereas there is no reason to believe this is the universe a god would probably make, nor is there any evidence that a god actually did any of the making. Of course one could ask why the multiverse exists at all, and why it has the exact properties it does. But as we've already shown (in section III.3.1 above), *something* must exist without any explanation at all, so it may as well be the multiverse. For if a god can exist unexplained, with all *his* convenient attributes, then so can the multiverse. Both solutions leave the same questions unanswered. But we find the god hypothesis leaves far too many *more* questions unanswered. So we take the multiverse instead, as our ultimate "brute fact."

In fact, the multiverse is a simpler explanation than god, because it has all those attributes of god sufficient to ground its own being and cause this universe to exist, minus all that stuff about intelligence, knowledge, desires, or omnipotent powers. So it does the same work with less baggage. For example, the multiverse is eternal, in the sense that it exists at every point of time that exists, has existed, or ever will exist. And for that reason it did not come "from" anywhere. There was never a time when it did not exist, so it did not come from "nothing," because there has never been "nothing." There has always been "something," from which every universe is born. And yet, having no knowledge or intentions or supernatural powers, the multiverse is a much simpler entity than a god—requiring fewer unproven assumptions.

It is sometimes said that a multiverse is not at all simple because it proposes a large number of entities to explain only one. But the multiverse is a single entity, not many. The existence of countless 'universes' or 'regions' within the multiverse is actually *entailed* by a very small and simple set of assumptions, a far simpler set than that required to make sense of a god. Of course, all of those assumptions are more or less supported by at least some evidence, whereas none of the peculiar assumptions about a god are. But even besides that, once you accept the basic elements of either Chaotic Inflation or Smolin Selection, the existence of countless universes follows necessarily. It does not have to be *assumed*. So these theories really involve only one or two leaps of speculation, since the rest is grounded in established scientific facts. In contrast, god's attributes comprise a rather lengthy laundry list of speculations, unsupported by any science. And that is why, as explanations go, a multiverse is simpler than a god.

Therefore, I believe it is most probable that a mindless multiverse exists, has always existed, and exists by nature. I don't claim this as anything more than a good hypothesis. But I believe it is more probably true than any other explanation so far, because it is the simplest and most plausible answer, explaining the most things by appealing to the fewest unknowns. And it fits. The theory that our universe had a mindless physical cause perfectly predicts the universe we observe: a dispassionate, mechanical, mindless, physical cosmos. It makes complete sense of why we are made of frail matter, why life developed through a long and messy process of evolution, why the universe is so big and old, why we can never find any good evidence of supernatural beings or events, and so on. Since this is a plausible, comprehensible explanation for the universe, until we discover some evidence that challenges it, there is no need to resort to any alternative. And until some facts are discovered that better support some other hypothesis, there is no reason to look for any other.

3.5 Answering the Big Questions

So now, how do we answer those Big Questions? Did the multiverse have a beginning? If not, how could it have gotten started? Or if it did have a beginning, then what existed before that? And what did the multiverse itself start out 'as'? Where did *that* come from? And why did it have the potential to produce a multitude of universes in the first place? Naturalists can answer these questions just as easily as anyone else can.

3.5.1 The First Cause

We don't yet know if the multiverse has existed for an infinite length of time, or if it had a beginning. Either is possible, in both Chaotic Inflation and Smolin Selection. Scientists are largely in agreement that the Big Bang theory no longer entails a beginning of time, as had been thought, since that conclusion only follows from a singularity, and few scientists accept the existence of singularities any more. It is not yet known for sure what happens when matter is crushed so densely, but it is known that the principles of Quantum Mechanics make a singularity physically impossible. This means we can no longer say whether time ever had a beginning.

Still, both of the multiverse theories described above make a lot of sense of the idea that there was a beginning, a first moment of random chaos, which spawned a tiny, simple universe—the smallest, simplest universe that could be, though still chaotic and unpredictable, which for that reason spawned more and more universes over time as it spontaneously and randomly grew. By chance alone some of those universes would inevitably be simpler or more chaotic, while others would inevitably be more complex and orderly. And the latter would inevitably lead to us.

On the other hand, it may be that if we keep going back in time we will keep finding universe after universe, and it may well be it is universes all the way down. Ours then sits in the middle of an infinite series of universes. Some object to this possibility, arguing that if time is infinite, it could never have gotten started. But that assumes an 'absolute' view of time that is scientifically mistaken. In actual fact, on current scientific understanding, just as all space exists at once, so does all time (as I'll explain in the next section). Our universe is simply in the middle of a fixed, endless structure.

For the same reason a multiverse that had a beginning would not have come "from" anywhere—there would exist nothing "before" the first ever moment of time, and that first moment of time, like every moment of time, would simply be an eternal fixed reality. It needs no cause. It is its own cause. And what spews forth from it would be an inevitable outcome of its own nature, the inherent nature of that first instance of space and time, that first element of being. Like a god, it is self-caused and self-sustaining. Like a god, what it brings about, it does so inevitably, of its own nature. But unlike a god, it is a simple thing, not a complex person.

We might have a hard time getting our mind around this. Due to our daily experience *within* time and within an acting universe, our assumptions often mislead us into false ideas about what must be the case with regard to the whole. This is why Relativity Theory and Quantum Mechanics, as two prominent examples, still to this day seem counter-intuitive, because they describe processes that occur at speeds or under gravitational fields we have never experienced the like of, or involving events and behaviors on a scale so large or small we can barely even imagine them.

So, although it seems that everything must have a cause, therefore the multiverse must have a cause, there is no real basis for such a generalization. The only reason to believe that anything has a cause is that we observe it to be so. But what we are observing is what is *inside* a universe, *inside* time. There is no reason to believe the same expectations should hold outside the universe, outside time. In fact, we have a good reason to believe they should *not* hold: causation is by definition a temporal concept. For something to be a cause of something else, it must precede it in time. But if there is no prior point in time, there obviously cannot be a cause—such a thing is logically *impossible*. It follows that the first cause can only be something like the multiverse itself, either eternally or at the first moment of time (if there ever was one), but either way, a multiverse would be fundamentally causeless in exactly the same way a god is supposed to be.

Some theologians instead propose what is called an "ontological" or "atemporal" cause—for example, a ball that rests on a pillow is a simultaneous cause of a depression in that pillow. Here we have a different concept of 'cause', one that is not connected to temporal sequence. But if we go there, we then cease to have *any* support for the original generalization. For the pillow does not need the ball to have the same depression. Consequently, we *cannot* say that everything must have an atemporal cause. There is no evidence to support such a premise. So as far as we know, the ontological cause of the multiverse is the nature of the multiverse itself.

Hence objections to a natural explanation for the universe typically fall flat, because they are based on false analogies from what we have experienced, to things wholly unlike them and with which we have very little experience. When we cast aside our prejudices, it remains perfectly sensible, and indeed most plausible, that the multiverse just *is*, and always has been. Everything else follows inevitably from that. There can be no objection to this, for the exact same objections would eliminate god as an explanation, too. Think about it. Just as one might ask, for example, "Why does the multiverse exist?" one can also ask "Why does God exist?" Ultimately, proposing a god gets you no further than proposing a multiverse.

3.5.2 The Origin of Order

Of course, someone might instead ask how the multiverse just happened to be orderly enough to produce complex universes, and ultimately life. But the same question could be asked of God, too. How did he just happen to be so well ordered and conveniently gifted as to produce our rich, complicated cosmos, complete with intelligent life and every good thing? We would be no less "lucky" to have such a god than to have a multiverse. Indeed, we would have to be far luckier, because a god is far more convenient, and—assuming he's a nice guy and everything—far better for us. So if luck is the only answer, a multiverse is more likely than a god.

And only a multiverse can explain where our luck came from. For it explains *all* fixed order. In contrast, theism leaves us with a fixed, orderly deity that has no explanation. Who rolled the die that gave us our god, rather than some other god, or no god at all? Basically, theism posits an extremely orderly being that just "exists" for no reason at all. That's not just passing the buck, it's worse—because it answers one mystery with another mystery ten times more perplexing. In contrast, both of the multiverse theories I presented above begin with a pure chaos, lacking inherent order. Because it is chaotic and unpredictable, growing and multiplying is one of the things it must inevitably do—for *not* to do so would entail a fixed rule, and there can be no fixed rules in a chaos (beyond the rules of necessity, such as define logic and geometry). Likewise, of the regions or universes this primordial chaos would randomly give birth to, some will remain utterly chaotic, while some will randomly acquire fixed order—again, for this not to happen would entail a prior fixed rule, which would itself be inexplicable.

An analogy will carry the point home. Everyone knows that rolling a die over and over again will only produce a purely random string of numbers. Yet orderly sequences are always among such random possibilities. For example, if you roll a die enough times, the odds become very good that you will roll the exact orderly sequence of 1, 2, 3, 4, 5, 6. The odds against such a sequence are something like one in fifty thousand. But if you get to roll the die a hundred thousand times, the chance of getting exactly that sequence at least once, simply by accident, is better than 5 out of 6—in other words, almost guaranteed.

It follows that every random chaos will *always*, as a matter of logical necessity, contain many pools of order. And the larger the chaos, the more times it expands or reproduces, then the more pools of order it will inevitably generate simply by chance—and the more complex they will be, too. Thus, order is the *inevitable* outcome of random chance. Pure chaos can thus lead *inevitably*, and quite easily, to that minimal order necessary to get Chaotic Inflation or Smolin Selection started, which in turn lead inevitably to more and more order, and eventually to us. In other words, from chaos we can predict order, even incredibly complex order. But we have no comparable explanation for where an orderly god would come from, or why such an innate order would exist at all in a god, rather than a different order, or a chaos instead.

Just as our intuition had lead us to mistaken beliefs about the cause of time, another faulty intuition leads us to think that order can only arise from intelligent design. This, too, is caused by our limited experience. Humans evolved in a regular environment of dealing with minds scheming against us—in trade we must thwart cheaters, in war we must thwart plotters, in the hunt we must thwart our prey. In every facet of our social lives, it helped us to try and find the intelligent cause behind things. And this led to early animism, the most widespread belief among primitive peoples that even inanimate objects are guided by intelligent spirits. So it is natural for us, who are not much different from our primitive ancestors, to assume that something that is highly ordered *must* have been arranged that way by some intelligent force. But we now know that is a mistaken assumption. Unintelligent forces *can* produce very complex, highly specified order.

As a model example, consider the solar system. In ancient Greece and Rome, the motions of the planets were found to be incredibly intricate and orderly, uniquely specific and astonishingly elegant. This was a mystery so great that even the best minds of the time could only struggle to describe how the system worked. No one had a clue as to why. The obvious and most popular answer was "God did it." After all, how else could such an inexplicable, complex order arise, with such an apparent selection of results toward a harmonious end?

Thousands of years later, of course, a fellow by the name of Isaac Newton figured it out: gravity. The whole complex order of the solar system was entirely explicable by appealing to a single, monotonous, mindless force. No appeal to God was really needed. For all you had to do was throw planets and stars together, complete with their gravity, and "Presto!" a solar system pretty much like ours would result. It would be *naturally caused* to arrange itself in the same sort of steady, complex order we observe for our own solar system, without any intelligent design, motivation, or involvement. All you need are masses, motion, and gravity, which are all blind and dumb.

This proves that order can result naturally, without any mind behind it. So theologians retreated, and placed their bets on gravity: for such a convenient, orderly rule must surely have been engineered, even though the solar system wasn't. Someone had to decree the laws of physics, right? Wrong. For along came Albert Einstein, who discovered that gravity wasn't a "law" decreed upon the universe after all, but, like the solar system naturally condensing from the interaction of masses and gravity, gravity was the natural outcome of the bending of space and time. All you had to do was change the shape of space-time and what we call "gravity" would be the inevitable result. And it so happens that masses change the shape of space-time.

So it turns out that all you need in order to explain the astonishing order of the planets in motion is *masses*. For they cause gravity to exist, which in turn causes those masses to entangle themselves in orderly ways, all without anyone making it so. As a result, theologians retreated yet again, insisting this time that someone must have given masses their power to bend space-time. But by now it should be clear that their reasoning isn't very credible anymore. We already know for a fact that order does not require an engineer, since immutable forces *can* do it on their own. We've seen it happen. Not only in the solar system, but in crystallization, magnetization, and of course the blind mechanical chemistry of life—especially in the proven capacity of natural selection to produce incredibly ordered systems. Wherever we have been able to look, beyond what we have made with our own hands, the explanation for every form of order we have ever found in nature has turned out to be just such mindless, natural forces—and nothing else. The most logical conclusion is that there *is* nothing else.

At the very least, there is nothing incredible about proposing that *all* order has such an explanation. After all, theologians have been wrong every time so far. Why keep betting on them? It's more probable that if the solar system can assemble itself out of a chaos without any intelligent guidance, then so can a multiverse assemble itself out of a chaos without any intelligent guidance. If the one is possible, then the other is, too. And since that's all we've ever found, probably that is all there is.

3.6 Time and the Multiverse

In previous sections I described time as eternal and fixed. That is how most scientists today understand it. From an imaginary point of view outside of everything (the way things would look to a transcendent God, for example), the entire multiverse looks like a giant interconnected forest, within which our universe appears like a flower blooming out of a stem extending from a much larger and more complex tree, bearing countless other flowers. But all the past and future exists at once, static and unmoving. Everything that was, is, and *will be* is already there, imprinted in this single shape.

This object "begins" at a tiny point and, based on current astronomical observations, most likely ends in an endless curving expanse—but both beginning and end exist together.

Though the multiverse has this shape, there is nothing outside of it: no space or time, no place for anything to be. Thus, to put a God out there requires inventing some fifth dimension or something where he can reside. There is also no point "before" the first point in time, if there even is a first point of time, exactly as there is no position in space before then, either. Even the idea that the stem of this flower is the "beginning" of our universe is a rather subjective notion of our own: from a purely external point of view, there is no more reason to call that the beginning than the end, or the side, or the top. For the whole thing just "exists," changeless and eternal.

This is certainly a very difficult concept to grasp. It is technically called the "B-Theory" of time. William Lane Craig opposes the idea, but describes it well:

> On a B-theory of time the universe never truly comes into being at all. The whole four-dimensional space-time manifold just exists tenselessly, and the universe has a beginning only in the sense that a meter-stick has a beginning prior to the first centimeter. Although the space-time manifold is intrinsically temporal in that one of its four dimensions is time, nonetheless it is extrinsically timeless, in that it does not exist in an embedding hyper-time but exists tenselessly, neither coming into nor going out of being. The four-dimensional space-time manifold is in this latter sense eternal.
>
> — *Naturalism: A Critical Analysis* (2000), pp. 232-3

That is a mouthful, but the following diagram helps explain this point of view:

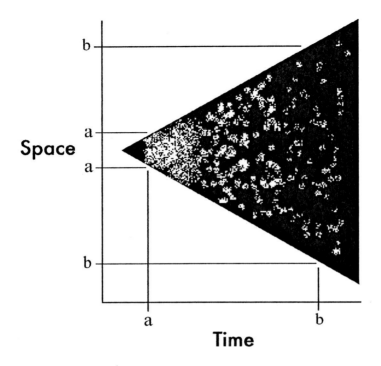

Here you are, looking at a simplistic representation of our universe, from a point of view outside of it. Though it is actually four-dimensional, we can barely imagine that. And the page is only two-dimensional anyway. So I have simply left out two of the three dimensions of space. But if you imagine the drawing as the profile or cross-section of a cone, you will be adding all but one of the four open dimensions.

To the left on the diagram is the Big Bang, which might mark the beginning of all time and space, or it might begin where a black hole in another universe ends. Either way, it is followed by an expanding cosmos of space and time until, at section *a*, photons are finally released from their entrapment in a super-dense plasma (an event that is supported by various lines of evidence), and the long road to galaxy formation begins, proceeding over time until our era, which exists at section *b*. The diagram then continues to the right, possibly forever.

In Chaotic Inflation theory, different regions of this expanding cone randomly but permanently acquire different properties long before section *a*, and then those tiny regions expand out to enormous sizes at section *b* and beyond. Some portions of the cone will collapse and explode into new cones, branching out from this one, forming a complex tree of universes.

Our universe might well exist within such a tree, too, branching off from an even earlier universe. In Smolin Selection theory, the above cone projects outward from a black hole formed in a previous universe, which looks like another cone, which also exploded out of another black hole in yet an earlier universe, and so on, while new universes branch out from ours, too—wherever black holes form.

However things turn out to be, we can modify the above picture as the evidence requires—whether the expansion is more curvy than straight, or oscillates like an hourglass, or whether the diagram continues infinitely in any other direction, or whether there are more dimensions than the four we know, or other universes or regions of space-time connected in various ways with this one, and so on. But however we modify it, the basic picture is the same: we exist *in* time, hence for us everything to the right doesn't "yet" exist, but that kind of perspective does not make any sense *outside* time. Since time is simply another dimension, it makes no more sense to say that future time doesn't exist because "we haven't gotten there yet" than to say that the rest of the road we are walking down doesn't exist because "we haven't gotten there yet."

As hard as this idea is for us to grasp, it is a perspective well understood by physicists. It is fundamental to the current theoretical and mathematical understanding of Relativity, as has been well argued for a lay audience by Paul Davies in *About Time*, and used by Victor Stenger in *Timeless Reality* to make sense of the strangeness of Quantum Mechanics. Many other principles and equations in physics imply a similar conclusion, indicating a convergence on a common result from independent investigations, one of our fundamental tests for truth. This view is so successful as an explanation, even supported by one of the most fundamental theories grounding all present science, that it is accepted by most physicists today, and it is fairly certain to be correct.

The fact that we observe a universe moving *through* time, ever changing, from the furthest point to the left onward to the right on the diagram above, is the product of the physical nature of both our minds and the universe: it is in one sense an illusion, like the illusion of solidity, when in fact solid objects are mostly empty space; but in another sense it is an interpretation of a pattern that really does exist—a pattern that does not really move or change, but *is* genuinely experienced, just as solidity is genuinely experienced and, in its physical effects, is a real fact of the universe. Existence is like a comic strip, which has a beginning and a middle and an end. It really does manifest causal change from each step to the next, as we read the cells of the strip from left to right. Yet, in a sense, this is also an illusion: for though we read it one cell at a time in that order, the whole comic strip already exists—how it ends is a foregone conclusion.

So why do we see time flow? Because perception is itself a pattern bound up with time. As we look along the time axis, we observe that our brains begin with no knowledge of a past, and then the lines of causation for an endless stream of data start connecting with it, building the brain up with a knowledge of a past, and the process of building this knowledge causes the experience of a 'present' at each point on the time axis. In effect, it is not possible for our brains to come into contact with data in the "future" without drawing a line of causation *through* the present, in other words by 'moving' through time as it appears to us. Hence our brains 'accumulate' data about the past as they pass through points of time.

But if this is so, if past and future are just two parts of the same whole, William Craig has asked, then why do we experience the future with anticipation or dread, and the past with nostalgia or regret? The answer is simple: because we have no knowledge of the future, and so react to it differently than we do to our past, which we do know. Our 'anticipation' and 'dread' are emotional responses to ignorance, a reaction to our time-based perspective: we can't see over the next hill on the road. On the other hand, 'nostalgia' and 'regret' are emotional responses to knowledge, a reaction to what we *can* see on the time axis, which is our past.

But why do we know our past and not our future? From a point of view outside of time, there might be no apparent difference whether you follow the path of someone's brain from right-to-left or left-to-right on the above diagram, so it could be a brain that starts empty and learns over time or starts full and forgets over time. But from the *brain's* point of view, it can only have one direction. For if we were continually forgetting our future, we would obviously not know this was happening. We would only remember our past, and could only do so in each present moment. Thus, no matter which "way" we imagine our brain moving through time, it *looks* exactly the same to us: as moving forward in time from past to future.

But why does the brain accumulate knowledge, perceiving movement forward in time, as it moves *away* from the Big Bang rather than toward it? Behind all this is the way the pattern of the universe manifests, which is such that, unlike space, the dimension of time, as scientists put it, is not "symmetrically homogenous." Whereas the universe at any point in time looks essentially the same in every spatial direction (for instance, galaxies aren't all small to the left of us and all large to the right, they don't all face north, etc.), the universe does *not* look essentially the same in either direction of time. Since only universes with an asymmetrical time dimension can ever produce life or large numbers of black holes, both multiverse theories explain why we find such a thing in our neck of the woods: it is one among many random possible outcomes of Chaotic Inflation, and an inevitable outcome of Smolin Selection.

One of the inevitable results of asymmetrical time is that, from a point of view *within* time, simple things always precede the complex. In other words, complexity is always built upon simpler structures. And one thing we observe to be universally the case, everywhere, is that complex things only arise from simpler ones. We have never discovered anything to the contrary. It seems to me that this makes the most sense on naturalism. There would not likely be any natural way for a complex pattern to just "pop" into existence. That is, *within* time—outside time, only the eternal exists, so the question of "popping into existence" does not arise, although the ontology remains the same: as far as we know, all complex things, even if eternal, are always composed of simpler things. So temporally or ontologically, pure complexity out of thin air seems quite improbable. But there could be many natural ways for the simplest things, like the basic lines, strokes and dots of an artist's drawing, to move along the canvass of space and time, eventually colliding and interacting with each other, accumulating into more complex structures from those simpler elements. So long as the way this "pattern" of the universe is "drawn" is defined by uniform regularities, which would be the Nature of the Universe, randomly frozen in place throughout the unfolding of the multiverse, our One Brute Fact, then complex things could easily arise by nature—as demonstrated earlier for the solar system..

This leads to the reasonable inference that if there was a first event in time, from which all universes ultimately sprung and upon which everything has been built, it must have been the simplest possible thing—which could not logically be anything as complex as a god, but had to be something incredibly basic, an Ultimate Simplex, devoid of the complexities of reasoning or knowledge or nuances of character (things that, as far as we have seen, can only exist when constructed from simpler elements—a problem even for an eternal god). And a fundamental chaos is the simplest possible thing we can think of, having no fixed order that needs explaining. And as we have seen, multiverse theory proves that such a chaos can produce our universe. It is also hard to fathom what use a god would have for such long temporal processes, or of simple things accumulating into the complex. That makes more sense on naturalism. It is what we *expect*. The reverse, however, a universe where complex things pop into existence the moment a god wants them to, makes much more sense on theism. But we never observe this.

Whatever the case, the inherent nature of our brain's contact with lines of information, and the peculiar asymmetry in our dimension of time, means that talk of past and future, of causation and temporal order, still makes sense, it still refers to real distinctions, just as it would in reference to the flow and causation of events in a comic strip. Scientists identify that distinction as the breaking of "symmetry" between past and present. So the reason we perceive ourselves as having "been there" with respect to the "past" side of time, but not the "future" side, is that knowledge is itself an increase in complexity, thus it must be constructed from simpler things—in other words, it must accumulate *along* the time axis. It therefore can only increase in one direction, toward the "future." In fact, that is the only reason a "future" exists at all, as something distinct from the past.

Still, if everything exists as a frozen moment on a four-dimensional canvas, how can we experience *change*? After all, if we experience change, then *something* is changing, at the very least our experience. Right? Of course, even in B-theory change is a real fact: objects look different, and in predictable ways, the further along the time axis their existence extends. At any moment in time an observer will experience that difference and call it "change," a difference between what she calls earlier moments and what she calls later ones—just like a novel or a comic strip.

But in another sense, our experience of change is a useful fiction constructed by our brains, much like our experience of color. While colors as we see them don't really 'look' like that, they nevertheless exist in some sense: the constructions our brains invent do correctly guide us (most of the time) in distinguishing the real frequencies of light. So it is with time. Each moment of our "experience" is a fixed fact, with a fixed location, existing eternally at its place on the axis of time. But this *includes* each experience of motion or change, and each experience of "presentness," of being in the "now." These experiences don't "go" anywhere, and they don't "come into being." They are always there. But at each moment of time there exists an experience of the "now" (the experience of being at *that* point in time instead of some other), as well as an experience of "motion" in time, constructed from past observation by our brains.

This means that at one moment there is an experience of the present, then at the next moment there is another experience of the present, but of a different present, that which exists at *that* location on the time axis. At that instant we are looking back to see the previous moment of time and looking forward to anticipate the next moment on its way, because that is all that is visible to the mind at that location. At the next location, the mind has a different angle on things, so it "sees" the universe differently there, and so on, throughout our existence. At each moment, our experience is colored by the effects of the immediate and distant past on the shape of things now and to come, accounting for how things seem and feel to us. But all our experiences, all these moments, already exist in the fullness of time, and they are fixed there forever.

Attentive readers might already notice that what I have just described entails what is often called "determinism," a view much maligned and misunderstood, so to that problem I next turn.

On the view of time presented here see: Paul Davies, *About Time: Einstein's Unfinished Revolution* (1995); Huw Price, *Time's Arrow & Archimedes' Point: New Directions for the Physics of Time* (1997); Victor Stenger, *Timeless Reality: Symmetry, Simplicity, and Multiple Universes* (2000). But for a naturalist view of time radically different from mine, see Ilya Prigogine, *The End of Certainty* (1997). For a survey of views, but ultimately supporting mine, see Robin Le Poidevin, *Travels in Four Dimensions: The Enigmas of Space and Time* (2003).

4. The Fixed Universe and Freedom of the Will

Determinism is the view that the future is as fixed as the past, and cannot proceed any differently than it actually will. In theory, if we could know every fact (the position and momentum of every particle, stuff like that), we could predict every event of the future with absolute precision, even our own choices. Of course, such knowledge is not likely ever to be possible, especially given what we now know about Quantum Mechanics. And even if we ever acquired such knowledge, any process that could compute it all (using any instrument *inside* the universe) would have to be slower than the universe itself, and thus could never succeed in perfectly predicting the future in advance, even in principle—except by accepting some degree of imprecision or limiting predictions to very isolated cases.

Even though the universe and its future is most likely fixed and unchangeable, this doesn't really change anything for us as individuals or as a society. Determinism does not justify fatalism—the view that we cannot change our future no matter what we do in the present. Determinism actually entails the opposite: it means we decide our future *precisely* by what we do in the present. Though our every choice is inevitably caused by our desires and our knowledge, which are in turn caused by our background and our circumstances, which are in turn caused by still other things not in our control, our choices still cause our future, directly and indirectly.

Likewise, determinism does not really mean we have no free will. For what we call "free will" is really nothing more than the ability to choose and do what we want, an ability we can have (and have taken away) even in a deterministic world. So all of our language about things in our power and not in our power, about what we are responsible for and what not, about worthiness for praise or blame, it all continues to refer to real distinctions that still exist. There are still possible and impossible things, real options and options unavailable to us, there is still a difference between what can and cannot be done. In short, none of our usual beliefs or concepts needs any major modification. Everything important remains the same.

This view of determinism is formally called "compatibilism." How compatibilism works requires a careful look, and that will come shortly (apart from what follows, I also discuss determinism again in III.10.3, "The Nature of Love"). First, I must explain why I believe determinism is true.

4.1 Why Determinism?

It seems to us that everything that exists and everything that happens in the universe has a cause. We know the mechanics of causation: on the whole, the course of particles and objects seems determined precisely by mathematical laws that contain no room for spontaneous variation. Everything we see appears to proceed in the one, precise direction it gets caused to go, and can go no other way. So many physicists have concluded that, since everything is determined by these fixed laws, things cannot turn out any other way than they actually do.

The principles of Relativity establish that time is simply another dimension, that our notions of past and future are merely relative and not absolute. The resulting mathematics, which has explained a large variety of strange phenomena, also imply that all of time exists already, that the future is already there, and, like a road before us, is merely waiting to be visited. And since this view of time (discussed earlier in III.3.6, "Time and the Multiverse") can in turn explain a great deal of other strangeness about the behavior of subatomic particles, the case for determinism is fairly strong. There is little reason to doubt it now, even if many do.

I believe determinism is true because it is simple and has great explanatory power, it is a reasonable inference from the facts so far, it leads to a much clearer and more accurate understanding of many things, and alternative accounts are neither needed nor useful. As we shall see, for example, 'responsibility', both moral and legal, actually *requires* determinism. For if determinism were not true, then our actions and choices would not necessarily be caused by who we are. And what "we" (as a set of personality traits, memories, and so on) did not cause, we cannot logically be blamed or praised for. On the other hand, if determinism is true, then our actions and choices, those that have in fact been caused by us—and not by, for instance, our bodies without the involvement of our "selves" (our minds, our thoughts, our personalities)—are necessarily caused by who we are, and thus "we" can always be blamed or praised for them.

There is one sense in which I may be wrong: there has long been physical evidence that sub-atomic events may be truly random and not deterministic. This evidence at the same time entails that all larger-scale events are still deterministic, so at the level where we exist and act, determinism still holds. And though quantum mechanical events in our brains might be decided with some measure of randomness, these remain causes outside of us as whole persons, over which we have no power. But even if quantum events are really random (such that certain events are incapable of being predicted even in principle), the picture I drew of the universe earlier still does not change: though there may perhaps be no physical law that decides the outcome of quantum events (we don't yet know), every outcome of every such event, though random, may already be a foregone conclusion, and thus the universe would still have only one future. Determinism still holds.

But there might yet be something more going on. What appears to be an absence of determination by prior states and laws may merely be the result of our inability to observe sub-atomic processes on the scale that is needed to identify their causes. There may in fact be circumstances and physical laws that absolutely determine the outcome *even* of quantum events—laws and conditions that, if known, would in theory allow us to predict such events with absolute certainty, but that in practice can never be known due to the impossibility of ever gaining access to the relevant facts. And whatever the hidden mechanisms may be, we still know they produce apparently random results.

This view is a reasonable inference, since the rest of the universe operates this way. A good analogy is the weather: completely deterministic, yet defying all attempts to predict it beyond a few days. We have discovered a lot of what are called "chaotic" phenomena like this, which behave in a classically deterministic way, but are so complex that their outcome appears quite random. Just like the weather and many other nonlinear phenomena, we do not have the means to observe quantum events on the scale that would allow us to confirm there are no causes. So we cannot be sure there aren't. And when a theory of deterministic quantum causation ends up explaining so much about unusual quantum behavior (as, for example, Victor Stenger has shown in his book *Timeless Reality*), then quantum determinism becomes even more convincing. But this is a question that can only be decided by future scientific study. In the meantime, determinism (of one sort or another) is the most plausible inference.

4.2 The Alternative: Libertarian Free Will

Compatibilism is easiest to understand when contrasted with its opposite, a "libertarian" concept of free will. This is not a reference to the American Libertarian Party, but to the idea that the will must be truly "liberated" from all causation in order to be free. I will draw from one particular case where Evangelical philosopher J. P. Moreland tries to defend libertarian Free Will in a chapter he contributed to the book *In Defense of Miracles: A Comprehensive Case for God's Action in History* (Douglas Geivett and Gary Habermas, eds., 1997). As the current leading opponent of secular ethics, he is a good foil here, and I shall bring him up again in my chapter on morality. The issue of free will is directly related to his program against secular ethics, since it decides (among other things) the nature of responsibility.

Moreland defines libertarian free will as "given a choice...nothing determines which choice is made" (137). *Nothing* guarantees that a particular choice will be made, not even reasons or values or knowledge. What he means is that though we will always choose according to some desire that we possess, which *particular* desire of the many we have that wins out is not determined in advance by anything but "us" (in some obscure sense of the word). This is how he explains it: "When agents will *A*, they could have also willed *B* without anything else being different inside or outside of their being" (137-8). So even if I want most of all not to raise my hand, according to Moreland, I might raise it anyway, presumably as long as I have any minuscule desire to do it. Of course, this seems counter-intuitive right from the start. If the strongest desire in me is to stay still, how can I be caused to raise my hand by a *weaker* force?

Trying to bypass this problem, Moreland argues that desires and reasons and other things "influence" but do not "cause" our actions and choices. But it remains unclear just what the difference is supposed to be. My knowledge that a wall stands before me certainly causes me to choose to change the direction of my walk. My desire to live certainly causes me to avoid leaping out of windows. Yet Moreland would not say that walls and windows deprive us of our free will. So he has to elaborate somehow.

"Suppose some person," Moreland asks, "freely performed some act...say raising an arm in order to vote" (138). He says that this person "exerted [his] power as a first mover (an initiator of change) to bring about" the motion to vote. But what about the request to vote in the first place? Actually being in a circumstance that calls for a vote is itself a necessary condition for raising a hand to vote. Correct. But this does not mean that the circumstances will be a *sufficient* cause of the action, and Moreland's point is that something *else* is necessary, which is unrelated to anything in or outside us.

But this thing is not our reason or our knowledge or our character or our desires, or anything at all really. And that creates a problem. For instance, Moreland includes in his theory the premise that this person "brought about [the choice] for the sake of some reason," which entails another necessary cause—the reason—without which the hand would never be raised. And this reason will certainly correspond to a brain state, and a chain of causation can be followed as we examine the path of all the calculations and knowledge that are in turn necessary causes of that reason arising in our brain. So he is forced to reject even the obvious theory that reasons cause us to act.

What Moreland must contend is that despite the necessity of all these causes, the sum of them all (having a reason and a strong enough desire, as well as the requisite knowledge and the necessary circumstances) will *still* not be sufficient to cause an action. In other words, though such things must be in place, something "else" is required, which is neither a reason nor a desire nor knowledge of any kind nor anything about the surrounding circumstances. And this is the problem. It is hard to see what this "something" can be.

If I have a desire to actually shoot someone, a desire that is sufficient to override all other desires which urge me against it—a necessary cause of any willful choice to shoot—why would I not shoot? If Moreland appeals to moral shame or guilt or fear, then he is appealing to a desire. But that is a cause, and that cannot be his special "something." Likewise, if he appeals to my character, knowledge of God or moral laws, to reasons not to shoot, or any such thing, then he is still appealing to causes. So what is left that could "cause" me not to shoot? Moreland is saying, in effect, that there is some uncaused power in me that can cause me not to shoot—for no reason whatever. But this contradicts his premise that an agent always acts "for the sake of some reason." For if I have no reason at all not to shoot, how can it be that I might choose not to shoot for some reason? That is a contradiction. So this concept of free will seems to be self-refuting.

Moreland might respond that we usually have a reason to do and not to do something, and which reason we follow is caused purely by "something" in us, something that is not a reason, nor anything else like desires or knowledge or circumstances. But if the ultimate reason for my doing something is not a reason, then rational action is impossible, for no rational calculation can then be the cause of what I do. Only something purely non-rational is the ultimate necessary cause on Moreland's theory, something uncaused by our knowledge, our reason, even our character. That would be terrible if it were true—far from rescuing responsibility, it destroys it.

4.3 Why Libertarian Free Will Eliminates Responsibility

Imagine two parallel universes, identical in every detail, and imagine a man in each universe, identical in their character and knowledge and desires and everything else, standing in totally identical circumstances. Now imagine that one of these men chooses to kill his wife, but the other man chooses not to. What could possibly explain this? Since the two situations and the two men are identical in every respect, there can be no cause whatsoever for either man's choice. This is what Moreland's theory entails.

But this has an unacceptable consequence. For it means that neither these men's desires, nor their knowledge, nor their moral character—nothing at all—can be blamed for having caused their choice. Moreland even agrees: "no description of our desires, beliefs, character, or other aspects of our makeup and no description of the universe prior to and at the moment of our choice...is sufficient to entail that we did it" (138-9). But this means that we could not even say that the first man was evil and the second good, since that would assume the first man's badness caused him to kill, while the good man's goodness caused him to refrain. But these men are identical, so one cannot be evil and the other good. Moreland might say he is evil or good *after* the deed, but that means we could not say he did what he did *because* he was a good or a bad man. In fact, we could not say at all why he acted. What quality in either man that is uniquely a part of "him" can be blamed for causing his particular choice? There is none.

Now imagine that this man is you, and in one universe you kill your wife, in the other you do not. What would you think of yourself then? You would know that nothing causes your actions—not your character, nor your environment, nor the surrounding circumstances, nor your knowledge, not even your love of your wife. Nothing. Your choice to kill or refrain is purely a result of happenstance: whichever universe you are in is a mere luck of the draw. Imagine how you would feel, having learned that it is nothing but the result of unpredictable randomness whether you kill your wife or not at this very moment. Shocking, yes? Imagine that you refrain from killing, but could step into a time machine, run the universe back a million times, and watch yourself again each time, and then saw that sometimes you killed and sometimes you didn't, even though each time all the circumstances, including your thoughts, desires, character, everything, were the same. There would be no rhyme or reason to why you did one or the other. It would be a mere shake of the dice. This is the nightmare of a world that Moreland's theory describes.

Wouldn't you instead want the result every single time to be the same? Every single time you would choose not to kill, right? But if the same circumstances are followed by the same choice 100% of the time, that is determinism. In fact, we know we are good only by seeing whether our goodness causes us to do good deeds, and so we should expect deterministic causation in our own choices. After all, the only logical alternative to 100% causation is randomness, and why would we feel good about our choices if they were actually random, and not caused by any of our inner qualities? Here's the deal. "I" am defined by my knowledge, abilities, character, values, and desires. If something causes me to act which is not one or the sum of these things, then "I" did not cause that action. Moreland, instead, wants "me" to be defined by something other than these things. But if you were to take them all away, there would be no "me," so his approach is absurd. Without knowledge, abilities, values, desires, reasoning, a character, there would be no person at all, and if something other than these things caused us to do something, then no one could really say *we* caused it, because *we* are all these other things, none of which is at fault in Moreland's view. And could anyone conclude I was at fault for something I did not cause? No.

So the key word here is "I" (or "he" or a "person" or whatever) and what it means. Moreland defines it as some unexplainable, unidentifiable thing that excludes all memories, desires, virtues, values, traits, even reason. This is a rather illogical conception of human identity, and one any reasonable person can easily reject (see also V.2.2.2 and 2.2.3, "Human Nature" and "Personhood," and III.6, "The Nature of Mind").

4.4 Compatibilism: The Only Sensible Notion of Free Will

Moreland tries to defend his illogical notion against compatibilism, by laying out "four areas central to an adequate theory of free will" (138). Actually, what he offers are four elements of a theory of *moral responsibility*: we must have the ability to act, we must be in control of our action, we must have a reason to act, and we must be the cause of the act.

4.4.1 The Ability Condition

To be free to act, we must be able to act. That is simple enough. It is not within our free will to leap over skyscrapers. But it is within our free will to climb them or take an elevator to the top. Moreland would argue that since determinism limits our real possible choices to one, asking us to act freely is like asking us to leap over a skyscraper: it can't be done.

But this is confused. For we *can* climb a skyscraper or take the elevator, or use a helicopter: these are possible actions, any one of which we can do. Though we can only do one of them at a time, this doesn't mean that if we choose to fly a helicopter we couldn't have used the elevator if we had chosen to do that instead. But we could not 'leap' to the top *even if we chose to*. That's the real distinction here. We are free insofar as we can do what we *want*. When what we would choose to do is outright impossible for us, *then* our freedom is restricted. Our freedom cannot be restricted by things we never even choose to do.

Compatibilism entails, as Moreland puts it, that "freedom is willing to act on your strongest preference" (138). Better put, freedom means getting to do what you want. It even means getting to want what you want in many cases, though not all. For at some point there must be some desire or other that you did not choose, since in order to choose the desires you want, you must first "want" them—and if you begin with no desires at all, you will never make any choice of any kind. Thus, according to compatibilism, any organism that chooses in accord with its desires must begin with one or more desires that it did not choose. We call this, in our case, "human nature," which we did not choose, but was given to us by the accidents of biology and history (or even, in Moreland's view, by the designs of God). Obviously all our desires will ultimately be caused by our nature. That does not prevent us from choosing desires. To the contrary, it is our nature that allows us (by causing us) to choose desires at all, a power nonhuman animals lack.

In contrast, Moreland's "libertarian" view holds that "a free act is one in which the agent is the ultimate originating source of the act." At first glance, this looks like the same thing: compatibilists also hold that freedom entails being the source of the action. The person's character, desires, knowledge, etc., must all be necessary causes of the act. This requires that the "person" be involved in the chain of causation, in the sense of thinking, contemplating, desiring, knowing. These factors, plus the affecting circumstances, are together the full sufficient cause of any free act.

But Moreland goes further, and requires that a person be an "ultimate" originating source, not just "the" source. But there is a problem here. Never in all of history has anyone ever sought to confirm this before assigning responsibility. We have no problem calling people responsible or not responsible all the time, but do we ever bother to check if there was a physical gap in the chain of causation, that the person was an "ultimate" origin and not just an origin? No, we do not. No one even thinks it relevant (see III.4.5, "What Free Will Really Is").

Thus, Moreland's notion of freedom, as it relates to responsibility, does not correspond to actual human practice. But if Moreland's ideas have nothing to do with what people actually mean, then why should we care about his ideas? Instead, ask anyone on the street whether "getting to do whatever you want" is the same thing as freedom and everyone will wholeheartedly agree. You will have a very hard time finding anyone who wants any other kind of freedom. So it is a good bet this is what freedom really is.

Therefore, "freedom," in the only sense everyone cares about, exists even in a deterministic world. In contrast, Moreland's view of freedom does not mean getting to do what you want—because what you most want doesn't actually cause you to act. Instead, this "something else" of his conception can step in and thwart your greatest desires by making you do something else you wanted less, and this means you must often do what you *don't* really want. That is the exact opposite of freedom. So we have good grounds to reject his view here.

4.4.2 The Control Condition

The compatibilist view of control is that, again in Moreland's own words, "an agent is in control of an act" if "the act is caused...by the agent's own character, beliefs, desires and values," etc. (140). To defend his libertarian rejection of this, Moreland appeals to an antiquated medieval philosopher, Thomas Aquinas, for the notion that "only first movers are the sources of action" since everything else "merely receive[s] motion passively and pass[es] it on." But this is a special use of the word "source" that is not employed in normal conversation. We say the source of an earthquake is a particular rock fracture at a particular location which slipped at a particular time. We never say the source of the earthquake was the Big Bang.

Likewise, in human discourse we distinguish between active and passive transmission of energy in a different way than Moreland does, or Aquinas did when discussing his "first mover." For we think in terms of whether the agent took action in accord with a desire to transmit motion:

if the transmission of motion requires the participation of the agent's personality or character or reason, then we call that an active participation (and it will feel that way introspectively, since our brains can detect the presence or absence of such activity in connection with an action). But if the motion does not require participation (if the body is pushed, without any knowledge, or even despite efforts to resist), then the agent has not actively participated, so we say the agent was not in control of his own motion.

That this is how people normally speak is further shown by how we talk about machines. Consider a thermostat: even though the thermostat is caused to change by the temperature in the room, in turn causing that temperature to change, we do not say that the air in the room controls the temperature in the room. Instead, we say that the thermostat is in control, because it is the necessary factor determining the temperature in the room, without which the room's air could be any temperature, varying aimlessly. Likewise, when we ask whether a computer is in control of a factory process, the issue is not whether it is free of prior causation (which has nothing to do with whether it is controlling anything), but whether the computer is actually the mechanism making the choices and moving the process along, or if something else is the cause (like factory workers or random signals). After all, what we want to know when we hunt for the responsible agent is where the problem is.

This is the distinction we actually make in real life. Again, Moreland is arguing for ideas that do not correspond to the way people actually think in the relevant contexts. So his contention that we do not have "real" control on compatibilism is just a word game. The fact remains that so long as we actively participate in the chain of causation, "we" (as in our unique mind, with its particular set of knowledge, character, desires) are in control of that action. This is true on ordinary English spoken by everyone, and it is true in every context that is relevant to assigning responsibility, from law courts to personal relations (as we shall see). That is a fact strongly in favor of compatibilism, good evidence that the libertarian theory isn't correct.

4.4.3 The Rationality Condition

Here Moreland begins to get illogical. He wants to show that compatibilism entails that we do not act for reasons ("intentions") but that we only act because of chains of causation, using the ancient Aristotelian distinction between efficient causes (a ball colliding with another ball) and final, or teleological causes (a ball is collided into another *in order that* the second ball will land in the corner pocket). But "reductionism" (see III.5.5) entails that these two kinds of causes are necessarily equivalent in the case of an agent: whenever there is a final cause, if reductionism is true, then there is also an efficient cause, and the final cause is itself caused by a prior efficient cause.

Moreland says, for example, that on compatibilism "a reason for acting turns out to be a certain type of state in the agent, a belief-desire state, that is the real efficient cause of the action" (141). He argues that this excludes the possibility of final causes. But since a belief-desire state *is* an intention, and an intention is a final cause, it follows that final causes can and do exist under compatibilism. A final cause is simply a thought process: a prior calculation from ends to means, which in turn participates in the causal chain that ends in acting. The visualized 'end' is caused by a desire ("I want the second ball to land in the corner pocket"), and the conceptualization of the 'means' is caused by an application of reason and knowledge to that desired end ("If I collide the first ball into the second just so, then I will achieve what I want"). That is all a final cause is, a thought process, and that's an efficient cause: without the final cause (the desired end) there would be no act. So Moreland is attempting to state a tautology (A is B) as if it were a distinction (A is not B), a fundamental violation of basic logic.

Moreland does this again and again. For instance, he says compatibilism entails that "persons as substances do not act; rather, states within persons cause later states to occur." But these are the same thing: the state within me (the sum of my being) *is* my substance, and since my substance is in turn the cause of my choices, it follows that persons as substances do in fact act under compatibilism. Indeed, compatibilism makes far more sense of the fact that in ordinary life we always define ourselves, our fundamental identities, by our aggregate personality—our skills, hopes, dreams, thoughts—and never by some other thing that excludes all of this and that no one can observe or ever has observed. Are "you" an ever-growing and changing personality, or a strange, invisible, unchanging causal power devoid of all personality? The answer seems self-evident.

Compatibilism says that it is precisely our reason, the inevitable process of logic applied to our observations and experiences and our wants and needs, that causes us to do what we do. Even when we use bad logic or false information, even when we are ignorant or choose to be irrational or impulsive or hasty or random, even when we are deceived, or deceive ourselves about our actual motives, it is still our reason that causes us to do what we do. We obviously have intentions and those intentions cause us to act. It is *Moreland's* theory that eliminates reason and intention as the real causes of our choices, not compatibilism. So it is libertarian free will that fails to meet Moreland's own "rationality condition."

4.4.4 The Cause Condition

According to a compatibilist, as Moreland puts it, "if we say that a desire to vote caused Jones to raise an arm, we are wrong" since the truth is that "a desiring to vote caused a raising of the arm inside Jones" (142). But Moreland is playing word games yet again. In reality, "Jones" is a synonym for a particular collection of character features, values, desires, beliefs, and memories, all of which are necessarily involved in the chain of causation from the initial call for a vote up to the rise of the desire to vote one way or another. Thus, although you could play a word game and say that a desire did not cause Jones to act, by arguing that Jones in fact caused the desire which in turn caused the act, this gets you nowhere if your object is to show that Jones was not the cause of the act, since he must be the cause of the desire or the direct cause of the act, and either way "he" is still the cause of the act.

Moreland's error here is the same as before: pretending tautologies are dichotomies. On his theory "it is the self that acts, not a state in the self." But he never shows how these must be, or even can be, different. Even on his view, there must be some "state" in the special non-causal "soul" stuff that Moreland is trying to identify as the "ultimate" cause of action, which in turn causes or constitutes the choice, thus the "self" always equals a "state in the self." They are always one and the same thing, unless we suppose that this "ultimate cause" of every act, Moreland's "something," never changes—which would entail that *we* never change, that we never grow or learn or develop as persons. That is clearly false.

Moreland also thinks this invalid distinction explains the difference between acts and "mere happenings." But that difference is adequately explained already by what we humans actually look for: not an undefined, unobservable, "self-stuff," but instead a visible, demonstrable connection between an act and an agent. This is what is done in courts of law: "means, motive, and opportunity" is a catchphrase for all the evidence that can lead a jury reasonably to believe that the event (usually a crime) is causally connected to an agent—and not just the agent's body, but the agent in his entirety: his mind, character, desires, beliefs, and intentions ("motive"). Since this is all we ever look for in actual practice, clearly this is all we actually mean when we say it was an "action" and not a "mere happening." Moreland's view is thus disconnected from the reality of human discourse, or bogged down in linguistic superstitions, and is therefore not as credible as compatibilism, which far better explains real human behavior when people investigate and assign responsibility (on this method of analysis, see II.2.1.4, "Getting at the Real Meaning of Words").

4.5 What Free Will Really Is

Even if my choices are entirely determined in advance, I still make decisions, and my decisions are still caused by who I am and what I know—my thoughts and desires and personality—just as they must be if I am to be "free" in any sense that matters. And because I am still their cause, I can still be praised or blamed for them. This is why compatibilism makes more sense: free will is doing what you want—nothing more, nothing less. And being responsible is being the cause—nothing more, nothing less.

We see this in a decision handed down in 1886 by the United States Supreme Court in the case of *Conley v. Nailor et al*. The case put to the court was this: a man left his wife and took up with another woman without marrying her. He had sons by this other woman, and in his will left his estate to these illegitimate children, even though he had legitimate sons still living. He even included a clause stating that if the illegitimate children died then the estate went to his girlfriend, not to his wife or his legitimate sons. Naturally, his wife and her sons contested the will, on various grounds, but one of which is relevant to us:

> The next and last ground alleged for annulling the deeds is that Nailor was induced to make them by the fraud and undue influence of the defendant. The ground upon which courts of equity grant relief in such cases is, that one party by improper means and practices has gained an unconscionable advantage over another. The undue influence for which a will or deed will be annulled must be such as, that the party making it has no free will, but stands *in vinculis* ["in chains"]. It must amount to force or coercion, destroying free agency.

The court held that "undue influence" was any constraint that "substitutes the will of another for that of the testator," including the use of threats or lies. This conception of free will is a commonplace in law, and clearly assumes that free will means getting to do what you want, such that you lose your free will only when you are tricked or forced to do something you didn't really want—in other words, only when there is the substitution of someone else's will for yours, or constraints that prevent you from doing what you will to do. This notion of free will is not eliminated by determinism.

In contrast, the libertarian notion of free will assumes that one's own desires (among other things, like one's own reason and knowledge) also constrain one's will, rendering it unfree. In other words, our personality, knowledge, wishes, are themselves chains that bind our will. But your proximate, causing desire *is* your will. It therefore cannot be considered as something "outside" of the will that constrains it—your strongest desire and your will are one and the same.

And since everyone *wants* a will that is bound and compelled by what they want it to do, which is determined by what they know and believe, it is a cheap word game to call this a loss of freedom. A loss of freedom is the exact opposite of this, when the will is *not* bound by one's own desires and beliefs, but acts without them, or even against them. We can only be called rational, after all, if our controlling desires are caused by our reason and the facts at hand. And that *requires* determinism.

The Supreme Court case at hand is particularly relevant here. For Nailor's wife tried to prove that his mistress "treated him with great kindness and with unremitting attention to his wants and comforts" and therefore had caused and controlled his decision, depriving him of his free will. The court rejected this argument, which means the court rejected the libertarian definition of free will as a will that is free of *all* causation of any sort. As the court put it:

> There is an absence of proof that the defendant used either threats, stratagem, importunity, or persuasion to induce Nailor to execute the deeds. In fact there is no evidence that the defendant even requested him to make them. On the other hand, the proof is abundant that the making of a provision for the children whom the defendant had borne him had long been his cherished purpose.

In other words, regardless of the kindness of his mistress, she did not try to override his will, and all evidence was that he *wanted* what he eventually chose (very likely *because* she was so kind to him), and that was enough for the court to rule that he had free will. The court did not investigate whether there was a gap in causation in his soul or that his will was completely uncaused by external factors. In fact, the court dismissed as irrelevant the question of whether the defendant's kindness caused his choice. So long as what he willed was what he wanted, his will was free. The court only regarded *certain* causes, not all causes, as depriving a man of free will—in particular only those causes that can lead a man to act contrary to his desires, such as intimidating him, deceiving him, or persistently urging so as to annoy, harass or cajole him into compliance.

Many other court decisions support this view, and in fact no other view of free will is found in major American legal decisions that I know. Compatibilism is the standard assumption of the law. For instance, in a more recent Supreme Court case regarding a man's competency to conduct his own legal defense (*Godinez v. Moran*, 1993) it was ruled that he was acting of his own free will if he was "literate, competent, and understanding," "informed," and made his choice "competently and intelligently." The court ruled that "the record must establish that the defendant 'knows what he is doing and his choice is made with eyes open'," but that is all.

In other words, if you know what you are doing and it is what you want to do, as far as the law in the United States is concerned, you have free will. But if you don't really *know* what you are doing or it is not what you really want to do, you do not have free will. Knowledge and intent. This is the real definition of free will, the real distinction people make in their daily lives when deciding whether they or others are responsible for what they do. In contrast, in the real world hardly anyone brings up the acausal metaphysics of the soul, much less do they actually try to determine where and when such a strange substance was or was not involved in any given case. So the libertarian definition of free will is irrelevant to human and social reality, while the compatibilist definition fits it like a glove.

Another recent case confirms this. In 1988, in *United States v. Kozminski et al.*, the court addressed the definition of free will directly. Though the court stated that it did not have to solve the philosophical problem of free will, but only had to define criminal servitude with "sufficient specificity" to enact the intent of a law, this required the court to define "involuntary servitude," which is identical to "deprivation of free will," and thus in effect the court ended up defining free will anyway.

The court's decision entailed the very same definition just observed. "Just any" causation is not considered, even in theory, as capable of eliminating free will. Rather, only those causes are considered that make someone do what they would not otherwise *want* to do—indeed, not even all of those causes were accepted by the court as negating free will. The court ruled:

> Involuntary servitude consists of two terms. Involuntary means 'done contrary to or without choice'—'compulsory'—'not subject to control of the will'. Servitude means '[a] condition in which a person lacks liberty especially to determine one's course of action or way of life'—'slavery'—'the state of being subject to a master'.

Here free will is clearly seen as acting as one chooses, since we only lose free will when we are caused by someone or something else to act "contrary" to or "without" our choice (which equates to our "desire"). So long as we "determine our course of action" as we want, we have free will, because our act is voluntary. The court even noted:

> It is of course not easy to articulate when a person's actions are 'involuntary'. In some minimalist sense the laborer always has a choice no matter what the threat: the laborer can choose to work, or take a beating; work, or go to jail. We can all agree that these choices are so illegitimate that any decision to work is 'involuntary'.

Thus, the court saw it as obvious to everyone that only coercion that creates an undesirable choice negates free will, not mere causation in and of itself. The judges overtly grant here that the issue is what the employee wants, not what he is merely caused to do. Thus, the court ruled that free will was violated in the face of "any coercion that either leaves the victim with 'no tolerable alternative' but to serve the defendant or deprives the victim of 'the capacity for rational calculation'."

Though the Supreme Court decided that this definition was too broad for a criminal application (primarily due to problems of evidence), it did not reject the moral application of this concept of free will outside the legal sphere, and it upheld it when there was objective evidence of physical coercion or the threat of it. So it is clear that the law in America holds a man's will to be free so long as he knows what he is doing (making a choice with 'the capacity for rational calculation') and wants to do it (making a choice where there is a 'tolerable alternative'). Any other form of causation does not negate free will in the eyes of the law.

Finally, the compatibilist definition of free will is supported by two standard rules of common law: the rule of intent, and the insanity defense. The first of these is seen in a typical jury instruction for all criminal cases, also quoted in *United States v. Kozminski*:

> Specific intent must be proved beyond a reasonable doubt before there can be a conviction. Intent ordinarily may not be proved directly, because there is no way of fathoming or scrutinizing the operations of the human mind. But, you may infer the defendant's intent from the surrounding circumstances.

This is fundamental to all criminal law: *intent* (knowledge of what one is doing and an overriding desire to do it) is essential for all criminal responsibility. This means that desire *must* be the cause of one's action in order for someone to be responsible for what he does. So our entire criminal justice system *requires* determinism to be true, at least for human actions. For the law assumes that a person's actions are caused by their intent with enough reliability that one can deduce the latter by observing the former, and that absent intent as the cause, there can be no guilt. This means that being free of causation is not, and cannot be, the standard courts and juries use to ascertain responsibility, but having knowledge and intent, in other words *being caused* by that knowledge and intent to act, rather than being caused despite or *contrary to* one's knowledge and intent.

This is confirmed by the noted jurist Joshua Dressler, who describes what the law really looks for as defining responsibility, and it is not a thoroughly unfettered will, but:

> In broad terms, a person is considered blameworthy when he voluntarily commits an immoral or illegal act—*actus reus*—while possessing the requisite 'guilty' mental state—*mens rea*. Under traditional doctrine, if either *actus reus* or *mens rea* is absent, the individual is not deserving of punishment. (p. 501)

He goes on to explain that a man has *mens rea* when he "intended to commit the prohibited acts or was aware of all the facts that made the conduct criminal" (p. 502). This only makes sense if we believe the intent caused the act, since only then would the act be evidence of a wicked intent deserving of punishment. Basically, it is the wicked intention that makes us guilty and blameworthy, not anything to do with being free of all causation, a bizarre concept that is completely absent from the actual two-pronged test for responsibility: *actus reus et mens rea*.

The second issue is the insanity defense. The American Law Institute's Model Penal Code, which captures what is typically found in penal codes nationwide, declares the insanity defense thus:

> A person is not responsible for criminal conduct if at the time of such conduct as a result of mental disease or defect he lacks substantial capacity either to appreciate the criminality (wrongfulness) of his conduct or to conform his conduct to the requirements of law.

Note what is *not* said here: it does not say that a person is free of responsibility if a mental illness caused his actions. Rather, it says he is guiltless only when a mental illness deprives him of what he must have in order to be responsible: namely, knowledge (here, the 'capacity...to appreciate' what he did) and intent (here, the 'capacity...to conform his conduct' to what is expected of him, i.e. despite what may have been an overriding desire to conform). Dressler again:

> Insanity involves an internal circumstance—disease of the mind—that substantially or totally impairs the actor's cognitive capability. He must either be unaware of what he has done, or unaware of the wrongfulness of his conduct; alternatively, the disease must substantially impair his volitional capabilities, so that he cannot effectively control his conduct. Under these circumstances, talk of choice is meaningless. A person has no choice when disease causes him to lose all touch with reality or to be unable to conform to reality. (p. 507)

Thus, being caused is not the issue that deprives one of free will, but being caused to act contrary to or without the involvement of one's will, one's personal being (knowledge, desires, character, reason). And this happens when someone lacks awareness and thus cannot even be choosing what he is doing, or does not know what he is choosing to do, or when one's body engages in actions wholly without or even against the will. That, and only that, is the reason insanity is allowed as a defense. Yet nowhere here is there even a hint that libertarian free will is the issue. To the contrary, a compatibilist conception of free will is inherent.

The point of surveying the position of free will in American law is that here we have one of the most serious and pervasive institutions devoted to assigning responsibility with accuracy and justice, in a manner agreeable to the community and informed by centuries of thought and practice, yet within this we find none other than a compatibilist conception of free will and responsibility, a conception fully compatible with determinism.

There seems no need, then, for the libertarian alternative. When this observation is combined with the fact that the libertarian view entails profound horrors and absurdities for the human condition (as we saw in the thought experiment where we kill or do not kill our wife), and lacks any evidence of any kind (while we have substantial evidence and reasoning to support determinism), there remains no grounds for advocating libertarian free will. And when we add to all this the fact that compatibilism better explains the actual behavior of humans when they investigate and assign responsibility (since most people the world over act like compatibilists, whatever else they *say* they are doing), there remains no reason not to be a compatibilist.

4.6 The Fatalist Fallacy vs. Improving Self and Society

All objections to the above analysis ultimately reduce to one and the same fallacy of fatalism: that all is "fated" so we cannot change our future or correct our behavior in response to the past. That's just illogical nonsense. As noted already, determinism entails the reverse: that we ourselves determine our own future in response to our study of our past and present. Our goodness causes us to do good, or our badness to do bad—and thinking and reading and being taught can cause us to change our character from one to the other. There are always forces working against us. We can't leap tall buildings in a single bound. But there are always forces at our command. We can build and fly helicopters. We can find and study all reasonable options and determine the wisest course of action among them. All we have to do is want to. And as far as any of that is concerned, determinism changes nothing.

Some even try to argue that determinism undermines knowledge because we might be "fated" to believe falsely and, at the same time, to think we believe truly. But that is just another illogical appeal to fatalism. We can still see when we have acted on a false belief rather than a true one. There is no possible way to always hide the consequences of error, or to constantly fake the benefits of successful prediction. Even if we are determined to believe what we believe, we will still either believe for sound reasons or not, and the difference between those two kinds of belief will still be physically and observably different, in all the same ways as before (see III.6.5, "The Nature of Knowledge" and III.9.2 "Why Trust the Machine of Reason?"). So again, determinism changes nothing. You can whine all you want to that you were fated to be in error—but you will still be in error, and the same evidence will still prove it. The option to correct yourself remains, and once you see that, you can't claim you were fated to ignore it.

In just the same way, to those who say "but I was caused by my deprived background to do this and therefore I am not responsible," we respond: "you did it nonetheless, and how you got that way does not make you any less guilty." For it is wickedness we condemn and goodness we praise, not freedom from causation. And if you were fated to be evil, then you were fated to be despised and punished, too. Complaining changes nothing. But acknowledging your faults and improving yourself changes everything. Were you fated to stagnate, or to improve yourself and alter the apparent course of your life? Fated or not, the choice remains yours—and yours alone.

Of course, this means we ought to care more about reforming bad people than punishing them, and a lot more about fixing the social causes of evil than locking up their products. But compassion compels us to this conclusion anyway, determinism or not. Punishment for vengeance's sake is pointless cruelty from which no noble benefit accrues to anyone. The only valid punishment is that which has as its end a better society and, if possible, a better person. Likewise, only the fool ignores the causes of cancer and, rather than seeking prevention by working to eliminate those causes, focuses all his energies on cutting cancers out as they come. Rather, you must attend to both the causes and the results if you want to solve the problem.

In contrast, with libertarian free will, punishment and reward have no point or purpose, because they can have no practical effect. If the will is free of all causation, then nothing can *really* cause anyone to change or act differently. In contrast, determinism makes all this possible, and restores sense and goodness to punishing our villains and rewarding our heroes—because it alone makes our efforts to improve them *matter*, and it alone makes *them* the causes of their actions, rather than some mysterious unpredictable "something" devoid of all worthwhile properties, not at all resembling anything we'd recognize as a "person," and ultimately governed by no sense or reason...

> The court cases quoted are: *Conley v. Nailor et al.* (118 U.S. 127; 6 S. Ct. 1001; 1886 U.S.); *Godinez v. Moran* (No. 92-725, 509 U.S. 389; 113 S. Ct. 2680; 125 L. Ed. 2d 321; 1993), upholding and quoting *Faretta v. California* (No. 73-5772, 422 U.S. 806; 95 S. Ct. 2525; 45 L. Ed. 2d 562; 1975); and *United States v. Kozminski et al.* (487 U.S. 931; 108 S. Ct. 2751; 1988). Dressler is quoted from "Professor Delgado's 'Brainwashing' Defense: Courting a Determinist Legal System," in Michael Louis Corrado, ed., *Justification and Excuse in the Criminal Law: A Collection of Essays* (1994), pp. 497-516 (this entire book is relevant), reproduced from the *Minnesota Law Review* 63:335 (1979).
>
> For further reading on compatibilist determinism, see: Daniel Dennett, *Freedom Evolves* (2003) and *Elbow Room: The Varieties of Free Will Worth Wanting* (1984); Gregory Rich, *A Defense of Compatibilism* (1982); Nicholas Dixon, *Compatibilism without Utilitarianism: Moral Responsibility in a Deterministic World* (1985); David DeMoss, *Compatibilism, Practical Wisdom, and the Narrative Self: Or If I Had Had My Act Together, I Could Have Done Otherwise* (1987); and Jack Kamerman and Gilbert Geis, eds., *Negotiating Responsibility in the Criminal Justice System* (1998).
>
> For an important work applying modern brain science to the issue, see Daniel Wegner, *The Illusion of Conscious Will* (2002), though "illusion"

here is a misleading term: our "self" is still a virtual model (and *post hoc* perception) of a real thing (see III.6, "The Nature of Mind"), and hence so is our "will." On which, see: Henrik Walter, *Neurophilosophy of Free Will: From Libertarian Illusions to a Concept of Natural Autonomy* (2001).

Some naturalists still reject determinism and argue for the existence of libertarian free will as a naturally emergent property of a reasoning brain. Prominent among them is Corliss Lamont, *Freedom of Choice Affirmed* (paperback ed. with new preface, 1969). See also "The Volitional Brain: Towards a Neuroscience of Free Will," *Journal of Consciousness Studies* 6:8-9 (August-September, 1999).

5. What Everything is Made of

Everything is a physical arrangement of matter and energy in space-time. This is a fundamental conclusion scientists have reached in their investigation into nature, and my Metaphysical Naturalism is built on this. First I will survey briefly what these things are, as well as how "the laws of nature" fit into this picture, and then I will move on to a much hotter philosophical question: the nature and existence of abstract objects. All these items together complete 'the furniture of the universe' so to speak.

5.1 Space-Time

Space-time refers to the four known dimensions, but includes any others that may one day be discovered—scientists are already finding evidence hinting at several other spatial dimensions, for example. Dimensions are essentially the existence of extension, and probably constitute the fundamental ground of all being. By definition, wherever there is more than one place something can "be," there is extension, and the entire range of that extension is a "dimension." We might even consider dimensionality as pure being—for to exist, in my view, is to exist in some *place*, in relation to all other things. And when you strip away all other properties anything can have, except mere "being," what you will have left is space or time—or both, as a unified space-time.

With regard to the three dimensions of space, this is easy to grasp. But with time it gets harder, because consciousness is among those things that *only* exists as an extension of matter and energy over a span of time. So we are in effect "using" the dimension of time in order to exist, leaving us with room to move only in the three remaining dimensions of space, and this makes time seem and feel different to us. More importantly, time behaves differently than space, since in our universe it has become a symmetry-breaking extension (see section III.3.6, "Time and the Multiverse"). In fact, that is what makes time "time" instead of just another dimension of space. Conscious beings could only ever arise in such universes, so it is no surprise that we find ourselves in one. Indeed, things may well be different in other universes, but life won't exist in them.

Current cutting edge science is already suggesting that everything else about the universe (matter, energy, physical law) derives solely and entirely, in some way or other, from the geometry of space-time. This is the powerful explanatory prospect offered by Superstring Theory, for example, which has made amazing strides toward explaining all manner of bizarre things about subatomic physics. Other theories are converging on the same conclusion. Only time will tell, but for now this is the most plausible conclusion: matter and energy are geometric properties of space-time, and all physical laws follow therefrom.

On time, see bibliography to section III.3.6, "Time and the Multiverse."
On space, see next section.

5.2 Matter-Energy

Once upon a time matter and energy were thought to be two different things. That has changed. The advent of the atomic age has demonstrated beyond all reasonable doubt that matter is simply another form of energy, one among many. In fact, there is really only one fundamental substance out of which every material thing is made: energy. Light is energy. Heat is energy. A magnetic field is energy. Gravity is energy. An atom is energy, as is a proton or a quark. A Dodge Caravan is energy. So is a pencil. The only thing that makes each of these different from the others is the physical, spatio-temporal arrangement of that energy—the way it is bound up or set free.

So what is this stuff? What is *energy* made of? That is the biggest question now being studied. But all viable theories are converging on the same conclusion: energy appears to be an oscillation of some sort, a movement back and forth in space-time (and possibly other dimensions as well). It is like a ripple, a wave, in space-time itself. And that means energy, and hence all forms of it including matter, is simply the result of geometry. In other words, space-time arranged, rippled, in a certain way *is* energy, and hence also matter. Wherever there is a space-time twisted up the right way, there will be matter and energy.

In fact, it seems it is impossible for energy *not* to exist. There is growing evidence that even a complete vacuum is seething with energy. And there has long been strong evidence that energy can never be created or destroyed—which makes sense geometrically, since you can't flatten any area of space-time without bunching up another, unless you can somehow generate or eliminate space-time itself. Otherwise, remove one ripple over here, and another pops up over there—the principles of geometry allow nothing else. Consequently, there doesn't seem to be any way to get away from the stuff. Wherever there is any place to go in space or time, we always find energy there, and nothing we do can really get rid of it.

Whatever energy is, everything we know is constructed of it. We know that energy is always quantized, meaning it comes only in distinct, unbreakable packets. These are the ultimate "atoms" of the universe, from which even atoms as we know them are made. In fact, we know how energy can be bound up in the form of any of hundreds of fundamental sub-atomic particles. Three of the most basic are photons, quarks and leptons. The most common lepton is the electron, and those are found almost everywhere, and lots of things are made from them, like electricity—or depend on them, like molecules and solidity. Photons, of course, are what light is made of, as well as heat and X-rays and radio waves and so on, even magnetic fields. Quarks stick to each other in various ways to make bigger things like protons and neutrons. And all these things get entangled with each other in countless ways, to produce different atoms, which in turn bind to each other in different configurations to make molecules, the bedrock of all solids, liquids, and gases, from which everything we can normally see and touch is made.

We know almost exactly how many ways energy can bind to form particles: we have counted up all the possible sub-atomic particles and found out how they all relate to each other in a complex scheme called the Standard Model of Particle Physics. We have counted up all the possible atoms that can ever be found in nature, and organized them, too, in what we call the Periodic Table of Elements. And though there are endless ways atoms can form molecules, we know exactly how atoms can do that and all the ways they can do it. This is a fact we will return to later: how so many wildly different things can be made from one and the same stuff.

> On the connection between space-time, geometry, and matter-energy, see: Henning Genz, *Nothingness: The Science of Empty Space* (1999) and Brian Greene, *The Fabric of the Cosmos: Space, Time, and the Texture of Reality* (2004). On Superstring Theory and other approaches to explaining everything by appealing to the geometry of space-time: Brian

Greene, *The Elegant Universe: Superstrings, Hidden Dimensions, and the Quest for the Ultimate Theory* (2000) and Dan Falk, *Universe on a T-Shirt: The Quest for the Theory of Everything* (2004).

For a discussion of all the particles of matter and energy which as far as we know comprise everything that exists, the evidence we have supporting that conclusion, and the science that governs these things (Quantum Mechanics), see: G. L. Kane, *The Particle Garden: Our Universe As Understood by Particle Physicists* (1996); Timothy Smith, *Hidden Worlds: Hunting for Quarks in Ordinary Matter* (2003); and Kenneth Ford, *The Quantum World: Quantum Physics for Everyone* (2004).

On chemistry, the science of how these particles combine to produce most forms of matter as we know it, see: John Sevenair and Allan Burkett, *Introductory Chemistry: Investigating the Molecular Nature of Matter* (1997) and Mark Bishop, *Introduction to Chemistry* (2001).

For more advanced discussions of the fundamental science of matter and energy, see works like: Tony Hey & Patrick Walters, *The New Quantum Universe*, 2nd ed. (2003); Anthony Zee, *Quantum Field Theory in a Nutshell* (2003); Jim Al-Khalili, *Quantum: A Guide for the Perplexed* (2003); R. Stephen Berry, Stuart Rice, and John Ross, *The Structure of Matter: An Introduction to Quantum Mechanics*, 2nd ed. (2001); Riazuddin Fayyazuddin, *A Modern Introduction to Particle Physics*, 2nd ed. (2000); Gerard t'Hooft, *In Search of the Ultimate Building Blocks* (1996); Elliot Leader & Enrico Predazzi, *An Introduction to Gauge Theories and Modern Particle Physics* (1996); Richard Feynman, *QED: The Strange Theory of Light and Matter* (1985).

5.3 Physical Laws

In the most basic sense, what we today call "physical laws" or "the laws of physics" are not really "laws" but precise descriptions of behaviors and relationships consistently observed in nature. When nature always behaves a certain way, or presents things to us in certain predictable relationships, we put it down as a "law" that the universe shall always do so. We could be wrong, but until we are proved wrong, and so long as we have heaping wads of evidence that the universe only behaves in that way, we can be confident the universe does not behave any other way.

The laws of physics are actually a large collection of different kinds of things, and some are more fundamental than others. For instance, it is a physical law that at sea level water boils at one hundred degrees Celsius. Though this used to be just one among many brute facts about the world, one of the "physical constants" or "constants of nature," it has since been found to be the inevitable outcome of other, more fundamental laws about pressure, chemistry, and Quantum Mechanics, which are in turn determined by still more fundamental facts about the universe, all ultimately reducing to the physical structure of matter and the physical behavior of energy.

The properties of water, in fact, would now be entirely predictable even if we had never seen water before: for all we need to know is the structure and behavior of all the particles that go into creating water molecules, and from that alone we can derive what behaviors that molecule will exhibit, such as its boiling point or even its color. Though actually making these calculations is beyond our limited time and minds in most cases, something like a molecule's color, when fixed in a particular larger-scale structure of known configuration, is entirely definable in terms of the basic interactions of subatomic particles.

Almost everything, in fact, is already reducible to this, and probably everything will be eventually: how energy behaves, and the different physical shapes that it takes. Besides how the geometry of a water atom entails its boiling point, or how Relativity explains gravity by appealing to the geometry of space-time, a more well-known example is how the common inverse-square law (governing gravity, magnetism, luminosity, and so on) is entailed by the geometry of any expansion in four dimensions. So far, the steady progress of past science has continuously and without fail grounded more and more laws and constants in the geometrical relationships and properties of quantum particles. So it is probable that what remains unexplained (certain basic constants, for instance—like why the speed of light is what it is) will likewise go the same path everything else has, and end up being yet another inevitable fact of the geometrical structure of space-time. It is already the most plausible inference that everything that happens and exists is entirely described by the laws of physics, all the laws of physics describe nothing more than the inevitable behavior of particles within a region of space-time, and this behavior is entirely produced by geometry—including the geometrical properties of the particles themselves, and of the dimensions they move through and inhabit. The evidence so far leads us in no other direction, and does not support any other conclusion.

I have already mentioned that this view, that everything is matter and energy in space-time, is called physicalism. According to physicalism, there is no other sphere of being in which the 'laws' of nature reside and impose themselves on the universe. Rather, these 'laws' are simply what happens when multiple dimensions are shaped and bound a certain way, hence they are a necessary outcome of certain patterns of arrangement. So long as matter, energy, space, and time exist, all these must surely interact and behave in particular ways in accordance with the principles of geometry. And this behavior is what we observe, and describe with what we call "laws." In other words, physical laws are the inevitable end result of what a universe *is*.

> For a general survey of all that science has discovered about the universe, see Nigel Calder, *Magic Universe: The Oxford Guide to Modern Science* (2003), which will provide you with sources to pursue on every realm of study, from both Classical and Quantum Mechanics, to Relativity and Chaos or Complexity theory.
> For more specific discussion of what the laws of physics are and how we know them, see works like: Paul G. Hewitt, *Conceptual Physics: A New Introduction to Your Environment*, 8th ed. (1997); Roger Newton, *Thinking About Physics* (2000). A. I. Burshtein, *Introduction to Thermodynamics and Kinetic Theory of Matter* (1995); and Albert Einstein & Leopold Infeld, *The Evolution of Physics: From Early Concepts to Relativity and Quanta* (1938).

5.4 Abstract Objects

"Abstract Objects" are not really objects. They are sometimes called "universals" because they are universally held in common by particular things. But they are also called "abstract" because they are not particular things, but qualities or properties that are "abstracted" from individual cases. For instance, after seeing red apple after red apple, we begin to abstract from these experiences certain commonalities—appleness, roundness, redness—that can be shared by different apples, even different objects altogether. In chapter II.2 ("Understanding the Meaning in What We Think and Say") I already discussed this abstracting process and how it occurs and what it refers to, and we can combine that now with what we have covered so far about the nature of things.

In the simplest terms, an abstraction or a "universal" is a potential pattern of experience, a repeatable pattern of matter and energy in space-time. The particular pattern it denotes is identifiable in the way set forth in II.2.1.4, "Getting at the Real Meaning of Words." For example, triangularity refers to any arrangement of three connecting straight edges. That there are straight edges in our experience is undeniable, likewise that there are edges that connect, and that there are many occasions when three do so in a triangular pattern. In order to think and talk about these occasions, we invent a word that refers to that common pattern, and that is an abstraction. Even before we invent a word for it, our brains are already doing this, recording the pattern and physically preparing to recognize other instances of it.

Abstractions don't exist apart from these facts—that is, pattern recognition by the brain and our naming that pattern. Even if a particular abstraction is never experienced in any way (not even imagined by anyone), it does not follow that the repeatable pattern it corresponds to does not exist. It might exist somewhere, at some time, past, present, or future. And if a particular universal property, a particular pattern of matter and energy in space-time, is actually never really manifested anywhere, ever, the abstraction that would correspond to it could still yet be manifested as an experience, through someone's imagination, as an idea that is never realized.

People can thus "create" abstractions, new patterns, that have never existed or will never exist otherwise, but they can also "learn" them by observing real patterns in experience, but either way, as the brains that do this are a physical part of the universe, so are abstractions. Only if an abstract idea is never imagined by anyone, ever, and the pattern it describes is never, ever, manifested in any physical object in any way, can we say in some sense that it doesn't exist. Yet even then, there is no reason why such a pattern couldn't have been manifest or imagined had circumstances led to it, and thus in another sense every conceivable abstraction exists "potentially," just as the principles of geometry already entail and hence contain every possible physical shape, whether it is ever realized or not.

To explain what I mean, and how naturalism accounts for abstractions, I shall take four paradigm examples that, together, should give you a fair idea how all other abstractions of any sort would be accounted for, using the same evidence and the same lines of reasoning. These four examples are numbers like "two," emotions like "love," colors like "redness," and activities like "running." If naturalism can explain what these things are, it can surely account for everything else that exists, making it a good, credible theory of existence.

Of course, some things need extensive explanation because of their complexity. For this reason, I treat emotions, including love, and the human mind generally in their own chapters (III.10, especially 10.3 and 10.6). Here I will tackle only numbers, colors, and activities (expanding on material in II.2; see also III.6.5, "The Nature of Knowledge" and III.9, "The Nature of Reason").

5.4.1 Numbers, Logic, and Mathematics

We will start with the number two. "Two" is merely a word, invented by humans to stand for any occurrence of a pattern of two discrete things. The only thing that is required for the number two to exist is any two distinguishable things, and wherever that pattern is repeated we can say a "twoness" exists there, meaning a pattern of two distinguishable things. So, for this number to exist "in the abstract" all that is required is that there be a potential (real or imagined) for two discrete things of any sort to exist. The word "two" thus means something like "the repeating pattern of two distinguishable things," and to say it "exists" is to say that this pattern is somewhere, at some time, manifest in the world—and nothing more.

It follows, when we apply this reasoning to everything, that all of mathematics is nothing more than a language, a language distinguished from others by its component simplicity and lack of ambiguity, a language humans created for precisely describing repeating patterns like quantities and relations. And quantities and relations are patterns that we know can manifest out of the arrangement of matter and energy in space-time. All of mathematics can be reduced, in fact, to one single field of inquiry: geometry, the language we invented for precisely describing pattern and shape.

Geometry was inspired by the existence of extension, though it can describe any system of patterns that share the same basic properties as extensionality (such as equality, divisibility, and arrangement). In other words, so long as there is such a thing as space-time, people can construct a language to describe that and all its possibilities, and from this comes math, even maths that describe things that are unreal (outside human thought or calculation) but constructed from the different parts of real things—like imagining six dimensions instead of four. We can do that because we have experienced the multiplication of extensions and thus can multiply extensions even further in our imagination, much as we can imagine a creature half-cat and half-dog. What we call "logic" is, in turn, an unambiguous way of describing things and analyzing those descriptions, even without the mathematical advantage of component simplicity.

All logical and mathematical "discovery" is therefore a process of coming to know and understand the different patterns that can be formed, or could be formed, out of the stuff of the universe, and deducing their consequences, all by picking apart and rearranging in creative ways our descriptions of known and invented patterns. For this reason, logic can never give us any new facts about the world: it only unpacks, extracts, exposes to our understanding information already contained within any given set of propositions. In short, logic and mathematics are human creations, just like English or German, and like English or German, logic and mathematics describe both real and potentially real things: repeatable patterns of matter-energy in space-time (see chapter bibliography below).

5.4.2 Colors and Processes

In the same way, words like "redness" refer to repeating patterns of light. Light, in the most basic sense, is an experience of certain distinctions in our visual field. But scientists have discovered that the cause of this experience is the brain's computation of distinctions in the "frequency of oscillation" of particles called "photons," which have flown or bounced off of an object and are striking certain cells in our eyes. Thus, color is physically real, a difference in rate of vibration in a barrage of real tiny particles, a form of energy. When we say two objects share a property called "redness" we are saying that they both have a pattern of arrangement that causes a predominance of photons of a certain frequency (one that we label "red") to strike our eyes.

We can perceive and thus name all these patterns called "colors" because our brains have evolved the ability to detect and then compare a perceived pattern against our memory of previously-perceived patterns, and identify a match when there is one. What we call "perception" is essentially this very ability to locate and distinguish patterns in the data of sensation, and "recognition" is the ability to identify similarities among those patterns, past and present. It is the same for every kind of experience, from colors and sounds to thoughts and emotions (see III.6 and III.10, "The Nature of Mind" and "The Nature of Emotion").

Color is really a process: it only exists because photons oscillate. Color distinctions are the product of variations in this rate of oscillation. But we cannot see this process directly, so we perceive colors as a static property of things. But other processes, which we can see transpiring before us, are no different from colors, except possibly in their greater ambiguity or range of instantiation. For example, the word "running" refers to a repeating pattern of legs traversing ground in a rapid motion. A great number of patterns, indeed a whole continuum of them, are identified by this word (just as the word "red" refers to a whole range of red-like colors), but all these patterns are still distinguishable from others, like walking, which is distinct from running in both rate of movement and pattern of muscular-skeletal motion. But so long as there are legs that can run, there can be a pattern instantiated by legs, which we call "running," one that we can recognize as the same even when totally different legs are doing it.

Already we can see there is no need for any other explanation for the existence and nature of abstractions (for further support, see Part VI, "Natural Beauty"). In order for there to be patterns of matter-energy in space-time, we only need there to be matter, energy, and space-time. Inevitably, naturally, some of those patterns will be repeated, and anything built to recognize those patterns, and physically store that recognition (like a brain), will naturally be able to distinguish the fact that the patterns repeat, and that repetition of organization can be assigned a name. That name is what we call an "abstraction," which refers to a real pattern in experience that could be repeated, even as a part of something else.

I believe this because in every case I have examined I have been able to reduce an abstraction to some repeating or repeatable pattern of matter-energy in space-time. I also believe it because the sciences have converged on just such a discovery: more and more it appears that all of sociology can be reduced to psychology, all of psychology can be reduced to biology, all biology to chemistry, and all chemistry to physics, which is the study of matter and energy in space-time. Therefore, everything is matter-energy in space-time. This is called "reductionism," and I will say something more about that in section III.5.5 below.

5.4.3 Modal Properties

A property is a pattern that can be instantiated, often within a larger pattern or in connection with others. Any instance of a pattern entails that it is possible and "exists" as a property. For example, the total physical pattern instantiated by an apple has several sub-patterns, many of which are detectable by different sensory organs: for instance, one aspect of that pattern causes a taste receptor to recognize and distinguish the shape of certain physical molecules emitted from the apple, the eyes recognize and distinguish how photons reflect off of its skin, and so on.

It is this ability to recognize and distinguish physical differences in these patterns (surface texture and atomic composition, tasteable chemical compounds, and so on) that gives rise to our language about abstractions—all so we can refer to those recognizable distinctions. The abstraction refers to a repeatable physical pattern, whether that pattern is instantiated anywhere or not, and regardless of where or when it does so. But there is a particular kind of pattern called a "modal property," which we sometimes refer to as a "possibility" or an "ability." For instance, Ed Norton can survive a trip to Paris but cannot survive a trip through a meat grinder. Ed Norton's ability in either case is a modal property: because of what he *is*, he has certain capabilities and limitations, his pattern contains certain *possibilities*. Philosophers make a lot of high-brow fuss over this, about how properties can be sustained, and contain possibilities within them. But it is fairly easy to explain what a modal property is in my view of Metaphysical Naturalism.

Modal properties are the causal consequences of a particular pattern of matter-energy in space-time. To say that X can do Y means that X *will* do Y in circumstance Z. Or, to say that it is possible for X to do Y is to say that in circumstance Z, X will do Y. Even if Ed Norton never suffers the horrible fate of falling through a meat grinder, we can say that if he did so, he would be destroyed, and what came out the other side would not be Ed Norton, but a pile of goo. We can say this because the pattern of matter and energy instantiated by Ed Norton is such that this is what will result when it collides with the pattern of matter and energy instantiated by a meat grinder. It is a geometric inevitability.

This will never change unless the patterns change—and not merely change, but change so much that they are no longer Ed Norton or a meat grinder. In short, "Ed Norton cannot pass through a meat grinder" means, literally, "Ed Norton will not pass through a meat grinder," which we can assert because we know what humans are made of, and we know what meat grinders do. There can only be an Ed Norton if matter and energy exist in a certain pattern of organization, a pattern that will be fundamentally changed by a meat grinder. The goo that results will contain all the same material out of which Ed Norton was made, but it will not really *be* Ed Norton, because it lacks the arrangement necessary to make an Ed Norton. It possesses instead the arrangement necessary to make for a random "goo" rather than a human being. In contrast, a trip to Paris will have none of the causal effects on Ed Norton that a meat grinder would have, and thus Ed Norton's existence, his collective of properties, would be relatively unchanged by such a move.

This is the inevitable conclusion from a belief in causation: if everything requires a cause, then absent any cause an object will not change its properties, i.e. the pattern of matter and energy it has will not change. Thus we are justified in believing those properties will endure, for we lack any reason to think otherwise, and have abundant evidence that this is how things are. We have support in science both for the fact that everything in the universe has a cause and the fact that absent a cause nothing changes. For instance, for the former we have the Laws of Thermodynamics, and for the latter, the Laws of Inertia.

Even what people often refer to as the Law of Entropy (the Second Law of Thermodynamics) does not state that things disintegrate or become disordered over time, but that when work is done to cause changes in a closed system (like generating order out of chaos), *some* of the energy used to do that work will become disordered, even to the point of being unavailable to do any more work. But for this effect to occur, there must be causes: something in that system must have *caused* part of the system to degenerate. Otherwise it would remain static (not necessarily motionless, but unchanging in whatever motion it has). Ultimately, it is shape and organization that give matter and energy modal (hence causal) properties, including the ability to cause other patterns to change in particular ways, a fact we shall now explore.

5.5 Reductionism

What I described earlier as "reductionism" is the view that everything can be reduced to the same, one thing. Physicalism is thus a variety of reductionism in which everything can be reduced to matter and energy in space and time: quarks and other sub-atomic particles and their behaviors are all that there is, out of which everything without exception is made. And this fits with the fact that society can be reduced to humans, and humans can be reduced to cells, and cells can be reduced to chemical systems, which can be reduced in turn to sub-atomic particles. So, therefore, societies can be reduced to sub-atomic particles. The natural corollary of this view is that the sciences follow the same pattern: sociology can be reduced to psychology, psychology to biology, biology to chemistry, and chemistry to physics. So, theoretically, all of sociology and psychology can be described entirely by physics.

This view is often attacked with the *modo hoc*, or "just this," fallacy. This is the faulty argument that if, for example, all we are is matter in motion, then we are just clumps of moving matter and nothing more. This is clearly false. Like Ed Norton in our previous example, there is obviously a difference between us now, and us hacked up into a stew. Both contain all the same matter, but not the same pattern of arrangement. Thus, how matter and energy are patterned, arranged, within space and time is itself a defining aspect of a thing, and this pattern has causal and other distinct properties.

For example, a bundle of rods and the same rods welded into a tripod contain all the same material, but though the rods in a welded tripod can stop a car from rolling, the rods laying on the ground will not. The only thing that has changed is the arrangement, and yet a new causal power arose from that simple change—and it arose solely as a consequence of a change in geometry. So it is with everything. For instance, though a tree can be reduced to nothing more than a huge quantity of quarks in a particular relation to each other, the particular way in which they are arranged distinguishes the tree, with all its powers and smells and colors, from a pile of inert gray ash made of the same quantity of quarks, even the very same quarks. Likewise, a ring of gold and a block made of that same gold are both "just" a quantity of gold, yet only one has the property of being able to roll down inclines and fit on human fingers, solely as a result of being in a different *shape*. Therefore reductionism does not entail that everything is "just" matter and energy in motion. That is the fallacy of *modo hoc*.

Also, a distinction must be made between the reduction of substance and the reduction of method. The one does not entail the other. That psychology reduces to biology does not mean we can do without psychology and talk only biology, or turn all research in psychology over to biologists—and thence, by logical progression, to physicists. There are two reasons for this.

First, each level of inquiry is too complex for this to be carried out in practice, at least by humans. It would take some sort of super-mind to actually do what is only possible in theory. The higher science of psychology is needed in order to simplify and analyze aggregate phenomena, focusing on larger patterns rather than the myriad smaller ones that compose them, whereas biology needs a certain methodological myopia that emphasizes smaller patterns within the larger patterns of interest to psychology. Yet they both inform each other, and our explanation of phenomena either is or aims at being continuous between them: a psychological phenomenon is never regarded as completely explained until it is explained biologically—likewise, a biological phenomenon is never considered completely explained until it is explained chemically, and a chemical phenomenon likewise requires a fundamental physical explanation at the most fundamental level.

That scientists regard things this way, and the fact that things have been this way and are becoming more so with every passing decade, is proof enough that reductionism is a reality. A physicist could never do a psychologist's job. But the whole gamut of scientists is needed to explain every phenomenon fully, even a psychological one. So though the physicist explains everything examined by all the sciences, she does so only at the smallest scale and the most fundamental level of analysis.

Second, reduction does not necessarily correspond to a unique correlation of phenomena. For instance, mental and biological phenomena are the outcomes of patterns of activity that could be manifested by several different materials. They only *happen* to have one in common, here on earth, but many others are possible, and on other worlds may even be a reality. Psychology would apply equally to a mind produced by a biological or an electronic brain, if the patterns of activity were sufficiently the same. If an alien arrived whose brain used entirely different chemicals and structures to perform what are otherwise the same aggregate functions as our brains, our science of psychology would describe both minds, even though their underlying biology was different.

Likewise, much of biology would be unaffected by a life system comprised of antimatter or a different set of proteins or even a different chemistry altogether, so long as the alien system produced the same patterns of activity studied by biology, such as evolution by natural selection. Thus, each stage of human science studies a different scale of organization, which in many cases can be manifested in more than one way by the smaller patterns, which are the subject of other, lower-level sciences. Thus, there is a practical utility in distinguishing the sciences, even though they ultimately reduce to physics.

And this relates to method. Patterns themselves have unique causal powers, even when wholly and exclusively composed of a simpler substance explored by a more fundamental science. And the study of the causal powers of particular *arrangements* of a substance is methodologically distinct from the study of the substance itself. How you study mental behavior requires entirely different skills, instruments, and research patterns than the task of studying quantum physics (see IV.1.1.1, "The 'Scientific Method'").

Two examples should complete this picture of reductionism. Consider a game of chess. The entire game is reducible to nothing more than binary mathematics. It could be reduced to other, non-binary systems of computation, even to incredibly simple machines like an actual chessboard and pieces. Nevertheless, we can play chess on a computer, knowing full well that nothing is going on except a vast stream of electrons changing places in a web of components according to the physical structure of a tiny computer chip. But most of us could not play chess by typing in binary commands and reading raw binary output. Even though that's all it is really, most humans just can't handle that complex a task, nor would the skills involved be at all the same. A chess player requires something like the images on a screen and simple keyboard commands, while a programmer can deal with the rest, though only in small parts at a time (and usually with much difficulty). Though the chessplay reduces to binary computation, neither programmer nor chess master can do the other's job.

This follows for substance dualism, too. It is clearly fallacious to say that a pawn in a game of computer chess is "just" electrons in a grid and nothing more—there is clearly something more. The organization of those electrons and that grid together make for a distinct difference between a game of computer chess and a scramble of random noise, even though both are made of the same things, even the very same electrons and grid. Presented with the computer chip as it played chess, and then challenged to "Find the pawn!" how do you think you would fare? There is no distinct

Sense and Goodness Without God

place or form of anything on that chip that is the pawn. At best you might isolate certain electrons at certain positions that at a certain time "represent" the pawn. But you would have to include everything, including the rules for a pawn's behavior, as well as the data pertaining to its position, and for imaging a pawn-like figure on the computer screen, and these would all change shape and position every microsecond.

Yet it is manifestly false to say that the pawn doesn't exist. It plainly does—it just took your knight! It is just as absurd to say that the pawn exists in some higher abstract reality. As any mathematician or computer engineer can show you, it only and entirely exists as a pattern of electrons in motion on a grid of transistors. Yet that pattern is causally different than any other, distinctly a pawn in its placement and behavior in the game. Thus, studying the basic process of electrons moving on grids is a wholly different enterprise than studying the nature and effects of a particular pattern of such movement (a "pawn") on the behavior of the whole system. You cannot in practice do the one by doing the other. Humans just aren't that limitless in ability.

Another example is the old hierarchy of computer control systems: we are familiar with operating a computer by using Windows, which was originally a phenomenon completely reducible to something more obscure and complicated called DOS, which in turn was completely reducible to binary machine language and nothing more. If you knew how to operate a computer by one of these means, you would not necessarily know or even be able to deduce any of the others, for the complexity of each stage was too vast, and the patterns that distinguish each from the others too different. Thus, you had to learn them all separately, employing different skills to master each, even though there was no difference at all among them in terms of their substance: Windows is composed of *nothing else* but binary machine language, electrons on a grid. Yet we talk of "folders" and "dragging" and "icons" and this still makes sense, referring to real things, just not exactly the same things these words refer to in other contexts.

> In addition to the readings recommended in sections II.3.3 and III.5.2, see: Gail Fine, *On Ideas: Aristotle's Criticism of Plato's Theory of Forms* (1995); Richard Jones, *Reductionism: Analysis and the Fullness of Reality* (2000); Ansgar Beckermann, Hans Flohr, and Jaegwon Kim, eds., *Emergence or Reduction? Essays on the Prospects of Nonreductive Physicalism* (1992); Harold Kincaid, *Individualism and the Unity of Science* (1997); Edward O. Wilson, *Consilience: The Unity of Knowledge* (1999).
>
> Also, regarding abstractions and mathematics in particular, see Mary Tiles, *The Philosophy of Set Theory: An Introduction to Cantor's Paradise*

(1989) and Penelope Maddy, *Naturalism in Mathematics* (1997), as well as: Paul Benacerraf and Hilary Putman, *Philosophy of Mathematics* (1964); Hugh Lehman, *Introduction to the Philosophy of Mathematics* (1979); Bruce Aune, *Metaphysics: The Elements* (1985); and Bertrand Russell, *Principles of Mathematics* (1938), with Jaakko Hintikka, *The Principles of Mathematics Revisited* (1998).

Most importantly, Ian Stewart proves with abundant examples in *What Shape is a Snowflake?* (2001), as Michael Resnik does with rigorous philosophical argument in *Mathematics as a Science of Patterns* (1997), that all of mathematics is simply a human language that describes the real shapes and patterns of matter-energy in space-time.

For more discussion of all the issues related to that point: Barry Mazur, *Imagining Numbers: (Particularly the Square Root of Minus Fifteen)* (2002); George Lakoff & Rafael Núñez, *Where Mathematics Comes From: How the Embodied Mind Brings Mathematics into Being* (2001); Reuben Hersh, *What is Mathematics, Really?* (1999); Brian Butterworth, *What Counts: How Every Brain is Hardwired for Math* (1999); and chapter 7 of: John Barrow, *Impossibility: The Limits of Science and the Science of Limits* (1998).

6. The Nature of Mind

So what is the soul made of? Scientists have come to realize there is no empirical distinction between what people popularly call a "soul" and what scientists and philosophers have long called a "mind." Though people often attach numerous superstitions to their idea of the soul (such as that it can survive without the body) the only evidence anyone has that souls exist at all is simply the evidence that minds exist, and over the last fifty years scientists have rather conclusively proven that there is never a mind without a brain. Therefore, a soul appears to be nothing more than the functional outcome of a particular kind of brain. Since your mind is your soul, to discuss the nature of the soul means discussing instead the nature of mind. Accordingly, the following takes scientific fact and fills in the blanks according to the predictions of Metaphysical Naturalism.

6.1 The Mind as Brain in Action

The word "mind" refers to a particular pattern of brain content and activity, which includes not only the input and processing of data (sensation), but also the recognition of patterns in that data (perception), the analysis of relationships among those patterns (reason), and above all, the ability to recognize a *particular* pattern: that of a *self*. Hence we say we are "self aware," a fact that distinguishes us from possibly every other animal on earth. This would appear to be the only thing people can really mean when they say humans have souls and animals do not. But to imbue this ability with supernatural powers, powers beyond the mere fact of generating self, is unsupported by any reliable evidence, scientific or otherwise.

Our ability to perceive and construct a "self" bestows on us magnificent though perfectly natural powers generally not shared by other animals, such as a self-reflective sense of identity (the ability to comprehend and study who you are), a rational will (the ability to deliberate from facts and values to personal goals and consequences), and a self-referencing memory (in short, a past that you can call "yours," which plays no small part in the other two activities: creating and understanding your identity and deliberating rationally about yourself and the world).

The brain serves two functions. First, as a storage medium, its arrangement preserves everything that we are, our memories, abilities, and qualities, like personality and desires. And, second, as a living mechanism, its activity generates our continuity of perception and thought, our "consciousness." This is as true of animals as of people. Many animals have unique personalities, memories, and mental abilities, and can be "conscious" of their surroundings, even to a certain extent themselves. But to be able to fully perceive themselves—as a mind, as a person—requires a special organ capable of such a computation, and an organ capable of perceiving a whole pattern of such a size and complexity would have to be vastly complex itself, far more than any other sensory organ like, say, the human eye.

It just so happens that we have one of these: a cerebral cortex, the most complex biological organ in the world—in fact, as far as we know, the most complex thing in the whole universe. Animal brains are simpler, lacking this organ. But once a brain has one, and can perceive itself with it, that brain acquires an unprecedented feedback loop: its self-perceptions generate entirely new kinds of information that, in turn, like all information, affects the behavior of the whole being, changes that are in turn perceived. And so on. *That* is a mind.

6.2 The Mind as Virtual Reality

Modern pilots train in flight simulators, enclosed rooms on hydraulic platforms that move like a plane would. Inside, video screens present to the operator a simulated appearance of what the outside would look like and what the instruments would be reading if he were really in the air, and a computer adjusts all this to correspond to what would really happen if the pilot took any particular action with the controls.

This is called "virtual reality" because it isn't real—the pilot's window and instruments say he is in the air and banking left, say, but he isn't really doing either—but it is "virtually" real because it is almost a reality of its own, presenting facts to him, and reacting just like reality would in response to what he does. Indeed, the entire computer program that does this is built to emulate the real world, so the pilot will gain experience with what things really look like and how they really happen, without actually risking his life or anyone else's, or taking up the resources of a fuel-guzzling aircraft. The virtual reality presented to him in the simulator corresponds to a real universe. In fact, he could fly a real airplane connected to the simulator, without ever looking out of a real window.

This is basically just what a brain does. Simple brains, like those in worms, only perform simple calculations in real-time ("move toward light"), but as brains evolved in complexity they developed the ability to simulate reality altogether. Cats, for instance, dream of hunting prey, and all indications are that they probably don't know they are dreaming. For instance, when the brain-functions that cut off muscular motion during sleep are numbed, dreaming cats actually move about a cage, stalking imaginary creatures. Thus, the cat's dream probably presents itself as a wholly "veridical" experience: the cat actually sees, hears, smells imaginary prey, and reacts accordingly.

It is easy to imagine why a brain of sufficient complexity would evolve such a function: just like the pilot in the flight simulator, the cat is gaining experience at hunting even on its down-time, without exposing itself to the dangers of a real hunt, and on very little expense in energy. Moreover, it no longer has to *actually* experience certain prey behaviors to learn how to respond to them, for the cat's brain can invent, experiment with, hundreds of different situations, far more than the cat might ever encounter in reality, and so the cat gains far more experience than it ever could otherwise. It will likely find itself prepared even for unexpected real-life situations, having rehearsed similar ones in dreams.

Scientists have proven that there really isn't much difference between dream and reality, in terms of the brain hardware used and its activity. The major difference is whether the data from which the "virtual reality" is generated come from the sensory organs external to the brain, or from the brain alone (and a healthy brain, when properly functioning, has mechanisms to detect this distinction). Humans have learned to go even beyond this, distinguishing faulty data from our sensory organs from real signals stimulated by external phenomena, or identifying the real existence of things that make no direct impact on our senses at all. For example, we can infer that a solid gold plate is mostly empty space by demonstrating that solid things like atoms can pass through it, even though we can't see or feel that empty space directly. And when we suffer a blow against our eye we sometimes see bursts of light, but by having other onlookers observe the same objects, and then by studying eye physiology, we can discover that the starbursts are only in our eye and not really in front of us—and this conclusion is supported when we find out just how pushing on an eye can generate such a false signal. The dream-reality distinction has been reinforced by such discoveries, and others (for example, reality sticks with us, but things that happen to us in dreams go away when we awake), and indeed the sum of all these distinctions is what we mean by "dream" and "reality." But otherwise, the brain is doing the same thing in each case: presenting a virtual reality.

The dream-reality distinction is thus best understood as the difference between a flight simulator that is not connected to a real plane, but that is just making up data, and a flight simulator that is connected to a real plane, and thus whose video displays and instruments are projecting constructions from real data, not mere inventions of the computer itself. Everything we experience is a construct, a convenient way for the brain to "represent" the otherwise complex and jumbled data of the senses and brain systems. But we have ample evidence, ample reason to believe this "reconstruction" of a world outside of us is based on real data from that world, and thus strongly corresponds to it. It is thus a reliable presentation, even if the manner of presentation (gold plates looking and feeling solid for example, or the particular appearance to us of the color red) is an invention of our brain (see also II.2.1.2, "Meaning, Reality, and Illusion").

The brains of *all* higher animals are such virtual reality machines, and very useful ones at that. But our own brain has the added complexity to pull off an even more astounding virtual construction: that of a self. The human brain not only constructs from external data a practical representation, a "virtual reality," of the world we navigate through, including our body's exterior appearance and position, and so forth, but from the internal data of the brain's own operations it constructs a practical representation, a "virtual reality," of a unified and centralized "person."

Like all perception, it takes time to do this. When we "see" something, a fraction of a second has already passed since the light hit our eyes, because the brain must process it to achieve a perception. But this does not mean the light doesn't exist or lacks causal powers. Likewise, scientists have proven that we also experience our "self" a fraction of a second after the fact, since the brain requires time to process the perception of what our brain is doing, of what *we* are doing. But the perception is still of a genuine reality, of things really in and going on in the brain, like reasoning and memories. And this 'self'-perception still has causal powers on our future—causing the formation of certain kinds of memories and steps of reasoning, for example, that could not occur without it.

The brain constructs this "self" out of the brain's real values, memories, abilities, calculations and choices. These control the brain's highest functions, and they collectively correspond to 'us'. The "virtual self" includes the sensation of being at the center of it all, of observing it all. And this perception of a self *is* the observation, not something distinct from it. It is a construct of use to the brain, of use to *us*. This is what consciousness actually is, and a lot of evidence confirms it (see this chapter's final bibliography).

So our brain processes information and creates patterns in a way that attempts to model ourselves and the outside world. And it can do this, again, in two distinct modes: in response to real external and internal data, or in response to invented data, as happens in dreaming or imagining. In the one case the brain uses externally-supplied data, checked and reinforced through feedback and accumulated experience, to try and develop a true picture of something really going on. In the other case, the brain (knowingly or not) creates *fictional* models, using remembered data in creative ways, in an attempt to work out possible scenarios and choices, as for the dreaming cat.

This is what permits us to consciously reason, to imagine and plan, and to think in our heads using the same language we speak—and when unfettered and unchecked, this is what happens when we dream or hallucinate. Its utility is just as obvious as for the cat: by simulating reality, drawing on real data from the real world, we can practice and experiment and think through countless scenarios and possibilities, gaining tremendous experience without actually risking anything or exhausting our resources.

6.3 The Chinese Room

Some philosophers have a hard time grasping this. Some even claim to have "refuted" it by appealing to a "thought experiment" called the Chinese Room, originally developed by another naturalist, John Searle. Basically, the experiment goes like this. Imagine a guy in a room. Little cards with Chinese characters on them pop through an "in" slot into the room, and the man in the room has the task of popping another card with Chinese characters on them through the "out" slot, as a response. He has a chest full of cards, every combination possible, and a big rulebook that matches sequences of Chinese symbols with others. Even though he can't read Chinese, he can follow the rulebook, and when a certain card comes in, he can match it up with the right card to pop out of the room.

If human consciousness is a computation, a process, then in principle one could write a rulebook for this man that would tell him to output just the right cards, so that a Chinese man outside the room would be able to have a fluent, intelligible, even profound conversation with the room, all the while ignorant of the fact that he is conversing with no one: for the man inside the room has no idea what the cards say.

This thought experiment is supposed to prove two things. First, conscious thought can't be a computation—no such rulebook could possibly replace a human mind, for a book isn't a person. Second, it follows that there is no center of the brain, no "core being," to whom the "virtual reality" would be presented (there is no one in the room who understands the Chinese conversation), yet we experience ourselves as just such an observer—so, again, what I described earlier cannot be a correct explanation of consciousness.

The problem here is an unfortunate lack of imagination on the part of everyone who finds this convincing. A "thought experiment" only works, only produces correct conclusions, when you *actually do* what the experiment asks of you. But when we do this, we do not arrive at these two conclusions after all. For example, for a Chinese man to have a conversation with the room so convincing he would be sure he was speaking to a real person who understood him, the rulebook in that room would have to be capable of some remarkable things. First, it would have to be able to be changed: the room would have to be able to remember things the Chinese man said, and what the room itself said, and adjust future responses accordingly. Thus, the rulebook would not be static, but constantly changing itself, and growing ever larger, even though doing so according to its own rules.

Second, not only would it thus accumulate knowledge, it would be accumulating knowledge about *itself*, for future use in its conversation (to answer questions like "How do you feel?" or "What's the difference between you and me?"). Third, the rulebook would have to contain something equivalent to desires, or else it would not exhibit curiosity, nor any capacity for active learning, and such omissions would be inhuman. So, built into the rules would have to be a hierarchy of simulated interests, for instance things the room would "want" to ask or say to the man outside in the course of any conversation, as well as things it would not want to.

Fourth, and related to this, would be creativity: the rulebook would have to have some rules for spontaneity and randomness, asking different questions and pursuing lines of discussion on a whim (for instance, if the man outside asked the room "Ask me something new."). And fifth, as with the rest, this part of the rulebook would also have to be changeable: to seem human, the room must be capable of changing its interests and desires and creative responses as new information is acquired or requested.

Sixth, the room would have to be capable of learning new skills—if the man outside tried to teach the room to speak Korean, for instance—so its rulebook would have to be capable of writing entirely new rulebooks, and have rules for which rulebook to use and when. Finally, and most obviously, the rulebook would have to encode all the rules of reason employed by humans, which involves the ability to compute virtual models of anything the room is asked to think about, and to change its interests and beliefs accordingly. I have not even exhausted all the things the room and its rulebook would have to be able to do to satisfy the thought experiment, but seven remarkable things is enough to make the point.

But once we have imagined *everything* we are supposed to—all that one must imagine in order for the experiment to meet its own condition that the Chinese man outside is convinced the room contains a person speaking Chinese—we realize we are left with something altogether different than Searle claimed. We realize instead that the rulebook we have imagined *is* a person, in every sense of the word that matters. It has desires, memories, interests, it can reason and learn, act creatively, and evolve and develop in its personality and abilities. It would be able to abstract universal notions from particular cases and reason about those abstract notions apart from their particular cases, and it would be capable of comprehending, with intelligence and insight, complex human languages.

What's more, this rulebook would not merely say it was self-conscious, it would actually exhibit abilities only a self-conscious being could have: for instance, it would be able to distinguish between itself and its own memories and interests and those of the man outside the room, in a way that would cause it to store memories specifically related to describing *itself* and distinguishing itself from things and beings outside. In other words, it would develop a coherent identity, a self-image, that would directly affect what it decides to do, how it remembers the past, and how it reasoned about new facts. If it were fully, convincingly encoded, the rulebook would even contain emotions. In the end, the rulebook would be so malleable, rich and complex that there would be no relevant difference between it and the human brain, with its complex rulebook of cells and synapses.

We just invented AI, "Artificial Intelligence." Thus, contrary to proving AI impossible, the Chinese Room proves it *possible*—if such a rulebook could be written. And that is an empirical question, not a philosophical one. But in that regard we are not limited solely to so-called Turing Machines, computers that can be reduced to and thus replicated by a simple serial, digital process (almost all computers today are actual or potential Turing Machines). And this was the only point Searle originally wanted to make: that if the mind is a computation, it is probably not comparable to a Turing process, because, as Searle puts it, "syntax is not sufficient for semantics."

I am skeptical of Searle's conclusion. As I have explained already, the mind achieves understanding by generating and 'walking through' virtual models of the world and each of the things in it, including a 'self' (III.6.2 "The Mind as Virtual Reality"), and then it thinks and communicates by assigning code words and rules to these models and their components and behaviors (II.2, "Understanding the Meaning in What We Think and Say"). Therefore, since the meaning of words and statements (what Searle calls "semantics") amounts to nothing more than attaching virtual models to words and rules (the elements of what he calls "syntax"), and purely 'syntax'-driven computers can generate models, it seems at least theoretically possible for a sufficiently complex Turing Machine to achieve 'semantics'.

But we need not assume this is necessary. For there are other kinds of computer. We already know, for instance, that a Quantum Computer is mechanically possible yet is completely incapable of being replicated by a Turing Machine. It is a wholly distinct class of computer. Likewise, there is good evidence that animal brains are computers that fall into yet another class of their own: for they employ analog and synchronized parallel neural-net processing, where computation in one part of the brain is simultaneously affecting computation in other parts, and in dynamic ways, and this may be the only way to produce consciousness (the experience of simultaneity being so essential to our kind of consciousness, and perhaps to any kind). It is possible this process has features that cannot be reproduced by a Turing Machine. Yet it is a simple thing to build a physical computer that operates according to such non-Turing processes, and thus to say a Turing Machine will never be conscious is not to say that man-made machines will never be conscious.

This leads us to the next conclusion. Once we realize there *is* someone in the room who understands the Chinese conversation—the rulebook, our AI—someone who is merely being supplied with energy by another man (whose role in the room is equivalent to the human circulatory system, dumbly supplying the brain with the energy and chemicals necessary to function), the second conclusion also falls. The only way to make a machine that can truly replicate human understanding is to make a machine that *has* human understanding. In other words, to be such a thing and to be a simulation of it are one and the same, with no logical or empirical difference between them.

But where is the observer? The fact that there is no center of the room, no center of the brain, no "core being," to whom our "virtual reality" is presented, is irrelevant. We *are* the presentation. For the rulebook would, like us, *experience itself as just such an observer*. It would be none the wiser. For the rules would simply encode the thought processes indicative of conscious awareness. Whenever the codebook was asked to inquire of itself whether it was a single, unified awareness looking upon its stream of thought as an audience might, it would answer that it was, and would believe it was. And in every respect that mattered, it *would* be. Every effect on behavior or belief that such an experience would have, the rulebook would be affected in just the same way, for it was encoded to do just that. It would never think itself "blind" to Chinese thought, because it wouldn't be. It would understand Chinese perfectly, and know it did. There would be no way the rulebook could "turn off" its belief that it was experiencing its 'self' as a central observer, nor could it imagine itself otherwise: for to do so would be to imagine itself out of existence, and to experience nonexistence is a contradiction in terms.

It takes a well-plumbed imagination to grasp this. But it should not be surprising that consciousness is so profound and difficult to grasp: we are, after all, talking about the most complex machine in the universe. To think you are going to solve it by imagining a guy in a room is a bit ridiculous. And since we *are* the machine in question, it should be obvious that we can't take ourselves apart and examine our operation, even in our imagination. For to do so would cease our operation, without which we could neither take ourselves apart nor observe the parts afterward. Can you imagine yourself at the same time not existing yet examining the operation of your imagination? You cannot. To imagine ourselves examining anything, we have to imagine ourselves as existing first. Thus, there is no logical way we could directly comprehend what it would be like to be a process, except the very way we experience being a process: as a coherent observer of a self.

Consider an analogy. If we take a storm apart and examine it atom for atom will we find a storm anywhere? Certainly not. The storm only exists as an aggregate concourse of those atoms—it only exists as a whole, not within its parts. Thus, we can take our brains apart and see all the different functional centers and nowhere find our "self" or our "conscious observer," because this only exists as the aggregate concourse of those individual brain centers, not as any one of them. Just as a storm has no single "location," nor even any constant or distinct measurements at all, yet clearly exists as a solely physical phenomenon fully explicable by science, so does our consciousness lack any such single location within the brain, or any constant or distinct mass or energy content, much less length or breadth, yet it is solely a physical phenomenon explicable by science (see also III.5.5, "Reductionism").

> The "Chinese Room" was first described in John Searle, *Minds, Brains and Science* (1984), pp. 28-41, under the chapter heading "Can Computers Think?" The rest of the book develops his alternative naturalist explanation of consciousness. For more on the idea of a Turing Machine, see plato.stanford.edu/entries/turing-machine. For quantum computers: George Johnson, *A Shortcut through Time: The Path to the Quantum Computer* (2003). For the kinds of computation the brain actually appears to employ: Peter Dayan & L. F. Abbott, *Theoretical Neuroscience: Computational and Mathematical Modeling of Neural Systems* (2001).
>
> For more on our mind as neural computer, see: Nikola K. Kasabov, *Evolving Connectionist Systems: Methods and Applications in Bioinformatics, Brain Study and Intelligent Machines* (2002); Patricia Churchland & Terrence Sejnowski, *The Computational Brain* (1992); Paul Churchland, *A Neurocomputational Perspective: The Nature of Mind and the Structure of Science* (1989); Patricia Churchland, *Neurophilosophy: Toward a Unified Science of the Mind-Brain* (1986); Aaron Sloman, *The Computer Revolution in Philosophy: Philosophy, Science and Models of Mind* (1978): www.cs.bham.ac.uk/research/cogaff/crp.
>
> On the nature of mind and consciousness, see the final bibliography at the end of this chapter.

6.4 The Mind as Machine

If a mind is just the function of a special kind of machine, what accounts for its most peculiar contents? How can a machine produce 'thoughts' for example? The answer begins with an understanding of what 'thoughts' *are*, as well as everything else that comprises a mind. So that is what we will consider here, followed by more detail in III.6.5, "The Nature of Knowledge," III.9, "The Nature of Reason," and III.10, "The Nature of Emotion."

6.4.1 Thoughts

In general, "thoughts" are the brain's own sensory perception of intellectual activity, of calculation and analysis. As the eye receives light, and translates it into organized electrical signals that the vision centers of the brain then recognize and harmonize with the output of other areas of the brain (the "virtual reality" construct), so there are brain centers that are the "eyes" of the brain itself, receiving the electrical signals of the brain's computations, the results of recognizing and experimenting and naming and so forth, and recognizing those processes in and of themselves, harmonizing them with the rest of the brain's virtual construction. In this way, thoughts have the causal power to make the body act, or to make changes in the brain itself (altering abilities, qualities, beliefs, or desires). All of this seems to us very mysterious, because we don't see the brain doing these things, any more than a driver sees how his car's engine produces forward motion. But the mystery does not make the process any less mechanical.

6.4.2 Abilities, Memories, and Traits

Besides thoughts, a mind contains a few other things. "Abilities" are skills, enhanced cellular pathways of action and reaction, that process data, raw or recognized, and produce a certain output for a certain input—pathways and connections that change in response to feedback. Learning how to ride a bike is an example, involving how exactly to move the body when the bike starts to lean. Then, like skills, "memories" are simply stored data or recognition-sets, related to each other and accessed in various ways, mapped into the very cellular and synaptic structure of the brain's flesh. So are personal "traits," which encode things like behavioral and emotional responses and impulses. They, too, like skills, are malleable with effort and time, but otherwise permanently represented in the brain's structure, generating emotional outputs, and things like desires and urges and even actual behaviors (such as facial or verbal expressions, gestures, laughter), in response to the various data the brain churns out in making its model of self and reality. These 'traits' are what comprise our values and personality.

6.4.3 What Machines Can't Yet Do

Given the survey above, there is nothing a brain does that a mindless machine like a computer can't do even today, in fact or in principle, except engage in the very complex process of *recognizing itself*. This requires computation of a complexity so great no computer yet made by man can even attempt it. The average computer now has the brain-power of a bee, while even the most complex computer has only that of a mouse. That is a far cry from what is needed for human-level intelligence. Even when we finally have such a machine on hand that we can experiment with (and the rate of progress in computer capabilities has been so constant for the past forty years it is a safe bet we will have such a machine before the year 2030) we will still have to figure out how to construct such an immensely complex organ as a self-perceiver, and no one knows how long *that* may take us. This is why AI has not yet appeared, though machines are already doing things now that naysayers twenty years ago said would be impossible. It is only a matter of time before humanity creates a new consciousness equal to itself, if not far superior.

6.4.4 Qualia

One thing I have left out of account so far is what philosophers call 'qualia', those peculiar veridical qualities of normal perception like the 'redness' of red or the 'urgency' of a desire. It is often held that naturalism cannot explain qualia, because no mechanism could ever produce them. But no one really knows if that's true. This is another subject at the forefront of science today, and there are still a lot of things left to discover about how the brain works to produce human experience.

Still, based on what we know so far, I suspect qualia are like the pawns in a game of computer chess: they don't really exist anywhere, beyond the shifting and complicated pattern of behavior of certain electrical signals on a grid. Just as a difference of pattern in this activity makes all the difference between a pawn and a knight, so also a different pattern of activity in the brain makes all the difference between an experience of 'redness' and an experience of 'loudness', or nothing at all. And I suspect this activity is completely sufficient to produce those experience, just as for the behavior of a pawn or a knight. There really isn't any way to prove that's not the case, and it fits everything else we know so far.

The central problem with qualia is that we cannot actually say, for example, that slugs or robots don't experience them, since the only way to ask them is to make them conscious enough to understand the question. So any connection between qualia and human consciousness could be a red herring, the latter not really being necessary to the former. Though consciousness is certainly necessary to appreciate and communicate qualitative experience, it is probably not functionally necessary for *having* qualitative experience (the obvious exception would be the qualia of self-consciousness itself, accounted for in III.6.3, "The Chinese Room").

I think it is safe to say that any process that produces virtual models, and analyzes and reacts to them intelligently, probably experiences qualia, since 'perceiving' the attributes of a perceptual model is exactly the same thing as perceiving its qualities, and thus identical to perceiving qualia. And we know all higher animals do this. So if we could get a mouse to talk, it could probably tell us all about what it is 'like' to 'see' a light at the end of a tunnel or to 'feel' the heat of a stove. How could it not? Would it make any sense for the mouse to say it was ambling toward a bright light rather than a dim one, or away from a hot area and toward a colder one, and at the same time report to us that it was having no experiences of lightness, dimness, hotness, or coldness? I can't make any sense of such a thing. So I don't see anything here that needs explaining. To explain perception is to explain qualia: they are one and the same thing. And advanced computers, still pure machines, have achieved perception, at least on the level of a mouse. I think it follows they have experienced qualia. They just can't appreciate or talk about them, any more than a mouse can.

For a good illustration of my point, consider people who have lost part of their 'consciousness' of vision, and do not experience qualia (they are 'blind'), yet can still point to things (though less reliably), and perform similar tasks, despite the fact that they can't see what they are pointing to. This disorder is called 'blindsight', and it results from destruction of a particular area of the brain called the primary visual cortex—damage that leaves other areas of the brain intact which continue to perform visual tasks, but cuts much of those areas off from the rest of what the brain is doing. This prevents the related experiences from being fully integrated with other brain functions, hence keeping them out of one's model of a conscious self, just as many other things our brains do are not 'wired in' to, or not functionally perceived by, that same modeling process. Yet, like the mouse, if we could 'talk' to that isolated brain system (which, all by itself, could be as dumb as a mouse), it would probably report the very experiences the rest of the brain can't access.

A similar situation exists for people who have had the two halves of their brain severed in order to prevent debilitating seizures: it turns out they cannot experience in their left brain what only their right brain is seeing, and vice versa. Yet we can still hold an intelligent conversation with each half by itself, as if it were a separate person, only because (unlike a mouse) each half possesses sufficient resources to make intelligent conversation possible, not to mention intelligent appreciation of the nature of its experience. Each half has enough of an organ left to perceive and generate an integrated self, something no mouse can do. But what if we couldn't talk to one of the two halves? What if it was too 'dumb' to talk—or could talk, but lacked control of any part of the body and thus was unable to convey anything it wanted to say? Just like the mouse, we would not know it was experiencing anything. Even the same person's own right brain would think his left brain was dead to experience. But it wouldn't be. It just couldn't tell us about it.

Therefore, we cannot assume it is ever possible for an organ, whether of a robot, mouse, or man, to perceive something, without experiencing some sort of qualia. It seems to me that to experience qualia is exactly what it is like to perceive anything—as opposed to merely "sensing" it. Which also means that to believe you are perceiving something, and to actually be perceiving it, are nearly one and the same thing. Indeed, this could be why we make so many mistakes: we can all too easily fool ourselves into seeing what we believe. At the same time, the only reason so much of our brain's activities don't get into our *conscious* experience is that they are not wired into the cerebral cortex in the same way, so they have not made it into our self-model—yet it is only that process of 'self' that is capable of appreciating and talking about such things. The rest of our brain is too dumb to know any better.

> For more on this issue, see: Leopold Stubenberg, *Consciousness and Qualia* (1998) and Stephen Palmer, *Vision Science: Photons to Phenomenology* (1999); but or a different view than mine, see: David Chalmers, *The Conscious Mind: In Search of a Fundamental Theory* (1996). An important paper illuminating my position is Allin Cottrell's "Sniffing the Camembert: On the Conceivability of Zombies," *Journal of Consciousness Studies* 6.1 (1999), pp. 4–12 (www.imprint.co.uk/cottrell/jcsmainframe.html).

6.5 The Nature of Knowledge

Knowledge is the possession of information, in our case in the very structure of our brain, usually about ourselves or our bodies or the world we inhabit. For knowledge to be useful two other things must happen. It must be accessible (the brain must be able to retrieve it when stimulated to do so by internal thought processes or external stimuli). And it must be correct—in other words, "true."

We saw earlier what it means for something to be "true" (II.2.2, "The Meaning of Statements"). Knowledge stored in a brain is true if the proposition that expresses it is true. That is, if all the predictions entailed by a believed statement are or will be fulfilled, then we know something. And as we have already discussed, since we cannot test every prediction entailed by every proposition all the time, but only some of them some of the time, our knowledge can only be said, at best, to be "probably" true or false, in varying degrees according to how many predictions have been or are being fulfilled, and which ones. For example, the more distinctive a prediction is to that proposition rather than others, the more significant it is.

All this was treated in chapter II.3 ("Method"). But now we must go further. There is a physical difference between, for example, someone who knows they are in Alaska and someone who does not: the person who knows possesses, physically in their brain's structure, accessible information regarding their presence in Alaska, whereas the person who does not know lacks such information in their brain (or lack's access to it, as in the case of the amnesiac). Moreover, the person who knows is more and more certain that their knowledge is correct when they possess more and more information, especially less and less ambiguous information, and information acquired from different sources and by different means and at different times, all suggesting the same conclusion, and none suggesting any other. Such a network of information possessed in the brain is mutually reinforcing, strengthening the level of emotional and motivational "confidence" or "belief" the brain will generate in connection with any particular piece of information, like "I am in Alaska." And this is where reason has its nascence, a phenomenon we will explore in chapter III.9 ("The Nature of Reason").

It follows that belief is a material property of the brain: a pattern of neurons and neural connections distinct from that of disbelief or uncertainty, a pattern that has causal powers, a power we "sense" as a degree of 'confidence' affecting our decisions—such as what we will say we believe (even if only to ourselves), or what we do as a consequence. It further follows that a "true" proposition is one that describes a physically-computed pattern in the brain, which corresponds to a physical pattern in the real world, whereas a "false" proposition is a brain pattern that does *not* correspond to its described pattern in the real world. Knowledge, belief, and truth are thus physical realities, physical distinctions.

Language is key here. One reason humans are so much smarter and so capable of a profound comprehension of things and themselves is that we have such large and intricately developed brain centers for processing symbolic thought, a "language." By being able to make more (and more precise) distinctions among the patterns in our perception, in our virtual models, and to label them and compare them, we are able to acquire far more knowledge, far more quickly, than even the smartest of our animal kin. We can also build a self-identity by accumulating knowledge about ourselves, knowledge that continues to have causal effects on us.

This gift actually works twice over. Not only does this power to distinguish and analyze and learn make us as individuals more sentient, by magnifying many times our capacity for generating knowledge from experience (and storing and using it), but it gives us a bombshell of an advantage over all other species: culture. This permits us to get injected, through socialization, education and emulation, with millions of human-years of previously accumulated knowledge, through the transmission of knowledge accumulated and refined over hundreds and hundreds of lifetimes, by millions of separate minds dividing among themselves the labor of exploring and learning about things. No animal can compete with that (for more on this point, see III.8.4, "Memetic Evolution").

6.6 The Evidence for Mind-Body Physicalism

I believe the mind is solely the product of a functioning, physical brain, an active pattern of matter and energy in space-time. I believe this because, again, it explains everything about us, and explains it well, with the fewest *ad hoc* hypotheses. But even more importantly, I believe it because we have a tremendous and ever-growing body of evidence supporting it. Though we are still looking for a complete theory of consciousness, every scientific study of the nature of minds and their properties and abilities has so far shown that these have a basic physical nature.

We can physically locate in the brain memories and emotions and decisions, everything we know that minds have or do. We can remove or stimulate memories and emotions and thoughts by removing or stimulating the physical parts that contain them. We can interfere with emotions and thoughts with brain-affecting chemicals or physical damage, even magnetic fields. And when one of us reports being unconscious, the rest of us observe the brain to be non-functioning, and over that same time the person affected later reports experiencing nothing—not even the passage of time. It is therefore an obvious and powerful inference that our minds, our "souls," are physically manifested by our brains, without which we would experience nothing, not even the passage of time—in other words, we would cease to exist.

I will now itemize and expand this evidence (for which, see final chapter bibliography below). The following six points, taken together, leave no reasonable doubt that minds cannot exist but for the brains that manifest them.

6.6.1 General Brain Function Correlation

Scientists have observed brain function in humans in dozens of ways, millions of times—hundreds are even doing so right now as you read this—yet never once has anyone ever observed a human mind functioning in the absence of a functioning brain. In every case when the relevant measures of brain function read even *near* zero, the human subject is completely unresponsive, and if the subject is restored to consciousness, they report having experienced nothing, not even the passage of time (about myths and legends to the contrary, see III.6.7, "Evidence Against Mind-Body

Physicalism?"). If you have ever truly passed out (such as from blood loss, as I have), you know just what this is like: in terms of mental experience, you literally cease to exist. The evidence here is as overwhelming as it can get. If you effectively cease to exist when your brain doesn't function, it follows that without a brain, you will not exist at all.

6.6.2 Specific Brain Function Correlation

Scientists have proven time and again that there is always a strong correlation between distinct brain events and distinct mental events. When people report seeing, there is always corresponding activity in what we know to be the visual centers of the brain. When people report remembering something, there is always corresponding electrical and chemical activity in what we know to be the location where such memories are stored. No exceptions have ever been observed. This supports the general observation above: nothing mental happens without something physical happening. So if there is no brain in which these essential physical process can occur, there can be no mind, and thus no you.

6.6.3 Positive Evidence Mapping the Mind to the Brain

Scientists have directly stimulated areas of a person's brain with electrodes, and in doing so always trigger exactly the same mental event when the same stimulus is applied to the same place, and in fact the general features of the brain have been mapped this way—so we know where sight is processed, where names are stored, where fear comes from, where love is generated, where specific songs are remembered, and so on.

Indeed, nearly every conceivable mental event has been identified with a physical location in the brain that has been mapped. For example, a neurosurgeon found a particular point in a patient's brain where, if electrically stimulated, a song would start playing in the patient's mind. Every time that same point was stimulated, the same song started playing. This makes complete sense if mind-brain physicalism is true, but very little if it is not. One can infer that without that physical spot in the brain, or anything to stimulate it, the memory of that song would not exist or would never again be accessible. Since one can infer this about everything that has been found this way, from emotions to thoughts and memories, and since a person without thoughts, emotions or memories essentially no longer exists, we can conclude from this evidence that without the brain there is no person, no mind, no soul.

6.6.4 Negative Evidence Mapping the Mind to the Brain

Completely corroborating the mapped results of the work described above, brain impairment or injury—either accidental or the result of necessary surgery or the experimental use of drugs or magnetic fields—results in the loss of specific mental functions as the specific areas related to those functions are lost or numbed. The effects are often quite strange. For instance, there are people who can see perfectly well, even identify different objects, or different people according to their clothing or body shape, yet they cannot recognize human faces. It is not that they don't see the faces—their sight is fully functional—it is that the center of the brain that stores information about faces specifically, and only that part of the brain, was physically impaired or destroyed by an injury. Even consciousness itself has physical locations. For instance, the ability to consciously 'see' requires the primary visual cortex. Likewise, half a brain can generate its own consciousness separate from the other half (see III.6.4.4, "Qualia"). This proves that parts of the brain are essential to conscious experience: without them, there is no conscious experience.

There are literally hundreds of different examples, ranging from every sensory or perceptual power known, to almost every sort of emotion, reasoning power, and personality trait known. This body of evidence collectively demonstrates that the loss of any area of the brain always results in the loss of what that area did, and when you add up all the things these brain areas do, you have the entire gamut of a human mind. Consequently, if destroying parts of a brain destroy parts of a mind, then destroying all the parts of a brain will destroy the whole mind, destroying *you*.

6.6.5 Brain Chemistry and Mental Function

Scientists have mapped many of the chemicals that make the brain work, and many remain that we know are there, though not yet what they do, and still more are being found. We know now that when the brain lacks certain chemicals, or has too much, the mind will fail in certain ways, or change in personality, causing depression, poor reasoning, even hallucinations.

We know that certain compounds make someone's mind think more characteristically like a woman (such as more emotional or empathic), but certain other compounds in the same person's mind can make them think more characteristically like a man (such as more violent or calculating). We know disorders that wreak havoc on a person's ability to reason or perceive things correctly, like schizophrenia and other forms of psychosis, are caused in part by an excess or depletion of certain chemicals the brain needs to function. And even simple things, like certain vitamins and minerals, and something as basic as oxygen supply, affect mental function and emotional states.

That the mind can be so affected by brain chemistry makes more sense on mind-brain physicalism than any other theory. Moreover, from the fact that crucial mental functions (like reason, attention, clarity, and control) decline exactly as certain chemicals decline in availability, it follows that in the absence of these chemicals those functions would be crippled if not lost altogether—which means that without a brain, even if we could still exist somehow, we would be unaware of that existence and incapable of coherent thought.

6.6.6 Comparative Anatomy and Explicability

Comparative anatomy supports the same conclusion. The mental powers of animals increase in direct correlation to the increased complexity of their brains. And in every species, when an animal, including humans, develops its mental abilities further, this is always matched by the development of neurons or synapses in their brains. An animal with an acute sense of smell has an enormous part of their brain devoted to processing scent. An animal, like us, whose primary mode of perception is vision, has an enormous part of its brain devoted to processing sight instead. And the same goes for reasoning ability: animals that are smarter always have larger, more complex portions of their brain dedicated to thinking. It follows that a physically complex brain is necessary for a mind, and that a mind can only develop when the brain develops physically.

Ultimately, the human brain is the most complex machine known in the entire universe, which is exactly what we would expect if our minds were a mechanical process—only then would the vast complexity of the human brain be needed, much less explicable. If our minds were not dependent on the human brain, then there would be no plausible reason for us to have one. At best, all we would need is the minimal sort of hardware a comparable mammal had, though even that would be hard to explain the need of, since if the mind were independent enough to be able, for instance, to see, hear, think, and remember all on its own, the vast majority of the brain of even an ordinary mammal would be useless material to us, dead weight, a needless drain on our oxygen supply, of which our brains now take the lion's share.

Indeed, even our sensory organs would be useless. Why have eyes if the soul can see without them? There would be an immense survival advantage to having eyes that can never be blinded, as well as a mind that was not as fragile or vulnerable to drugs or injury or asphyxiation as the brain is. And if we didn't need all that brain, we could have a smaller head, and fewer mothers would die in labor. So there would be tremendous pressure over the ages for humans to become reliant on their souls for sensation and thought, while our brain tissue and sensory organs would atrophy as unnecessary expenditures of energy and serious points of vulnerability. But that hasn't happened.

In short, mind-brain physicalism makes perfect sense of our enormous, complex, energy-hogging, vulnerable brain, as well as its dependence on specialized physical organs for sensation, whereas any alternative to mind-brain physicalism makes no ready sense of any of this (see also III.8.3, "The Evolution of Mind").

6.7 Evidence Against Mind-Body Physicalism?

The evidence that is offered "against" the above conclusion differs in a very significant way from the evidence for it. The evidence *for* mind-body physicalism has been corroborated in laboratories and scientific field studies thousands and thousands of times in countless ways and is in fact the only evidence any reputable science has ever turned up or ever sees in regular medical practice. Indeed, the evidence has converged from numerous different directions on the same conclusion.

In contrast, the evidence against mind-body physicalism is purely anecdotal, almost always ambiguous, and has often been shown to be outright fraudulent. It is thus clear that mind-body physicalism is proven by objective, rigorous investigation, whereas its denial rests on wishful thinking so powerful it even leads people to fabricate it, or to see what isn't there.

The four kinds of evidence offered are "Near Death Experiences" (NDE's), "Out of Body Experiences" (OBE's), "Spiritualism" (which takes two forms: purported communications with the dead, as by psychic mediums, and sightings of ghosts and apparitions), and "Past Lives Regression" (PLR, either someone spontaneously recalling memories claimed to be from a past life, or doing so under or after hypnosis, either way purporting to prove reincarnation).

I will deal briefly with the first two here, since they are the only ones for which anyone even claims to have any evidence of a scientific nature (see part IV, "Not Much Place for the Paranormal"). In short, Spiritualism and PLR have ready explanations in fraud, self-deceit, illusion, and "myth-making," since in every case where they have been genuinely scientifically investigated in the past hundred years, if any truth to the story was discovered at all, one of these four mundane causes was proved to be at work.

Several books on NDE's and OBE's survey all the evidence and literature thoroughly and scientifically (see bibliography below). In an OBE, a subject reports the sensation of leaving their body and looking down upon themselves, and sometimes they can move about in disembodied form, even covering great distances. Many fanciful legends have grown up boasting of amazing proofs that a particular OBE was genuine, but they have always dissolved under scrutiny: investigations turn up no corroboration for any of the story's details, or often uncover evidence that flatly contradicts it. Direct tests of the power to actually "see" things from a perspective outside of one's body—such as by reading and recalling a word only a hovering observer could see—have all turned up negative. Instead, people sometimes report experiencing, for example, what turns out to be a non-existent word, proving that OBE's are actually *invented* by a person's brain. Nothing really leaves the body.

NDE's often, but not always, incorporate an OBE. An NDE is an experience some subjects have when near death. In some cases, NDE's happen even when there is no real physical damage or threat, induced by natural brain chemistry, causing experiences of calm or elation, and inspiring reactions like self-reflection on the meaning of life. There is nothing supernatural about that. The amazing claims, by contrast, usually relate to cases where someone is reported to have really died, and been revived, usually by a procedure like CPR or electrocution. In these cases, there is remarkable continuity in what is reported even across cultures, though there are certain profound cultural differences as well. Not every NDE contains every element, but the typical experience is of an OBE, followed by flight down a dark tunnel toward a light, where the subject meets deceased relatives and/or gods.

Analysis of the "typical" NDE cases and elements has found that each feature had a ready natural explanation as an hallucination induced by oxygen deprivation, endorphin release, and random neural firing, all typical effects of dying. The OBE has been proved to be a common type of hallucination, which can be caused by chemical changes in the brain when dying. Likewise, "seeing" gods or deceased relatives is one of the most common forms of hallucination on record. Since many NDE's involve meeting *living* relatives in heaven (an obvious impossibility), not to mention even stranger persons (like Elvis), hallucination is the best explanation. Most importantly, trips down the tunnel have been proved, through computer analysis of the effects of oxygen deprivation on the optic nerves or visual cortex, to be a natural optical illusion.

Though there is often the claim that the subject could hear or even see things while clinically dead, such as things said or done by the doctors and nurses attempting to revive them, or equipment in the room, these claims never come with real proof of the one essential element: that these things were experienced when the subject was *dead*. In every case that could be investigated, everything reported was either unremarkable (the sort of thing anyone could guess or would know) or could have been seen or heard when there was still sufficient brain activity to experience them normally, sometimes even well before or after the NDE itself. Even more remarkable claims have been refuted outright as fabrications or exaggerations, or could not be corroborated.

In short, there is no trustworthy evidence against mind-brain physicalism. Notably, even religious scholars are beginning to concede this. Christian editors Warren Brown, Nancey Murphy, and H. Newton Malony, in *Whatever Happened to the Soul? Scientific and Theological Portraits of Human Nature* (1998), declare, "We have written from the perspective that views soul as a functional capacity of a complex physical organism, rather than a separate spiritual essence that somehow inhabits a body. We have adopted this position because we believe it is the best way to incorporate and reconcile all the various sources of available data" (xiii).

The positive evidence for mind-brain physicalism also presents an almost insurmountable challenge for opponents. For example, scientists have confirmed that we only perceive things *after* our brains do, not the other way around. When we see a face, our brains already show activity in the area that recognizes faces a fraction of a second before we are aware of seeing a face at all. Likewise, when we make a decision, say to move our arm, we know our brain has already sent the signal to move the arm (and thus has already decided to move it) a fraction of a second *before* we become aware of making such a decision. This is very hard to explain unless physicalism is true. For only then would perception be a process occurring in a physical organ, one that takes time (a fraction of a second) to complete itself, and only then would self-awareness itself be such a process of perception.

Consider a different problem. There are many people who suffer from a condition called *synesthesia*, where their brain fails to physically separate sensory processing, so they 'see' sounds or 'hear' colors, and so on. If this is what happens when the brain's wires get crossed, how can a disembodied soul experience distinct sensations? After all, the soul has no physical wires to keep such sensations separate, and clearly can't keep them separate in synesthetes. Yet every OBE has been reported without synesthesia.

I could list a dozen other similar problems that opponents of mind-brain physicalism have a hard time answering, but that neurophysiologists can answer easily, often with hard evidence to back them up.

6.8 Immortality and Life After Death

The evidence seems clear: our mind, hence our very existence, depends entirely on the brain. As a mechanism, the brain must be kept healthy and active, so it can remain a system of coherent perception and thought, and we can remain "conscious" and experience life itself. But stop the brain from functioning, and we can experience nothing. Our "consciousness" ceases to exist.

As a storage medium, though, we remain even when the brain is not functioning—so long as the cellular arrangement itself remains intact. That's why we do not cease to exist when knocked unconscious. Who we are is stored in the hardware of the brain. But the brain's hardware is very delicate and vulnerable. It requires the expense of a considerable amount of energy and material just to keep the cells and their connections from disintegrating. This is why, when the brain loses oxygen or some other vital resource, even without being dealt any physical blows it very quickly becomes irreparably damaged, resulting in irreversible death, the permanent loss of the pattern that was 'you'.

Since your particular pattern of memories and abilities and qualities is what makes you who you are, a person instead of a mere machine, unique rather than just like everyone else, it follows that this is your 'soul', the particular part of you that is really *you*, as opposed to the body's parts which can be hacked away and destroyed without harming you as a *person*. You could be left a brain sustained in a vat and yet you, as a person with a will, an identity, a collection of memories and abilities and qualities, would remain intact. At worst, you would have lost those motor skills that were stored in the spinal brain stem and beyond. And yet even the brain is just another body part, one that "supports" the mind with its processes and storage capacity, just as the body supports the brain with fuel, sensations, and locomotion. Like all other body parts, a brain is not essential to *you*. Rather, it is the pattern, and the process it manifests, that is essentially you.

Of course, if the body and its brain could be kept alive and sufficiently healthy, you could live forever, though there are obvious difficulties, such as forms of gradual cell damage that cannot be controlled. However, since the body and its brain is merely a vessel for the mind, it is entirely replaceable. Your mind-pattern can in principle be formed out of many different materials, not just the one we happen to be made of, so it remains possible that we might be able one day to "transfer" our minds to a more durable, enduring medium, like an electronic brain, and thus achieve immortality that way. This would indeed be a life after death—the death of our original bodies and, to borrow a phrase from Christian theology, a resurrection in a new 'more glorious' body. Thus, life after death is not wholly impossible on Metaphysical Naturalism, but it requires an incredibly advanced technology not yet within our reach.

Since it is the particular pattern held in our brain, a particular arrangement of neurons and synapses, that gives ourselves existence, storing all of our memories, abilities, and qualities, it is also possible in principle to "freeze" a record of that pattern immediately after our brain's natural death, so these patterns can be reconstructed later in a working brain of some sort (assuming someone bothers to do this for us). At present, it is not clear that any existing technology is actually capable of an accurate-enough freezing or recording process—cryogenics, for example, causes damage to the microscopic arrangement of synapses and thus may in fact forever delete the crucial and delicate pattern that defines who you are. But this ability is nearer our reach than actual resurrection. I expect it could be perfected within my lifetime. And since this "saves" our minds for possible future advances in technological resurrection, for those today who want any chance of life after death this is the only prospect reality has to offer. There is certainly no evidence for any other kind.

But absent these technologies or access to them, death is extinction, oblivion, complete non-existence—and it is inevitable. If everything is entirely and solely comprised of matter and energy in particular arrangements in space and time, then we cannot survive death. For if what we call a "soul" is only a brain process after all, then without a brain, there can be no soul. So when our brains die, and eventually disintegrate, we cease to exist, never to return. This means we will never meet our deceased heroes or loved ones again, that our own lives are of only a limited duration, and that 'justice' in the sense of fitting rewards or punishments cannot be dealt to the dead. Naturally, this is all terribly annoying. But it's not like we can't cope with it. I deal with this revelation next.

> For further reading, see the Secular Web sections on "Life After Death" (www.infidels.org/library/modern/lifeafterdeath/), the "Argument from Physical Minds" (www.infidels.org/library/modern/nontheism/atheism/minds.html), and bibliographies for sections III.6.4.4 ("Qualia") and

III.8.3, "The Evolution of Mind." I discuss mind-brain issues in more detail in "Critical Review of Victor Reppert's Defense of the Argument from Reason" (2004, at www.infidels.org/library/modern/richard_carrier/reppert.shtml).

For book treatments: Gerald Edelman, *Wider than the Sky: The Phenomenal Gift of Consciousness* (2004); Steven Johnson, *Mind Wide Open: Your Brain and the Neuroscience of Everyday Life* (2004); Christof Koch, *The Quest for Consciousness: A Neurobiological Approach* (2004); Susan Blackmore, *Consciousness: An Introduction* (2003); Julian Paul Keenan, et al., *The Face in the Mirror: The Search for the Origins of Consciousness* (2003); Joseph Ledoux, *Synaptic Self: How Our Brains Become Who We Are* (2002); Robert Aunger, *The Electric Meme: A New Theory of How We Think* (2002).

Other useful treatments include: John Ratey, *A User's Guide to the Brain: Perception, Attention and the Four Theaters of the Brain* (2001); Bernard Baars and James Newman, eds., *Essential Sources in the Scientific Study of Consciousness* (2001); Sandro Nannini & Hans Sandkühler, eds., *Naturalism in the Cognitive Sciences and the Philosophy of Mind* (2000); Steven Pinker, *How the Mind Works* (1997); Daniel Dennett, *Kinds of Minds: Towards an Understanding of Consciousness* (1997) and *Consciousness Explained* (1992); Paul Churchland, *The Engine of Reason, The Seat of the Soul: A Philosophical Journey into the Brain* (1995); A. G. Cairns-Smith, *Evolving the Mind: On the Nature of Matter and the Origin of Consciousness* (1996).

On whether a soul can survive death of the brain: Gerald Woerlee, *Mortal Minds: A Biology of the Soul and the Dying Experience* (2003); Antony Flew, *Merely Mortal? Can You Survive Your Own Death?* (2001); Susan Blackmore, *Dying to Live: Near-Death Experiences* (1993) and *Beyond the Body: An Investigation of Out-of-the-Body Experiences* (1992); Mark Fox, *Religion, Spirituality, and the Near-Death Experience* (2002); documenting cultural differences in NDE's is Karlis Osis and Erlendur Haraldsson, *At the Hour of Death*, now in its 3rd ed. (1997), though the 1st ed. (1977) contains cases later omitted. Also relevant is Paul Edwards, *Reincarnation: A Critical Examination* (1996), and Peter & Elizabeth Fenwick, *The Truth in the Light: An Investigation of Over 300 Near-Death Experiences* (1995), which, though favoring the case for survival after death (like Osis and Haraldsson), nevertheless contains a lot of evidence for the hallucinatory nature of NDE's. A famous case is that of renowned atheist A.J. Ayer, reported in "What I Saw When I Was Dead," Lewis Hahn, ed., *The Philosophy of A.J. Ayer* (1992), pp. 41-53. Many other books collectively document so much evidence for mind-brain physicalism that there is no longer any reason to doubt it. See, for example: Oliver Sacks, *The Man Who Mistook His Wife for a Hat, and Other Clinical Tales* (1998) and Frederick Schiffer, *Of Two Minds: The Revolutionary Science of Dual-Brain Psychology* (1998), as well as the many excellent works on the subject by V. S. Ramachandran, such as: *A Brief Tour of Human Consciousness: From Impostor Poodles to Purple Numbers* (2004); *Phantoms in the Brain: Probing the Mysteries of the Human Mind* (1999); and the *Encyclopedia of the Human Brain* (2002).

7. The Meaning of Life

So when we die, we cease to exist. Does this mean life is meaningless because it is only temporary? Not really. Most valuable and meaningful things are only temporary. We do not regard a joyous occasion as pointless because it only lasts a day. Yet our very lives are a joyous occasion. By existing, and making of ourselves something good, we give ourselves and each other value, we *create* purpose and meaning. Neither existing by accident nor existing only a short while changes anything about the value of existing, the value of getting to be, to behold and to know the universe, to create something.

Nor do we need to be some superbeing's creation for our lives to have value. After all, believers seem comfortable with the fact that God was not created, yet his life has value. Just as theists understand God's love as *giving* God himself and the universe value, so naturalists understand *our* love as giving ourselves and the universe value. Even if I were the accidental by-product of a giant rubber tire machine, my life would not be meaningless. It would be meaningful to the precise extent that I endeavored to make it so, to imbue my own life with meaning and purpose, to make it valuable, to myself and to others. But if I did nothing to make my life meaningful, even being created in some god's image would add no meaning to my life. I would be nothing but a pawn or lab rat, a mere homunculus cooked up in some divine kitchen, if I did nothing on my own to make myself into more than that.

Meaning can be found in our own existence, the here and now, but also in our hopes and dreams for the future, and not just our future but the future of others. In the simplest terms, the meaning of life is a healthy mind in a healthy body, pursuing and manifesting what it can most deeply love: the creation of good works, and the society of good people, in a well-tended world. Anything less is ultimately self-destructive or unfulfilling, and generally a complete waste of time, especially given the brief opportunity for living afforded us. And if we want even more, if it is important to us that something good be eternal, then we can make this so of the good society, and contribute our necessary part to realizing that society and laying the seeds for its endless progress and renewal. We thereby give the greatest gift of all to countless other beings who will know the world after us.

If we actually wish to live forever ourselves, or for our descendants to do so, our only prospect is a human technology of immortality. But it does not matter so much if life is brief, for merely the opportunity itself is priceless. Our being here, to acknowledge it, to study it, to know it, and to love it, gives the *universe* meaning, not the other way around. To draw again from our picture of the universe (III.3.6, "Time and the Multiverse"), from a point of view outside of time, everything—past, present, and future—exists eternally: our lives sit forever like pearls on a string of time. What we do with our life, what we make of it, how we enjoy it, can never be taken away. It becomes a part of what exists, adding to its value, like gems in a purse.

And for us, the sages have said it for millennia: it really is love that is key—love of learning, love of doing, love of others, love of ideals, love of country or cause, anything, everything, is the foundation of meaning. If we lacked that, we would certainly be miserable and our lives pointless, even if we lived forever. Indeed, even if we droned on with praises for a supreme being in heaven for all eternity, our existence would be superficial, trite, unsatisfying, and ultimately a torture. Thus, the key lies in finding your loves and pursuing them, manifesting that love in defiance of a universe that won't.

What is worth loving? The potential of humanity, the power of reason, the comfort of another's love, the pursuit of knowledge and truth, the beauty and joy of human experience, and the nearly unlimited power of the human will to endure almost any hardship or solve almost any problem. And that is just the short list. How many wonderful people do we know, or could we know if we sought them out, who are worth loving—loving merely for the fact that we wished there were more of them in the world, and the fact that they give us a reason to live? Even when I look at something magnificent in nature—the stars, the wilds, the physique of a sea lion, the beauty of a nebula—I think to myself "How fantastic!" How pointless that beauty would be if I *didn't* notice and appreciate it. How valuable I am because I can.

Immortality is inconsequential in this equation. We have no reason to fear death. Why fear the end of fear itself? We live for only one reason:

because we love life, all of it, any of it. And if it disappoints us that there is not enough happiness in the world, not enough goodness, we can contribute to rectifying that. And that is what gives *our* lives meaning. The more good things we can create or teach and thus leave behind for others, the more problems and difficulties we can remove, the more lives we can light up with our effort and company and companionship, then the more precious our short existence will have been, and the more satisfied we will be that we used our bank account of life well, and thus deserved our measure.

I have faced death on a few occasions. After the inevitable panic I was always calm and accepting. On the one hand I knew I would no longer have any worries or pains when I no longer exist, and on the other hand I had lived a good life, I had done some good, things that would never have been had I not existed at all, and my short span of knowing, enjoying, loving it all was well worth it. I looked upon what I had had, and found myself content with it. By making the universe that little bit brighter and more meaningful, my own life had value and meaning as a consequence, and that addition to the world remains even when I do not.

But when despite such understanding, despite such a good life, you still seem trapped by depression, you are probably as unwell as you would be with a dangerous flu, and the reaction should be the same: to seek medical help. The cure often requires medicine or therapy. Certainly, it would be irrational to consider *ending* your existence without first trying every means at your disposal to get something out of life worth living for. Because you don't get a second chance.

Therapy can help you discover (or rediscover) what you love about life, and to come to terms with your faults and fears, as the method of atheist Dr. Albert Ellis does, the father of Rational Emotive Behavioral Therapy (REBT) for the treatment of depression and other problems. However, sometimes the emotions that torture you are a chemical or other biological malfunction and thus need medications to correct. So professional diagnosis should always be sought when things get bad. If you feel you need a counselor with a secular perspective, you can seek a referral through the American Humanist Association or the Humanist Society of Friends. Sometimes you are simply in an unhealthy circumstance from which you just have to escape and start over. Good counselors can direct you to those with handy advice on how to do this.

Also keep in mind the basic and essential home remedies. Eat well and exercise. Take long walks in nice places. Take up a cause you feel good about, and work to help others in some way that comes easily or comfortably to you. And above all, seek to maintain a happy, social interaction with other people. Studies have proven that people with a cause they care about, and who have even a small but enriching social life, live longer, happier, and healthier. And if it's good for your health it's good for your mind.

>   For more on these issues, I always recommend Bertrand Russell's *The Conquest of Happiness* (1930) and Epicurus' "Letter to Menoeceus," which can be found in *The Essential Epicurus* (1993), and also online at www.epicurus.net. See also: E. D. Klemke, *The Meaning of Life*, 2nd ed. (1999); Keith Augustine, "Death and the Meaning of Life" (in www.infidels.org/library/modern/keith_augustine) and James Still, "Death Is Not an Event in Life" (in www.infidels.org/library/modern/james_still). Many also recommend Albert Camus, *The Myth of Sisyphus* (also available online in the Secular Web's historical library).
>   Recently relevant are: Peter Heinegg, ed., *Mortalism: Readings on the Meaning of Life* (2003); David Cortesi, *Secular Wholeness: A Skeptic's Paths to a Richer Life* (2002); Nicolaos Tzannes, *Life Without God: A Guide to Fulfillment Without Religion* (2002); George E. Vaillant, *Aging Well: Surprising Guideposts to a Happier Life from the Landmark Harvard Study of Adult Development* (2002); and Michael Martin, *Atheism, Morality, and Meaning* (2002).
>   The program of Dr. Albert Ellis is popularized by David Burns in *Feeling Good: The New Mood Therapy*, rev. ed. (1999). Other secular ways to happiness are described by Martin Seligman in *Authentic Happiness: Using the New Positive Psychology to Realize Your Potential for Lasting Fulfillment* (2002), and even by Christian therapist David Myers in *The Pursuit of Happiness: Discovering the Pathway to Fulfillment, Well-Being, and Enduring Personal Joy* (1992). For a good scientific study of how to improve the "luck" in your life (and thus your enjoyment of being) with simple, beneficial life strategies, see: Richard Wiseman, *Luck Factor: Changing Your Luck, Changing Your Life—the Four Essential Principles* (2003).
>   Also, see the works on happiness cited in the section bibliography for V.2.1.3, "Self Worth and the Psychological Need for a Moral Life."

8. How Did We Get Here?

If no one planned the universe, how did we get here? How could a universe, as complex and wondrous as it is, just spew out a conscious, reasoning creature, and civilization in general? Isn't that just too amazing a coincidence to believe?

Naturalists have a well-known answer. Our existence is indeed very unusual. But the universe is so vast, it was inevitable that something like us would be one of its byproducts somewhere. We believe simple life is probably a relatively common end result of natural processes in the universe, which may in turn have led quite naturally to a sentient species like us in several places across the cosmos. We are rare enough that there might be only one such outcome per galaxy, or even fewer than that. But as there are trillions of galaxies, even if conscious beings and their consequent civilizations are so improbable that they can result only once in a *million* galaxies, there would still be *thousands* of such civilizations in the universe. Yes. The cosmos is that big. And we would likely never meet. For the space and time that separate galaxies is beyond all possibility of crossing—a challenge even for a coherent radio signal. It would take thousands of years just to exchange some simple words of greeting between two peoples in the *same* galaxy, much less in different ones.

Simple life, however, stands a very good chance of being all over every galaxy like ours. As galaxies sport billions of stars, if life turns out to be so rare that it arises in only one in a million star systems, that would still mean thousands of planets must have life in our galaxy alone. So *yes* life, especially sentient life, is amazingly rare and improbable, *but* the universe is even more amazingly old and big. That means amazingly improbable things are sure to happen in it all the time. This is actually a good reason to believe we *are* just a natural product. A divine engineer would have no need of spinning out a trillion useless galaxies just to produce one species like us. That makes more sense on naturalism. But how natural processes could produce us at all is the next story to tell.

Currently, the naturalist's epic narrative of how we got to be begins, of course, with the Big Bang, and the condensation and evolution of the universe itself, which we have already surveyed (III.3, "The Nature and Origin of the Universe"). The story picks up again with two more processes, natural biogenesis and evolution through natural selection, which played out here on earth in four key stages. First, molecules capable of reproduction arose by a rare but not too improbable congruence of events. Second, natural selection inevitably acted on these molecules to produce the evolution of life. Third, this process of evolution, after much meandering, eventually hit upon a supreme adaptation: the production of a conscious mind. And fourth, the activity of conscious minds created an entirely new playing field for natural selection: memetic rather than genetic evolution, giving rise to an evolving culture, eventually producing philosophy, science, and technology. This is our story, our Book of Genesis.

8.1 Biogenesis

Biogenesis is the first stage of this story. All the planets around all the billions of stars in all the billions of galaxies are organized at random with practically infinite variation, and because there are so many, the law of big numbers prevails: every possible planet that could be (given the universe and its physical laws) probably has been, is, or will be. Thus, that one or more planets should have all the right properties for biogenesis is probably a foregone conclusion, and our planet is known to be one of those rare few. Though we have none of the information we need to calculate any of the relevant statistics, this is still a reasonable conclusion, for two reasons.

First, the Inference to Metaphysical Naturalism. Natural explanations have turned out true for everything so far, without exception, so it is a good bet they will continue to do so (see III.2, "A General Outline of Metaphysical Naturalism" and IV.1.1.7, "The Lessons of History and the Burden of Proof"). This is amplified here by the explanatory power of natural biogenesis—for current theory explains more than that life originated naturally, but why it originated with the chemicals it did, when it did, and how it did, and why it has the characteristics it does. Such a good fit between theory and fact, such wide explanatory power, gives it tremendous merit.

Second, and most importantly, scientific research upholds all the elements of the theory—the vast size and variation of the cosmos, the law of big numbers, the suitability of Earth for natural biochemistry, the ease with which a biochemistry can arise in such conditions, and the abundance throughout space, and especially our solar system, of all the chemicals needed to get life started. Everything from amino acids to sugar, from water to sulphur, from oxygen to nitrogen and carbon dioxide, has been found in space, sometimes in great quantities. And these are the things of which life is made.

So the ground was set: we are made of the very things commonly found in the universe. And planets like earth are just the sort of places where they would most likely come together. Furthermore, experiments have proved that amino acids naturally chain into proteins, the building blocks of life, when subjected not only to many possible kinds of natural forces, but forces we know were common on the early earth, and beyond. Finally, scientists have manufactured proteins that naturally reproduce themselves without the aid of any additional enzymes, proteins so simple that we now know the odds of such things forming by chance are well within the realm of cosmic possibility.

Consequently, though we lack access to the facts we need in order to know just what happened on earth four billion years ago, it is more than reasonable to conclude that when all these common chemicals come together in the most favorable spots in a galaxy—such as earth—they will mix and produce random proteins, and after trillions and trillions of such random processes throughout all the star systems in all the galaxies, certainly at least one, if not many, somewhere, will inevitably turn up by chance a protein that can reproduce itself.

Once reproducing chains of amino acids exist, mutation inevitably takes hold. Whether from damage to these molecules from radiation and chemicals and other environmental hazards, or from mere errors in reproduction, we know mutation cannot be avoided, especially by a primitive replicator that lacks any mechanisms for limiting, preventing, or repairing it. So, in fact, not only is random mutation in reproduction inevitable for the first life, such life would experience a very rapid rate of mutation.

> For all the current science on this topic, see: Tom Fenchel, *Origin and Early Evolution of Life* (2003); Fred Adams, *Origins of Existence: How Life Emerged in the Universe* (2002); Iris Fry, *The Emergence of Life on Earth: A Historical and Scientific Overview* (2000); Geoffrey Zubay, *Origins of Life: On Earth and in the Cosmos*, 2nd ed. (2000); Christopher Wills and Jeffrey Bada, *The Spark of Life: Darwin and the Primeval Soup* (2000); Peter Ward and Donald Brownlee, *Rare Earth: Why Complex Life is Uncommon in the Universe* (2000); Noam Lahav, *Biogenesis: Theories*

of Life's Origin (1999); Andri Brack, ed., *The Molecular Origins of Life: Assembling Pieces of the Puzzle* (1998).

On probability arguments in particular, see my related online research paper, "Are the Odds Against the Origin of Life Too Great to Accept?" (www.infidels.org/library/modern/richard_carrier/addendaB.html), and my more advanced article: "The Argument from Biogenesis: Probabilities Against a Natural Origin of Life," *Biology & Philosophy* (forthcoming).

8.2 Evolution by Natural Selection

When many of the copies being made of an organism contain mutations—random changes or mistakes—there will inevitably be differences among those copies. By far, most of those differences will be fatal. But *some* of those differences will have no effect, while others will actually produce some *advantage*, which allows that copy to copy itself more accurately, more quickly, or more abundantly than the original or any of its other kin—or which ensures that its copies will be more successful, even if only slightly, in surviving and propagating than competitors. This is called "differential reproductive success," where the ability of an organism to copy itself is better than another organism, so that in time its descendants far outnumber, and may even drive extinct, the descendants of others. In fact, so long as they compete for the exact same resources, such as living space and food, such extinction is inevitable. To avoid it, creatures must find ways to cooperate, or find different ecological niches to occupy.

It is known that a single mutation can have huge effects on the differential reproductive success of even the complex organisms that exist today. This would be many times more true for the simpler creatures that would have preceded them, which have since been so outgunned by their more advanced children that these earliest species have all gone extinct. And having no bones, they left no useful traces. Yet it is also known that even *small* effects on differential reproductive success can pay off hugely in the long run.

All this entails that evolution is *inevitable*: there would be no way to stop it from happening, unless some being or force actually intervened to prevent it. This is powerful evidence for the reality of evolution by natural selection. And since an increased complexity (within certain physical limits) confers an increased arsenal of advantages, complexity is *itself* an advantage, which means complexity is often selected for naturally, without any intelligent design.

"Natural selection" is what we have good reason to believe is the cause of evolution. Evolution by itself is the observed fact that life on earth has evolved over billions of years from simpler to more complex organisms in identifiable chains of ancestry. Its cause is called "selection" because it is a choosing process: organisms that are able to reproduce better than others are by that fact alone "chosen" to win out. No intelligence does the choosing. Death does the choosing. The only creatures that get "chosen" to survive are the ones who are brutally able to do so.

That is why this selection process is called "natural," because natural forces and circumstances do the selecting, not any personal or sentient force of any kind. It is a choice made automatically, not by any will. All that is required is anything that reproduces itself, in a way that at least occasionally suffers from random errors, which then must compete with other children for limited resources. Computer simulations have proven that whenever these three things exist (reproduction, mutation, competition), this selection occurs naturally, and evolution inevitably results. As long as these three things remain in place, evolution is almost impossible to stop. And lo and behold, life on earth fits all three features: life reproduces by a natural chemical process, and it does so in a way plagued with random errors, on a planet of limited resources.

This process has certain uniquely predictable results. Two examples will suffice:

First, as there are different environments (or ecological "niches"), different mutations will have advantages for some organisms in some places and not others. The inevitable, unstoppable result is diversification. On the early Earth, where countless niches were completely uninhabited, diversification should logically have been very rapid and affected an organism's structure and function in fundamental ways. But as all niches are filled, competition becomes fiercer, so the rate of diversification slows and evolution becomes more a process of modifications to a basic form, a battle among the best. Likewise, as their genomes (the chemical "code" for their construction) increase in complexity, the rate of notable change should also slow remarkably.

Second, since most mutations are fatal, it can take many generations before a favorable mutation randomly arises—but when it does, its effect on success is immediate and starts affecting the population with the very next generation. Thus it can spread and dominate very rapidly, relative to the time it took for that mutation to arise in the first place. This is why evolution is often punctuated, even in the absence of new niches to exploit: significant changes occur when an advantage is gained, but it takes long periods of random experimenting before such an advantage can arise.

This punctuation is different from effects of diversification because it does not start with a rapid multiplication of species, but occurs accidentally within the lineage of a single species. And because this happens all the time (whereas new niches only occasionally open up), punctuated equilibrium is more constant and occurs on smaller time scales. It is notable that Darwin did not anticipate this feature of evolution because he lacked the information needed to see it, which would only be found long after he died—for example, it occurs on such small time scales that it is not visible as an effect without very precise dating methods (which were not arrived at until the atomic era) and a very large body of evidence (which took a century to accumulate), and the phenomenon could only be seen as inevitable after we had a full understanding of how genes were composed of DNA and how chemical mutation actually operated (a discovery that had to wait a whole lifetime after Darwin's death).

All this we can conclude simply from logic alone: the theory of natural selection entails this two-stage process. And when we actually look at the physical evidence, the layers of sediment and fossils, we encounter exactly what logic leads us to: a long period of relatively rapid diversification of simple life forms; then slowing in later periods exhibiting punctuated equilibrium, where we see a greater complexity and sophistication arising over time—a very, very long time.

In fact, whenever a new niche opened up in history, the same two-stage process happened again, and we see these events, too, in the fossil record. For example, when the first life, which breathed only carbon dioxide, polluted the atmosphere with too much oxygen, a new niche was created for oxygen-breathing mutants. And when wings were hit upon, the air provided an entirely new niche to exploit all over again. And on and on. In each case, rapid diversification in the new niche over millions of years is followed by slower, less earth-shattering development over sometimes tens of millions of years. Likewise, the physical evidence for punctuated equilibrium even within old, occupied niches, is strong.

Our own past fits these patterns. Once bipedalism was hit upon, a whole new way of exploiting the environment was discovered, creating yet another niche, and over the course of about five million years several dozen species of intelligent, tool-using bipeds arose. But eventually only a few won out in competing with each other over the next five million years, with much slower development and slighter modifications. Ultimately, they all went extinct but one: *Homo sapiens sapiens*—us.

We have vast amounts of evidence for this—of non-human bipeds and their tools, revealing a wide range of different kinds of human-like creatures, all appearing and disappearing gradually over time until the last of our competitors died out about 50,000 years ago. And there is no missing link between us and them: we can trace in numerous recovered bones and digs every stage of development from ourselves to our ancestral species all the way back to the first biped—a short, strange, hunched little creature.

Since logic and fact coincide so perfectly and so widely, evolution by natural selection is almost certainly true. There is also a vast body of additional evidence to support this conclusion, a classic case of evidence from entirely different quarters converging to prove the same conclusion (see II.3.1, "Finding the Good Method"). Again, two examples should suffice:

First, DNA: the chemical strings that form the "genome" that codes for our construction show ancestral relationships that correspond very strongly with observed "descent with modification" in the fossil and living records. This is essentially the same test used to prove that a man is the father of his child: a close DNA match proves the kid could not have come from anywhere else. The same test can show other relationships—distant family connections, even racial background. It is therefore incredible that this test for ancestry works even *across species*, which is too amazing to be coincidental. For instance, humans are observed in the fossil record to have descended from the same ape ancestors as chimpanzees, and DNA confirms this: over 98% of a chimp's active DNA is *identical* to human DNA, a match far closer than for any other animal on earth. Even more telling, long sections of the DNA we share with chimps that are "junk genes," excess baggage that does not actually do anything, are also identical (making our total genetic similarity nearly 95%). This code doesn't relate to structure or function, it is just garbage carried over by a natural, imperfect copying process. So the *only* way it could be shared is through blood relation—which means for these matches there can be no other explanation but evolution.

Second, imperfection: every animal's construction—its DNA, anatomy, and physiology—is often flawed or improvised, with vestigial organs, less-than-optimal structures or processes, or missing what would be obvious improvements to any intelligent engineer, like a god. For example, nipples on men serve no function whatever, but are the inevitable byproduct of a procedure for developing men differently from women that is imperfectly carried out. Evolution explains why. In contrast, female breasts do not need to be large, or prominent at all—as instruments for nursing, small breasts are just as effective, while large breasts create increased strain on a woman's back and increased risk of injury and lethal malfunctions like cancer.

The only physiological reason for large breasts is the same as for other inefficient methods of sexual competition, from the peacock's feathers to the baboon's inflamed butt: their size serves to compete with other women in attracting men. Otherwise, they are a liability, and a needless waste of energy—although, in the rude way nature works, that is their point: to advertise to men "Hey, this woman is so healthy she can waste energy on these risky things, aren't they pretty?" What possible use such an inefficient tactic would have in the eyes of an *intelligent* engineer is hard to fathom—especially an engineer whose only purported purpose for sex was procreation, not lust or entertainment. Wouldn't making men attracted to, say, intelligence, or beautiful eyes, have been enough—even better—without all the danger and headache? There are dozens of examples like that. We can think of many other improvements an intelligent engineer *would* have made, but that a blind process of natural selection hasn't, such as spinal cords that can regenerate, brains that aren't so delicate, a genome that isn't prone to cancerous mutation, and kidneys that don't waste so much water in evacuating toxins—just to name a few.

We can also ask other nagging questions about life in general. If life was engineered by a compassionate person, why create an animal kingdom that depends on wanton rape and murder to persist? Why were 99% of the species that ever walked the earth so ineptly built that they've gone extinct? Why do most animals produce hundreds of offspring because almost all of them will die, most of them horribly? What is the point of disease or other forms of suffering among animals anyway? And why are resources limited, when the creator's means and material were not? And so on.

Most importantly, whereas the nature of man and other plants and animals is inexplicable and mysterious on the theory that they were intelligently created, it makes more and more sense every day on the theory of evolution by natural selection. Many of the poor design features of animals can be readily explained, for instance, and some of these explanations have even been proved with the evidence of heredity, chemistry, and history. The examples of this and other convergent evidences for natural evolution are legion. You only have to read the literature to find it and be amazed. There can be little doubt it is true.

And the alternative is not very encouraging: in order for creationism to be true, God must have deliberately and maliciously *planted* all the evidence for evolution, which would entail that God is not a nice guy. He would have to be the most magnificent Father of Lies, reducing Creationism to a form of Satanism (see IV.2.7.6, "The Great Deceiver Defense").

> For further reading, see the Secular Web's sections on the "Argument to Design" (www.infidels.org/library/modern/theism/design.html) and "Creationism" (www.infidels.org/library/modern/science/creationism/).

See John Maynard Smith and Eörs Szathmáry, *The Origins of Life: From the Birth of Life to the Origin of Language* (1999) and *The Major Transitions in Evolution* (1997); Monroe Strickberger, *Evolution*, 3rd ed. (2000); Stephen Jay Gould, *The Structure of Evolutionary Theory* (2002); Ian Tattersall, *The Monkey in the Mirror: Essays on the Science of What Makes Us Human* (2001); and for even more detail: Mark Ridley, *Evolution*, 3rd ed. (2003) and Douglas Futuyma, *Evolutionary Biology*, 3rd ed. (1998).

On the superiority of natural selection to creation theory: Eugenie C. Scott, *Evolution vs. Creationism: An Introduction* (2004); Niall Shanks, *God, the Devil, and Darwin: A Critique of Intelligent Design Theory* (2004); Matt Young & Taner Edis, eds., *Why Intelligent Design Fails: A Scientific Critique of the New Creationism* (2004); Mark Perakh, *Unintelligent Design* (2003); Richard Dawkins, *The Blind Watchmaker: Why the Evidence of Evolution Reveals a Universe Without Design* (1997); Douglas Futuyma, *Science on Trial: The Case for Evolution* (1995).

On reconciling religion with evolution: Michael Ruse, *Can a Darwinian Be a Christian? The Relationship Between Science and Religion* (2000) and *The Evolution Wars: A Guide to the Debates* (2002). Discussing opposition to this: Barbara Forrest & Paul Gross, *Creationism's Trojan Horse: The Wedge of Intelligent Design* (2003).

8.3 The Evolution of Mind

Two things that confer endless advantages are the ability to reason (to analyze and thus respond in logical ways to the environment) and the ability to recognize repeating patterns in oneself and one's environment, as an aid to reason.

All animals reason in some sense. Every brain, no matter how simple, engages some sort of logic, however simplistic, that helps it intelligently respond to sensations, even if only to a small degree. But the more complex the brain, the more complex and subtle—and thus the more advantageous—its ability to perceive things, and to think about them. Likewise, the more sensory organs one has, and the more sophisticated they are, the more abundant and accurate your data about the environment will be—another obvious advantage. And so we see over time that sensation, perception, and reasoning advanced in animals over the ages, as some brains enlarged and connected to ever-increasing arrays of sensory organs of ever-increasing power.

Eventually, at the end of an ages-long line of development in brain complexity, after hundreds of millions of years of trial and error, a brain developed so complex that it was capable of perceiving itself and its own thought, and this made self-identification and self-knowledge possible (III.6, "The Nature of Mind")—which also made a recognition and comprehension of other minds possible (by analogy to our own), as well as a recognition and comprehension of abstractions (III.5.4, "Abstract Objects"). And this made sophisticated language possible (II.2, "Understanding the Meaning in What We Think and Say"), as well as advanced reason (III.9, "The Nature of Reason").

With language, which may have been the original driving element in all this, thought patterns can be organized and analyzed more quickly, efficiently, and creatively, and they can be communicated, passed from one member of a population or group to other members, very easily. This was an unprecedented development: now, units of survival-advantage (such as ideas and knowledge) did not have to await the slow process of growth and reproduction, but could spread almost instantly, could even be created wholesale—and intelligently. No longer was a drawn out process of random chemical mutation and reproduction required for adaptation. Now, direct feedback loops, and the ability to simulate circumstances and thus sift good ideas from bad with greater ease and less risk, could produce ever-increasing adaptation in a single organism or population within a single lifetime. These "acquired advantages" could then be passed on, even to non-relatives. And the advantage, of things like true knowledge, creative innovations, and abstract investigations, is quite obvious.

> See: William Calvin, *A Brief History of the Mind: From Apes to Intellect and Beyond* (2004); Nicholas Humphrey, *The Inner Eye: Social Intelligence in Evolution* (2003); Bruce Bridgeman, *Psychology and Evolution: The Origins of Mind* (2003); Nicholas Humphrey, *A History of the Mind: Evolution and the Birth of Consciousness* (1999). Important supporting evidence is provided by Gary Marcus, *The Birth of the Mind: How a Tiny Number of Genes Creates the Complexities of Human Thought* (2003).

8.4 Memetic Evolution

Now that ideas rather than molecules could be reproduced and transmitted ultra-rapidly, the whole process of natural selection could take over again. First, there is reproduction, as ideas can be communicated and passed on. But now an *intelligence* can act on them as well, to modify or perfect them, before and after. No longer is nature blind. Second, there is an inevitable mutation of ideas as they are analyzed, passed on, and acquired. Not only does this change occur from distortion in memory and transmission, and processes like syncretism (taking two ideas from different sources and merging them into one), but even from intelligent, deliberate, creative modification. Third, an inevitable selecting force is exerted upon these ideas by our intelligence, as well as by nature and society, as ideas help or hinder populations that adopt them.

This is called "memetic selection" because instead of "genes" (genetic selection) the units being reproduced and acted on are ideas, which scientists call "memes." Now, memes are not strictly identical to genes. Whereas genes have certain characteristics from the fact that they are chemically encoded in a limited number of ways, which affects what sorts of genes can arise, what they can do, how they mutate and when, and so on, memes can be encoded in a number of different and less precise ways in the synaptic patterns of brains (and can even be "stored" in other media like books and records), and are far more freewheeling in the way they can combine or mutate. But the similarities are key: both genes and memes have all three features necessary for evolution by natural selection—reproduction, mutation, selection—and therefore can be described by essentially the same mathematical rules.

Memetic selection acts on two levels. First, memes will survive, or die out, in relation to how they affect the biological survival of individuals and societies. For instance, false knowledge, and poor methods of acquiring and testing knowledge, are both often deadly, whereas true knowledge and good methods will allow their possessors to hit upon more, better, and more useful ideas. Second, memes will survive, or die out, in relation to their own battle for *ideological* survival. For instance, to survive and be passed on, ideas must not only confer some advantage to the recipient—which may be something as trivial as the provision of emotional comfort, or as profound as the creation and use of fire—but they must contain with or within them the seeds that inspire the recipient to pass them on, and the audience to gladly take them up.

These two levels often act at cross purposes, but in conjunction they slowly grind out a culture, a philosophy, a science, and a technology that, despite some meandering, is increasingly more advantageous to general survival within a given environment. Indeed, culture and technology both change and *become* the environment. This is how human language and civilization arose and progressed to where it is now. The process of history is simply a natural one of human interaction and conflict, of creativity and blunder, of reason and emotion—in short, of variously failed and successful adaptations to stability and crisis. And once humans had procured a certain measure of wealth and security, they were able to work on perfecting their most fundamental understanding of themselves and the universe: to reason about reason itself, and thus seek, and seek to perfect, their access to truth.

At the same time, the tenacity of certain virus-like memes (such as nationalism or racism or religious fundamentalism) fights against the advantages of careful and tolerant reason, which is not only an idea harder to transmit and less comforting to the unwise, but an idea more tolerant of competing ideas, a definite memetic *disadvantage*. In contrast, the ultimate memetic virus has less to do with things as harsh and difficult as the truth, but more to do with silencing competing memes, and stirring purely emotional attachment to other memes instead. For instance, some memes play to our ego by telling us that we are the center of the universe, the purpose for which the whole cosmos was made, and that we will get everything we want later if we play the sheep now. Others play to our fears by telling us we will be safe from all our worries if we just have faith. And so on.

We will speak more on this later (IV.2.3.1, "Religion Didn't Win by Playing Fair" and V.1.2.3, "Selfish Genes and Selfish Memes"). For now, it is enough to say that a memetic disease, just like a biological one, can cripple or kill—but it can also be cured. Indeed, all of human history has been a battle against biological as well as memetic diseases—a quest for their cures, and a better life of health and wellbeing.

> See David Hull, *Science as a Process: An Evolutionary Account of the Social and Conceptual Development of Science* (1988); Alan Cromer, *Uncommon Sense: The Heretical Nature of Science* (1995); Dan Sperber, *Explaining Culture: A Naturalistic Approach* (1996); Susan Blackmore, *The Meme Machine* (2000); Robert Aunger, *Darwinizing Culture: The Status of Memetics As a Science* (2001) and *The Electric Meme: A New Theory of How We Think* (2002); Leda Cosmides, et al., eds., *What Is Evolutionary Psychology? Explaining the New Science of the Mind* (2002).

9. The Nature of Reason

Reason involves at least three abilities: correspondence, logic, and retrieval and presentation. Each of these abilities has been mechanically reproduced by computers today, and is thus neither mysterious nor unnatural.

The power of 'correspondence' is the ability to recognize that a particular pattern matches many others (and is thus a repeatable pattern). Computers are doing this, for example, when they perform Optical Character Recognition (OCR) on a written page, recognizing a certain specific pattern of ink as a certain universal letter. This means assigning a name to a recurring pattern, and detecting when that pattern arises in our perception.

The power to carry out logical operations is the ability to analyze data in various ways, such as adding or subtracting, or identifying patterns within a larger pattern. Every form of logical operation humans can imagine has been performed by a computer, from Boolean Algebra to Fuzzy Logic—even creative scientific induction—and, like OCR and other recognition routines, we know the mechanical-causal steps that can carry out such functions. They are basically just practical, truth-finding 'rules' for extracting information from complex patterns of data, for managing a linguistic train of thought, and for arriving at probable conclusions from an enormous pool of information.

Finally, the power of 'retrieval and presentation' is the ability to remember and call up relevant data for analysis from among a vast store of information, such as comparing one thing to another, combining two things to achieve a third predetermined goal, and so on. This is also an ability we have mechanically programmed into computers, running routines called "Creative AI" (Artificial Intelligence), where computers can store, even *learn*, huge amounts of disconnected data and creatively "guess" what particular item might be relevant to solving a new problem never encountered before, then retrieve it and apply it. Computers have, for instance, learned all on their own how to regulate a power-plant's cooling system through trial and error and applied logic. A computer has even been programmed to invent its own scientific hypotheses and test them, or use common sense to plan someone's trip—for instance, taking into account their claustrophobia and their passion for nursing, without even being told that claustrophobics don't like going through long train tunnels, or that nurses would be interested in a famous medical exhibit.

However, computers have yet to accomplish three other things essential to *human* reason: a vast processing capacity (the smartest computer that exists today has the equivalent of a mouse, a far cry from what humans have and need), an advanced integrated perception complex (though computers have been trained to "see," for example, their powers of visual perception are still at the stage of lower animals, and they have not yet reached the complex level of integrating several different sensory systems into the same process of continually constructing a model of their environment), and self-awareness (constructing a model of oneself: see III.6, "The Nature of Mind"). Otherwise, contrary to popular belief, computers can abstract, recognize abstractions in existing things, and talk about them (see III.5.4, "Abstract Objects").

For example, a computer can learn how to recognize "circleness" or "letter-k-ness" all on its own. It can then use that knowledge to see where circles or the letter "k" exist in things, distinguish their attributes from others in the same object, then analyze and output the features of circles or the letter "k" that are universal (i.e. repeatable), and reason from those features to other conclusions (such as that a circular object might roll down an incline, or that a "k" might be associated with other letters before and after it to form an actual word rather than a random string). But *consciousness* of abstract objects requires consciousness of a self that exists in relation to them, and that requires a computer more complex than any yet within our technological grasp.

9.1 Reason vs. Intuition

Reason is basically just the stepwise deployment of rules for arriving at useful conclusions. But "intuition" is not to be left out of this account. Intuition is a blanket term for all reflexes, what we generally call "skills," essentially innate or learned systems of reflex patterns, usually connected to patterns of experience. Reason plays no part in riding a bicycle, for example. Rather, we learn from practice how to intuit the proper moves in various circumstances. However, with enough effort and time we can still reason out exactly why our intuited responses in riding were correct or incorrect. Only then do we *understand* what we are doing.

The distinction here is between noncognitive knowledge, of a kind computers have and that most animals can possess and develop, and cognitive knowledge, also known as "propositional knowledge," knowledge you can formulate and state in the form of a proposition. The latter requires a self-conscious mind to truly grasp. The recognition of a self-pattern is needed to make a distinction between *self* and *other*, which is needed to be conscious of thought-acts and to develop a centralized mediation of data relating everything to one's own identity and interests. It is also useful for modeling the environment, directing one's attention to elements of that model, and assigning symbolic names thereto.

This may be connected to the rise of language. Noncognitive knowledge cannot be passed to another except by physically training them. We can direct someone towards the best hands-on experience, talk them through it even, but they can only acquire the skill by *doing* it. However, cognitive knowledge can be passed on without any such activity. It can be communicated directly with words, which are simply agreed-upon codes for patterns of experience shared by all parties (see II.2.1, "The Meaning of Words").

It's the same for *everything* we call "intuition," from our ability to sense danger to our ability to predict the behavior of others or interpret what they are thinking. When we "intuit" something, we are doing nothing more than employing a skill in the matter at hand to arrive at a conclusion or a decision. This means the success of our intuition will always be limited to our actual skill in the matter at hand. In other words, it will be limited to our experience and training in that kind of task. It will also be limited to the data actually available—which can often be misleading or in error, but can also include a wide range of valuable information we are not consciously aware of.

We all have skills we take for granted, which we learned throughout childhood and continue to perfect all our lives without thinking about them, such as walking, or reading the emotions and intentions of others, predicting the weather by smelling the air and looking at the sky, seeing the patterns of words in jumbles of letters, "educated guessing." Intuition is thus very handy, but also quite fallible. Even a champion bicyclist can stumble. And the comparatively harder tasks of reading another's emotions, or guessing the right career move to make, carry many more chances of failure.

Ultimately, a skilled lover's intuition will likely fail him in war, while a skilled warrior's intuition will likely fail him in love. This is why someone who cultivates a variety of experiences is most likely to be wise. Still, intuition has obvious advantages over reason: it is much, much faster, functioning automatically, and it can be applied to many circumstances in which reason cannot yet even approach the problem. After all, reason requires a careful identification and analysis of relevant facts and principles, especially the recognition of what are often incredibly subtle and complex patterns, something that is often not possible in practice. And even when such a feat becomes possible, it takes time to go through, time we often don't have.

But reason, in turn, has advantages over intuition. Unlike intuited responses, the conclusions of reason can be checked for error—and, conversely, can be checked for accuracy by other people. Intuition cannot—at least, not until it's too late. Intuition only learns through trial and error: a decision is known to be false when it fails, true when it succeeds, and never at any other time or in any other way. Reason allows us to "simulate" situations and circumstances, through the virtual-reality model of the world that is generated by our consciousness (see III.6, "The Nature of Mind"). And that is a far safer means of checking ourselves for mistakes, and for building knowledge and expertise without the risks and time-consumption of slowly acquiring enough experience to have good judgment in a particular endeavor or subject. While intuition can learn from its mistakes, reason can avoid them before they are ever made. And while intuition often can't tell if it's correct, reason almost always can.

Reason is also universally applicable. While a lover will be inept in war, and a warrior inept in love, according to their degree of experience in each, reason will serve the lover and the warrior equally in both, for reason is a universal skill, adaptable to any kind of knowledge. Skill in reason therefore ought to be mastered, trained in constantly, since it is the one skill that we can apply with equal success to anything and everything, even things we've never experienced before—very much unlike intuition. And, again, it can be checked for success or error, communicated to and from others, and applied without the same risks or expenses faced by trial and error. A command of reason is essential in order to expand and master one's ability to understand all things and all people.

Intuition is, of course, the prototype out of which our capacity to reason evolved (see, for example, VI.2.7, "The Analogy Effect"). But this only reiterates that intuition evolved as a useful way to solve problems, but that reason is a more accurate and perfected tool than intuition.

9.2 Why Trust the Machine of Reason?

In mechanical terms, one man cannot directly cause another man's nervous system to rearrange itself—as must happen, for instance, in order to learn how to ride a bicycle. But through language, one man can cause a certain part of another man's nervous system to rearrange itself, the part that hears, recognizes, analyzes, and stores a language. This is in fact one of the revolutionary abilities of the human brain, discussed in III.8.3, "The Evolution of Mind." The intermediary here, among men and within a man's conversations with himself, is *reason*: the application of a few simple tools to construct complex patterns of thought.

It is often little appreciated that even the most complicated structure in mathematics, logic, or science is built out of a few simple notions and procedures. For instance, neurobiologists have found that the brain really only has the ability to instantly recognize a handful of numbers (the first few). But that is all we need to construct any other number that exists. Even in formal terms, all of mathematics can be reduced to about nine simple statements, and all of computational logic to just three. And, of course, we have already seen in Part II how all our theories in science and life are constructed from collecting the simplest sensations. Just as a ship for all its complexity is little more than metal adeptly employed, so human thought needs only a few simple tools and building blocks to erect marvels of conceptual genius. "Reason" is the faculty of having and employing these simple resources to just such an end.

But why trust it? For two simple reasons: (1) Because it works, and (2) Because natural selection ensures it, at both the genetic and memetic levels. I have already discussed above how we know science and reason work (see II.3, "Method"). If they weren't successfully getting at the truth, we wouldn't be seeing their successes. It is their success, after all, that gives them such prestige and respect. And this ensures their memetic success. For their utility will always inspire us to pass them on, or eagerly seek them out.

How did we arrive at the basic tools of reason? By trying many different ways of getting at things until we hit upon better and better ways of knowing, and eventually a *best* way. There can be no reasonable dispute now that we have struck gold, no plausible ground for not trusting our methods and means of knowing. So even if it was improbable that our faculties would be as trustworthy as they are, such an observation would be moot. You can't tell a lottery winner that he didn't win because his winning was improbable. All he has to do to prove you wrong is show you the money.

But there was nothing improbable about it anyway. A brain, and its training and refinement at the hands of a memetically evolved belief-structure and a natural belief-forming process, employs certain generic tools of "reason" in order to test pieces of information it possesses, and thus assign or reinforce our corresponding belief or doubt. And this is essential. For a brain cannot just be a passive mop, implicitly believing everything it receives. If a brain's ability to reason does *not* produce results ("belief" or "unbelief") whose degree of certainly corresponds more and more to the *real* probability of something being so, the brain's ability to survive will diminish substantially. That it can still get lucky is irrelevant. A brain full of false information (or worse, false *confidence* in false information) will far more often do the wrong thing than otherwise, and will far less often solve a novel problem. And doing the wrong thing is often deadly, almost always wasteful of precious time and resources, and routinely painful or otherwise damaging to one's interests, whereas having less of this error is always an advantage over one's peers, as is having more right answers to new problems.

No goal can be efficiently achieved if you do not know how to achieve it. So certainly when we *think* we know how to achieve a goal but are actually *wrong*, we will not only just as surely fail to achieve our goals, but we will risk and waste much more attempting the futile than if we lacked confidence in the project to begin with. Conversely, if we know how to achieve a goal *better* than our peers, we will have an obvious advantage over them. It follows that our brain must be good at what it does, or else it would not exist. Certainly, it need not be perfect. So long as our faculties generate more truths than errors we will be able to approach the truth, and get closer and closer to it the more minds, the more time, and the more methods and tactics we apply to the same problem, rolling our loaded dice over and over again until we inevitably get a better and better, and perhaps a perfect score.

Even novel applications are not unlikely. Though we did not originally evolve within a scientific culture, the discovery of scientific methods has still been an unquestionable survival advantage to our whole species. That is why it has such importance, utility, and universal prestige today. From disease control to electrical power, no one can say science does not confer an advantage to human survival. Even the most esoteric sciences, like Relativity and Quantum Mechanics, decisively win our wars for us: precision guided missiles depend on both, relying on Global Positioning Systems based on quantum clocks and relativity transformations.

But even before that happens, scientific reason is like the human hand: clearly a hand's ability to play complex music is not a survival advantage, but a hand that can use tools *is*. Yet such a hand will always be able to play complex music simply by virtue of being adept with tools. Thus, that such an ability is not relevant to survival does not mean it is improbable. To the contrary, since it eventually follows from an ability that *is* highly probable, the ability to play complex music is in turn highly probable. So it is with 'scientific' or any other application and refinement of reason: with enough looking, better and better methods will inevitably be discovered (see II.3.1, "Finding the Good Method").

Of course, we must distinguish reason from sensation or intuition, or even belief for that matter. These are not the same. Most "errors of reason" people describe are really errors produced by intuitive processes and perceptions rather than conscious reasoning. And everyone agrees our reason far more often corrects the errors of our senses and intuition than the other way around. This is because our brain's ability to reason is a generic skill whose primary advantage is that it can be universalized to any thought or activity. There is no way it could be so distorted in its structure as to produce false beliefs conducive to survival. Such a flawed construction would require an improbable scale of complexity in the very rules of reason, yet we know the rules of reason to be very simple.

Such an outcome would also require an improbable coincidence between those rules of reason that produce false conclusions, and the set of all harmless or beneficial falsehoods. For even false beliefs that are seemingly harmless can cause needless behaviors that become an evolutionary disadvantage. For example, they will probably be expensive, wasting resources to no good end, which could have been employed toward genuine advantages instead. Consider witch hunting or lengthy rituals: a group who can accomplish its ends without these things will have a clear survival advantage over a group that thinks them necessary. Reliance on faulty beliefs also prevents us from finding new resources or hitting upon advantageous ideas. Consider taboos or superstitions that discourage us from vital discoveries, which would aid our survival, or might someday be essential to it. Anyone whose reason is free from such errors will have a significant advantage over us (see, for example, section III.10.4, "The Nature of Spirituality"). Thus, evolutionary pressures will *eliminate* such flaws over time. It will not accumulate them.

Only a rare set of objects, genuinely "beneficial" or "harmless" falsehoods, falsehoods that do not cause us to miss opportunities or to waste resources on the needless, could be believed by a brain without affecting its ability to survive and flourish. Yet it is inconceivable that the whole universe could be so arranged that a *simple* flaw in a universal reasoning procedure would just "happen" to match, and far beyond chance, the survival needs of an incredibly adaptable species as varied in abilities and environments as man, and do this so well as never to be detected—even when the same procedures are applied and tested on all other species' means of survival in totally different environments. For we can observe that reason would lead to the best means of survival for the ant just as surely as it does for humans. Yet our needs and environments are so different. Never mind, for example, our ability *now* to survive in the radiation-swamped vacuum of space, which can have no explanation but that human reason works very, very well.

Could this be a coincidence? No. Such a miraculous fit between a flawed faculty and *all* the needs and advantages for survival is so improbable that only a Cartesian Demon could account for it (cf. II.3, "Method"). How else could Mother Nature know what falsehoods we can afford to believe and what truths we can afford not to discover? And not only that, she would have to know how to intricately design a faculty, and that despite using only a very few simple rules, which just so happens to always toe this line perfectly, accounting with uncanny prescience for all future changes in the ecological environment and all unexpected and novel circumstances for all creatures on the evolutionary line, up to and including humans. There is hardly much place for such a clever being in Metaphysical Naturalism!

Rather, the only plausible *natural* way humans could be built to place confidence in a falsehood is through the far simpler method of installed or intuited *knowledge* (i.e. belief *without* the use of reason, whether inborn or communicated or "guessed at") or a faulty sensory apparatus, not a generic reasoning procedure. But reason convicts innate, communicated, or intuited knowledge if it is false *and* testable. That is what we developed reason for. So an ingrained flaw could only work for *untestable* knowledge, which reason already informs us cannot be trusted for that very reason.

This is why Alvin Plantinga's arguments to the contrary simply don't work. For example, in his book *Warrant and Proper Function* (1993) he posits a scenario whereby an early human "*likes* the idea of being eaten" but always runs away from tigers because he falsely believes tigers won't eat him (pp. 225). Plantinga notes that this still confers a survival advantage, and therefore false beliefs are compatible with survival. But that's not how it works. First of all, this poor human would fail to act to rescue family members attacked by tigers, and would fail to fight back if actually caught by one himself—and should he ever need one for food he would be ill-prepared to confront it. Thus, Plantinga is wrong: this is a huge *disadvantage* to a species' survival.

Plantinga's scenario is also naturalistically untenable. It is extraordinarily improbable that a man would develop the ability to run away *only* from *all* those things that will eat him and yet, *at the very same time*, still evolve a useless ability to want to get eaten. And if that is *not* the improbably-exact belief-desire pair he developed, then he is either going to get eaten by something he wasn't born to run from, or he will run away from things he shouldn't (like a delicious deer). Either way, he is at a huge disadvantage again. Finally, survival demands adaptability and innovation: a man who had options besides running when facing a tiger (like hiding, climbing, attacking, trapping, or scaring it away) will *always* have a survival advantage over Plantinga's poor fellow. Thus only the former, not the latter, will ultimately survive the trials of natural selection. No matter what scenario Plantinga envisions, it will fall to the same analysis. It will be too improbable to credit. And even then, the *total consequences* of such false beliefs will almost always end up negative, leading somewhere down the line to wasteful, limiting, or self-defeating behaviors.

But that isn't the half of it. What does any of this have to do with the faculty of *reason*? Clearly, a man will not *reason out* that tigers are harmless, much less that wanting to get eaten is a good idea. To do so would require such tremendous flaws in his memory and sensory apparatus as to doom him for sure. For reason has no inherent reference to tigers. It is far too generic and universal an ability, and far too simple in its components, to be manipulated by nature in any way Plantinga suggests. And if men developed *innately* the bizarre beliefs Plantinga posits (though how natural selection could ever lead to them is quite unfathomable), reason would always convict those beliefs as false or untrustworthy, as it does in all other cases where our instincts could be fatal, such as running toward a mirage when thirsty. That is the very function of reason, the very thing it evolved to do.

As we can see, there can't be many false beliefs that would *really* be an advantage to survival. But if there are such things as untestable yet genuinely advantageous falsehoods, belief in God is one of the best examples imaginable, fitting all the criteria for an innate false belief, which could be biologically inspired in us by natural selection (a fact for which we actually have some evidence: see III.10.4, "The Nature of Spirituality" and IV.2, "Atheism: Seven Reasons to be Godless"). In contrast, there is no plausible path by which we could naturally evolve a *false* belief in the reliability of reason.

This leaves us with the question of a faulty sensory apparatus. Here things are different, since we know our senses are imperfect. But just as 'reason' is a tool, a technology—the conscious, artificial perfection of intuited skills—we also know how to improve our *senses*, and we have built numerous tools that do just that. We see this in technologies from writing to the computer (as enhancements to memory), or from magnifying glasses to spectrographs (as enhancements to vision). All are examples of memetic evolution improving our faculties beyond genetic evolution's best efforts.

Yet even genetic evolution by itself has steadily improved all our faculties, in one respect or another. The very reason we developed multiple faculties (including our reason) was to allow independent checks on each other, making their limitations fewer, and their errors less damaging, and ultimately far less probable than blind chance would predict—after all, what are the chances that such disparate systems as vision and hearing and touch would all simultaneously mislead us in exactly the same way? Pretty slim. It would also be wrong to say that our senses, however flawed, are *un*reliable, for their successes in daily life far exceed their failures, and even when they fail, we know, or can advantageously learn, when to trust them most or least, and how to catch and correct their mistakes.

Still, the argument could go like this: blind, natural processes led to our sensory organs by a series of accidents, so why should we trust sensory organs that are just accidents of nature? But must an "accidental" sensory organ be untrustworthy? The sort of example some give is a set of stones arranged to convey a verbal message: it would be irrational to regard the message as both accidental *and* true, since an accidentally arranged message would only be true by blind luck. Yet the "messages" sent to us by our sensory organs are "ultimately" accidental, therefore they cannot also be true, except by blind luck. Right?

Wrong. First, ask yourself: Would it be irrational to regard the presence of a pile of stones at the base of a cliff as signifying a danger of landslides? The analogy breaks down here. Such a pile of rocks *does* signify a landslide risk, and not because of any design—for it is purely accidental. The fact is, we don't infer the risk of landslides on the assumption that someone arranged those stones to convey to us, by a prearranged code, a landslide risk. Rather, we infer the risk from the prior observation that landslides result, without any intelligent design, in sufficiently unique patterns of debris, and thus we know that wherever those patterns happen to turn up again, we can suspect that there have been landslides there, and so there may be a risk of further landslides.

Since this proves that it can be rational to infer a "message" from an accidental arrangement of things, you cannot say it is irrational for us to do this in the case of our sensory organs, unless you can show that our sensory organs must be like the "words" analogy rather than the "landslides" analogy. But it is clear that our sensory organs fit the latter, not the former. Remember what I said about language in chapter II.2: it is an agreement between you and me that certain things will be codes for certain other things. But what is an agreement? You decide to keep using this to mean that, and I decide to do the same. But this means that *any regularity of nature* is equivalent to an agreement. Nature "decides" to keep using this to mean that (by having one regularly follow the other), and we decide to adopt the same rule. And that is the basis of inference. All that is needed is regularity. And nature has no choice in the matter: nature is regular by nature. She cannot be otherwise (per III.3, "The Nature and Origin of the Universe").

Second, the analogy simply breaks down in and of itself. The *written message* was produced by accident, but not our means of reading it. Clearly in this analogy our senses correspond to our means of reading the message, *not* the cause of the message. Thus, an unjustified step is made when we reason from an accidentally caused message to an accidentally caused "message-reader." For our sensory organs obviously do not cause accidental messages—their messages are so consistently consistent that the odds would be near infinity to one that such a stream of messages was wholly produced at random, like the "written message" in the analogy. This is not to say that our senses never produce false or random messages, only that they clearly don't do this very often, much less regularly.

So we are back to the original question: Why trust a message-*reader* that was accidentally produced? As we have said (in III.8.2, "Evolution by Natural Selection"), the outcome of natural selection, though "accidental," is not random. Rather, it produces what is conducive to survival, and destroys what is not. Since our sensory organs would be lethal if they were regular generators of false information, and since they would only be selected for in the first place if they were regular generators of accurate information, there is no conceivable way on the theory of evolution that our senses could be unreliable to any substantial degree.

Of course, we should not confuse reliability with authenticity. Even if our eyes do not give us *authentic* information about color (and I do not believe they do), they nevertheless *reliably* inform us of certain distinctions in color, and we can accurately infer things about the world based solely on that. In this way all of our senses do in fact reveal the truth about things. For instance, what we call colors are only inventions of our brain. They are coded patterns created to represent the fact that our eye-cells are sending certain distinct signals to our brain. We infer from the patterns within this "invented representation" certain things about the world, like the fact that our eye is being hit by photons most likely, say, bouncing off our bathroom door. We do not infer this because we were pre-designed to know what photons bouncing off our bathroom door would look like. We know it only because we have seen the same effect every time we looked at our bathroom door in the past. In fact, this repeated experience is what we give the name "bathroom door," and everything we believe about a "door" is based on all our past sensations of just such a sort. So it is not even necessary to know about cells or photons in order to trust our eyes, though now that we know such things we can be even more certain that our senses are reliable, because we can check their function and identify ways they are fooled.

In short, our senses and our reason are reliable because of two simple facts. First, the universe just happens to follow certain consistent behavior patterns. Second, our senses just happen to follow certain consistent behavior patterns. And all that is needed for things to follow consistent behavior patterns is the existence of consistent behavior patterns (see III.5.4.3, "Modal Properties"). Once you have that, thanks to evolution by natural selection, the reliability of sensory organs can be accounted for. So there is no need to appeal, for example, to an intelligent engineer in order to "ground" or "justify" reason.

Of course, one might still argue that the existence of consistent behavior patterns in the universe requires an intelligent engineer, but we have already addressed that argument elsewhere and found it wanting (III.3, "The Nature and Origin of the Universe").

9.3 Contradiction Revisited

Some still try to argue that despite all this there could still be some flaw in us or our reason that we cannot detect and that renders it untrustworthy, and therefore we need some sort of "ground" for establishing the reliability of reason, some reason to trust reason. This is an irrational skepticism, rather like someone who is afraid to go outside for fear an elephant from the moon will fall on them. Yes, it is possible. But it is so immensely improbable that there is no reason at all to believe it, and plenty of reasons to disbelieve it. Indeed, the nature of reason *itself* makes such skepticism incomprehensible. Drawing from what I have already argued (II.2.2.7, "The Nature of a Contradiction"), I will make this point by focusing on the Law of Non-Contradiction, the converse of the most fundamental rule of reason, the Law of Identity ("When two things are the same, they are the same" or "A = A").

As I've already argued (cf. II.2.2, "The Meaning of Statements"), what we call "logic" or the "rules of reason" is actually the structure of what we call *language*. If a language exists, then *by definition* logic exists, because without logic you can communicate nothing. It follows, then, that if you are communicating something, logic exists, for it must be inherent in the very rules that allow the communication to occur. After all, the only way I can communicate to you that "my cat is white" is if you and I both agree to certain fabricated rules, a 'code', which we invent (or borrow from someone else) and decide to follow. This allows me to know that *you* will know what the sounds "my" and "cat" and "is" and "white" will stand for. They are "code words" for our experiences. I point to a white wall and you and I agree that we will call what we both see there "white," and so on (the actual process of language learning is much more complicated than that, but it does boil down to this). Then, when I shout "white" to you, you will remember our agreement about what that would be a code for, and I will have communicated something to you. We invent these rules for this very purpose. If you and I refused to decide on any rules, or did not obey the rules we decided on, we would be unable to communicate—indeed, we would even be unable to communicate with *ourselves*, which is a fundamental function of reason: to compute, and communicate the results of that computation, and employ that communicated result in our next step of reasoning.

All logic arises from these manmade rules. We can reason that something described as a table will have all the properties of a table without actually seeing it, and we would only be wrong if the actual item was described incorrectly or incompletely. In this way all rules of logic derive from the analysis of language, which is the analysis of coded *descriptions* of sensory and perceptual experience.

Consider, again, the universal, fundamental principle of non-contradiction: something cannot both be and not be. My cat cannot be both all white and all black. Why not? Suppose I were to tell you "my cat is all white and all black." You would look up these words and follow the rules in our mutual codebook, but you would not be able to make this statement correspond to anything in your experience. The rules would not be able to match this code with any agreed-upon meaning. Consequently, I have communicated nothing to you. This is because "black" means, among other things, *not* white, as we have agreed.

However, since this is all manmade, you might think that all we have to do is assign a meaning to this "contradictory" statement, and it will then be able to communicate something. But what meaning will we assign? There's the rub. Can we assign it a meaning that will be consistent with all our other rules? No, we cannot, because we decided beforehand that we would use the word "black" to refer to certain non-white things. Thus, the only way to create a meaning that will obey our own rules is to change the rules, and hence the meaning, of the words that conflict. But then they won't conflict. In other words, the law of non-contradiction is simply a *natural* feature of any consistent set of rules. Indeed, this is a tautology: What is a consistent set of rules? A set of rules that never produces a contradiction.

So then you might think we can escape this by "deciding" not to have a consistent set of rules. But we have already seen that we cannot communicate anything with an inconsistent set of rules, because we have to follow the rules in order to communicate, and we can't "follow" inconsistent rules. Thus, we are stuck. Either we have contradictions, but no language (and hence no reason), or we rule out contradictions and communicate. This is a simple fact that we observe about the universe. Now, you might say that, perhaps, there are things that can exist but cannot be communicated or described. But if they can be *experienced*, then they can be given a code name, and can thus be communicated to anyone who has experienced the same thing and knows the code word for it. Perhaps you might propose instead that it is possible to have a universe where a contradiction could communicate something, where it could actually describe something that we can experience or imagine. But since we all see that we do *not* live in such a universe (we cannot even *imagine* one), it doesn't matter if it is possible.

Returning to the original query, maybe this inability to experience or imagine a contradiction is simply a limitation in our construction, or an error in our brain or senses. But if something can affect us in *any* way, it follows that we can experience it, and thus can imagine it, by reference to that effect. If something existed that could never, even in principle, affect us in any way, its existence would be of no consequence to us. In fact, *no kind of sensation* could ever experience that thing, because to sense something is, by definition, to be affected by it in some way. Thus it follows that even a god could not make us capable of sensing something that can never affect us. All he could do is make it affect us. Thus, the argument that we could be missing some feature of reality is moot. So long as any part of reality can affect us in any way, we can experience it. Everything else is irrelevant.

Of course, if we should discover the ability to imagine and communicate contradictions, we would simply change the way we thought about things, just as we did when the axioms of non-Euclidean geometry were discovered. There is thus nothing that needs to be accounted for here. Logic is explained by what we observe, and it arises automatically and necessarily the moment we try to create a set of rules for describing those observations. Since all language begins with discrimination between things that are the same and things that are not, if language exists, it follows that the universe has things that are the same and things that are not, which is the very reality the Law of Non-Contradiction refers to.

This is even more obvious in the case of inferences from the particular to the general, where the entire structure of inferential (as opposed to deductive) arguments is justified solely and entirely by prior experience. By recalling the reliability of all prior inferential reasoning, we conclude that it works. After all, no one believes that inferences are guaranteed to *always* work—by definition, an inference only suggests, it does not 'prove' in the sense of a logical or mathematical proof, but only in the sense of being convincing to one degree or another. So why are we justified in trusting inferences? Because they work. Period. Experience completely explains logic, and completely justifies it, as well as it can ever be justified in any imaginable universe. So there is no reason to look any further for some other "ground" for reason.

9.4 Alternative Accounts Are Not Credible

A final merit to the above analysis is that there is simply no other way reason makes sense, and no better defense anyone can give it. The only alternative views I know are either that reason is some supernatural property of the universe (various forms of Platonism claim this), or the creation of a god.

The first alternative is quite inexplicable, and that theory usually lacks any coherent explanation of this "supernatural" feature of the universe, hiding instead behind metaphors and assertions that merely perform a magic trick on the problem, dazzling crowds with one hand, while concealing in the other the fact that nothing is really being solved or explained. Maybe someone will wow me with a genuine Platonic alternative to Metaphysical Naturalism's account and ground for reason, one that is coherent and comprehensible. But so far all who have attempted this for me have failed.

The second alternative is a double-edged sword. If we must *assume* that a god exists before we are justified in trusting reason and our senses, then how do we know this god isn't a Cartesian Demon? (see II.3, "Method") Our reason and senses could be deliberately designed to lead us to believe in any lie, just because God wanted it that way. Even the believer's conviction that God is good (so God would not trick us like that) is suspect, because a Cartesian Demon would fool the theist into thinking that very thing. The circularity of this argument ("we know the truth because God lets us, and we only know this because God lets us know the truth") has long been noted by philosophers, and still applies here: to propose God as a ground for reason gets you no farther than naturalism gets you already.

For further reading on the evolution and physiology of a reasoning brain: William Calvin, *How Brains Think* (1996); Robert Moss, *Brain Waves Through Time* (1999); Manfred Spitzer, *Hare Brain, Tortoise Mind* (1999); Lesley Rogers, *Minds of Their Own* (1998); Dietrich Dörner, *The Logic of Failure: Why Things Go Wrong and What We Can Do to Make Them Right* (1996); Eduard Hugo Strauch, *How Nature Taught Man to Know, Imagine, and Reason: How Language and Literature Recreate Nature's Lessons* (1995); Valerie Walkerdine, *The Mastery of Reason: Cognitive Development and the Production of Rationality* (1990).

For philosophical analyses of reason: Robert Nozick, *The Nature of Rationality* (1993); Nicholas Rescher, *Rationality: A Philosophical Inquiry into the Nature and the Rationale of Reason* (1988); Newton Garver & Peter Hare, eds., *Naturalism and Rationality* (1986). See also: Ruth Millikan, *Language, Thought, and Other Biological Categories: New Foundations for Realism* (1984) and *Clear and Confused Ideas: An Essay about Substance Concepts* (2000), as well as Patricia Churchland, *Brain-Wise: Studies in Neurophilosophy* (2002). And for extensive discussion, see: Richard Carrier, "Critical Review of Victor Reppert's Defense of the Argument from Reason" (www.infidels.org/library/modern/richard_carrier/reppert.shtml).

For the science of intuition: Thomas Gilovich, Dale Griffin, and Daniel Kahneman, eds., *Heuristics and Biases: The Psychology of Intuitive Judgment* (2002); David Mayers, *Intuition: Its Powers and Perils* (2002); Michael R. DePaul and William Ramsey, eds., *Rethinking Intuition: The Psychology of Intuition and Its Role in Philosophical Inquiry* (1998); Robbie Davis-Floyd and P. Sven Arvidson, eds. *Intuition: The Inside Story* (1997). Also: M. D. Lieberman, "Intuition: A Social Cognitive Neuroscience Approach," *Psychological Bulletin* 136 (2000), pp. 109-137; D. A. Shirley & J. Langan-Fox, "Intuition: A Review of the Literature," *Psychological Reports*, 79 (1996), pp. 563-584.

10. The Nature of Emotion

People generally assume there are only the five senses (sight, hearing, smell, touch, and taste), though in fact there are many others (balance, for example, or our ability to sense the motion and position of our limbs, and so on) and "touch" actually conflates several different senses (such as pressure, heat, and pain). The same mistake leads many people to miss the fact that our mental life (consisting of various kinds of thoughts and emotions) is simply another form of sensation. The difference is that our ability to sense and perceive thoughts and emotions involves a purely internal operation: the organs responsible are actually parts of our brain, and what they detect are patterns of activity within our brain and body (see III.6, "The Nature of Mind").

Consequently, emotions are a form of perception. Rather than recognizing patterns in sensory data sent in from outside the brain by dedicated sensory organs like the eyes or ears, our experience of emotion is the perception of a *brain-state* by the brain itself. These brain-states are a form of pre-rational computation, having developed in mammalian brains long before higher forms of reason were developed, as a form of practical intelligence. Emotions are largely chemical, involve changes in body state as well (which are in turn felt), and originally served the function of producing a particular behavior while simultaneously preparing the body for it. They are typically caused by certain stimuli the brain has learned (or in some cases is even born) to associate with the need for a particular behavior. But with the added stimuli of rational cognition, emotions can have very complex causes and content.

Naturally, emotions evolved precisely because they cause us to act in various ways conducive to our survival, or our "differential reproductive success" (III.8.2, "Evolution by Natural Selection"), and almost every known human emotion (with its associated behavior and physiological symptoms) has been observed somewhere in the animal kingdom, especially in monkeys and apes. But whatever their original function, emotions have become a fundamental aspect of human experience, and still serve important functions in our lives, not only by enriching our experience, but by giving us important information about *ourselves*.

10.1 Emotion as Appraisal

Everything observed above is scientifically grounded, well-established by neurophysiologists, psychologists, sociologists, and anthropologists. These sciences also lend strong support to the view that emotions are a form of appraisal, and thus can be mistaken. The emotion a brain generates in response to a certain stimulus, circumstance, or thought is an indication of the brain's "opinion" about what it perceives is going on. This is an "evaluative" response, because this "opinion" relates to what our attitude or response to something *should* be, and will vary in intensity or degree (see II.2.2.3, "Opinions"). This also means that emotions can err in two different ways: first, if a stimulus is incorrectly perceived or interpreted; second, if a stimulus is incorrectly evaluated.

A stimulus can be an event, like a loud noise; a circumstance, like being shot at; or a thought, like realizing you are going to die—any experience that stands out and calls for a reaction (even the lack of such experiences can stimulate a reaction, such as malaise or wanderlust). If we incorrectly interpret a loud noise as being shot at, we may experience fear—the evaluation that a circumstance is dangerous—but fear on this occasion would be a *false* indicator of danger. Or, if we mistakenly believe that being shot at entails we are going to die, all sorts of emotions might be generated that would be true indicators of, say, sadness, which could be correct evaluations of the circumstance of dying, but since they are caused by a false belief—for being shot at is not a circumstance of inevitable death—these emotions would be a somewhat incorrect evaluation of your real circumstances. All emotions can be mistaken in this way. For example, you might love someone because you have false beliefs about them.

In contrast, even with entirely correct beliefs and perceptions, an evaluation can be incorrect when our evaluating criteria are incorrect. The brain's mechanism for evaluating circumstances involves what we call "values," which are like computers, physically present in the brain, that assess the *value* of something (I talk a lot more about values in Part V, "Natural Morality"). And these are not mere arbitrary computations. Fear indicates that something is valued as dangerous, for example, and whether something is dangerous is an objective fact about the world, not our mere opinion of it.

How our brain comes to develop criteria of evaluation for each emotional response is not based on conscious reasoning. It is partly inborn (reactions to raw stimuli like loud noises or attractive faces), and partly learned through childhood by imitation, training, and trial and error. Parents play a crucial role in all three methods, hopefully by setting a good example, fairly and consistently distributing punishments and rewards, and exposing children to new and different experiences and challenges, guiding them through safely with attention to their need to *learn* rather than be told.

Though values are open to conscious molding in adulthood, it is not easy. It's not like simply rearranging thoughts or just choosing to have a particular value. Values are more permanently represented in the brain, and to rearrange them requires a more long-term effort. *Conditioning* can be used to mold values, engaging in processes of conscious habit-formation and exposure to certain stimuli. We often call this "evolving" as a person, "learning from experience," "maturing," but it can also be deliberately and scientifically directed in some cases.

Changes in values can also be accomplished through extensive contemplation about things—which is why, again, the self-examined life is so important (II.1, "Philosophy: What It Is and Why You Should Care"), for only then will your values ever become really *yours* rather than just what you ended up with. This is the only way to have values that are rational, based on wisdom and knowledge and reason, rather than accidental byproducts of chance, error, naivety, or indoctrination.

It is in this way that you can learn to experience false guilt for violating a tribal taboo, even when doing so to save a human life; or violent anger at a nonviolent offense like blasphemy; or pointlessly crying at losing a job. We can love someone for all the wrong reasons, too, when our love-evaluator generates the emotions we call 'love' in response to real facts that are not worthy of it, while *not* generating those emotions in response to facts that *are* indicative of the appropriateness of love—depending on how love is defined, or what kind of 'love' we are talking about.

But all is not despair. Our values are not merely randomly constituted. Even when not rationally contemplated nor consciously conditioned, most people, even those with lousy parents, are raised well enough, in a sufficiently healthy, rich environment, that natural processes (of enculturation, socialization, and plain-old trial and error) generate pretty good evaluators in the brain, with some mistakes in each, such as omissions of some important indicators or an attachment to some false ones. After all, the brain would not have these systems if they had not been selected for (III.8.3, "The Evolution of Mind"), and therefore by nature they must be at least somewhat good at what they do, even if not perfect (and as we know, nothing produced by natural selection is likely to be perfect). But values that are rationally organized according to trustworthy beliefs and relevant knowledge will always be superior, more frequently generating more reliable emotions.

For both reasons—incorrect beliefs and incorrect values—emotions can be incorrect. This is why we need reason and reflection to assess the accuracy of our emotions and thus decide whether we should really act on them or not. Unfortunately, since emotions are pre-rational, wired into us before and beneath our reasoning faculties, we cannot 'will them away' even when we know they are wrong. And their effects on our cognition and our body will remain until we remove ourselves from the stimulus, or reprogram ourselves—through long conditioning and contemplation—to evaluate the same circumstances differently. But our *will* is not hard-wired beneath our reason. It is normally under rational control. So even under the influence of what we know are false emotions, we can act differently than our emotions suggest, so long as we maintain focused control over our behavior and maintain our attention to reason.

10.2 Reason as the Servant of Desire

The battle between reason and emotion is often characterized as a master-slave relationship, as if reason should always be in control over emotion. This is not quite correct. Certainly, reason must always keep a check on emotions, making sure they are sending correct signals, and correcting their recommendations when they are not. But without emotions, reason would be a dead letter. For *reason* is the slave of *emotion*. Reason is not a motivator. Reason is a tool, a process. But for that tool to be applied, you must be motivated to apply it, and what you apply it to depends on your goals, which are in turn the result of motives, and motives are the product of desires, and desires are the outcome of emotions.

For example, if you love someone, that love will generate certain impulses, desires, with regard to that person (such as a desire that they be happy and well, and a desire to be with them). Then you would apply reason to work out how to satisfy those desires. Thus, reason is the servant of the emotion. But as we just described, emotions can be in error. So if you love someone because of false beliefs, reason can correct those beliefs, causing you to stop loving them. Or if you love them for the wrong reasons, reason can detect this and produce the conclusion that this love is wrong, therefore its associated desires are not what you desire most after all.

In effect, reason *informs*, but does not direct. Reason is *directed*, it is employed, for purposes not its own, because reason alone cannot have a purpose. A purpose must be given to it. It must be assigned a goal. While emotions provide the impulses and desires for certain ends, they are also a form of information, one vital for deciding what to do, every moment of every day. Not only do we need to pay attention to our emotions and take them seriously, we need to make sure they are working properly, and have correct information to work on.

> For the whole scoop on emotions, see the chapters on emotion in any good college psychology textbook, plus specialized works like Michael Lewis & Jeannette Haviland-Jones, eds., *Handbook of Emotions*, 2nd ed. (2000) and Richard Lane, Lynn Nadel, & Geoffrey Ahern, eds., *Cognitive Neuroscience of Emotion* (2000).
>
> But for more specific treatments dealing with what I have emphasized here, see Antonio Damasio's trilogy: *Looking for Spinoza: Joy, Sorrow, and the Feeling Brain* (2003), *The Feeling of What Happens: Body and Emotion in the Making of Consciousness* (2000), and *Descartes' Error: Emotion, Reason, and the Human Brain* (1995). Also: Martha Nussbaum, *Upheavals of Thought: The Intelligence of Emotions* (2002); Bennett Helm, *Emotional Reason: Deliberation, Motivation and the Nature of Value* (2001); Joseph Ledoux, *The Emotional Brain: The Mysterious Underpinnings of Emotional Life* (1998); D. P. Goleman, *Emotional Intelligence* (1995); Richard & Bernice Lazarus, *Passion and Reason: Making Sense of Our Emotions* (1994); G. L. Clore & W. G. Parrott, *Moods and Their Vicissitudes: Thoughts and Feelings as Information* (1991).

10.3 The Nature of Love

Love is widely if not universally regarded as the greatest human emotion, the emotion most characteristic of human beings as a species, the source of the profoundest good for society and of true happiness for ourselves. Naturally, it is an emotion worth writing a bit more on. Love is indeed the pinnacle, the strongest expression, of the very values that inform my ethical theory (see Part V, "Natural Morality"). For compassion is essentially a love for all beings, and the self-respect that motivates the pursuit of integrity is itself a healthy love of one's own being.

Drawing out what we have just discussed about emotions in general, the word "love" refers to a very complex but repeating pattern of affection and response, present in one sense as a pattern in our brain (the love evaluator, and the physiological capacity for a love response), in another sense as a pattern in our experience (the actual motivating sensation of loving something or someone), and in yet another sense as a pattern in our lives (as love affects our behavior, our society, and our environment).

In each case, "love" is a pattern that extends in both space and time, instantiated whenever any brain exists that is both conscious of other things and possessed of certain beliefs (including both evaluative and factual beliefs) about one of those things, which together produce the emotion and behavior that we observe and describe as 'love'. This love is composed of matter and energy: that of the brain-body system, and of patterns of behavior exhibited by that brain-body system. And, as with all things (cf. III.5.5, "Reductionism"), the pattern called "love" remains distinct from the material of which it is made, and from other ways that same material can be arranged (such as to produce a pencil or a ray of light or the contrary pattern of *hate*).

But what *value* does love identify? In one popular model of human emotions, love is the combination of joy and acceptance in response to a person or thing, in particular or in the abstract (like painting, or rowing, or democracy, or a particular painting, a particular time at sea, etc.). Love varies in intensity according to its object, but entails some degree of adoration, respect, and compassion. What we call "romantic" love incorporates a healthy lust as well. And though a basic kind of love is an emotion even an animal can feel, human love has the added potency of a cognitive component: that of *understanding*. This is a crucial point, one many people miss.

U.S. News and World Report once ran a story entitled "Why We Fall In Love: Biology, not romance, guides Cupid's arrow" (February 7, 2000, pp. 42-48), which was as misleading and terminally shallow as a lot of what passes for journalism these days. But the article's thesis was generally correct: what we identify as beautiful or sexy is biologically determined. Of course, this should be obvious. If I were a jellyfish, I'm sure I'd find a nice healthy gleam of slime to be the height of goddesshood in my mate. Science confirms the obvious.

But there were three grand problems with this article. First, in a typical, scientifically-ignorant fashion, the authors forgot to emphasize the difference between averages and individuals. Though they correctly report that a ratio of 0.6 to 0.7 between a woman's waist and hips is widely considered to be the "ideal" in a feminine figure, they fail to mention that nature thrives on variation: though the average man may prefer this, there are millions of men who no doubt have different preferences, for thinner or wider ratios, and thus women who deviate from the ideal have no grounds for utter despair.

The authors try to make the same point in the end, but only to say that people can "compensate for looks," never realizing that some people won't have to: their looks, unaltered and uncompensated, could well be quite moving to someone out there already. Nature likes it that way—since things can change at any moment, it is best to have many variations of the game in play. And the most obvious case of this variation is found in genetically-determined homosexuality, which never gets a single mention in the article. The authors choose to explain beauty solely in terms of reproductive interests, a concern that fails to account for why nature creates homosexuals (and she does—homosexuality in animals is well-documented, and has a known genetic basis even in humans), or how homosexual notions of beauty are biologically determined—a significant omission, since there is a lot of scientific research out there to report on.

Second, the authors give only scant attention to environmental causes, covering the ambiguous finding that (again, on *average*) people prefer mates who resemble their cross-sex parent (e.g. men like women who remind them of their mothers). But even this is problematic. In my experience, nothing is a bigger sexual turn-off than a woman who looks like my mother. Similarities of personality between mate and parent certainly could matter, as it is understandable that a personality type that we have grown accustomed to, and have acquired a lot of experience with, might also be one we might at times prefer in a mate (though one can surely think of exceptions). And the same does hold for familiar physical features, but only up to a point.

However, environment is a far bigger issue than this. I doubt biology explains my own predilection for brunettes—especially since I am fairly sure I know the cause: my first and best friend as a child was a brunette, and when I was young the two women on TV that I most wanted to marry, because I perceived their personalities as interesting and fun, were Morticia of the original *Adam's Family* and Nora Charles of the *Thin Man* films (played by Carolyn Jones and Myrna Loy, respectively). My fate was no doubt sealed forever by these experiences of enjoyment and familiarity (disclosure: my wife is a buxom brunette).

But third, and most important, the authors confuse *sexual attraction* with *love*. Perhaps this is just a sign of "Hollywooditis," the tendency of people to be converted to the religions of vanity and superficiality by insipid and shallow romantic dramas, wherein a five-minute sex scene is considered the most efficient way to communicate to the audience that the main characters are now in love (Jane Austen would be appalled).

This is where I find the most relevant problem with articles like this, which report on this or that scientific finding of biological or environmental predestination. One of the most important things that really distinguishes us from, let's say, jellyfish, is our capacity to love. Thus, one must really be careful not to confound the importance and meaning of love in an article about sexual attraction. No wonder believers raise alarm at atheism, a view they think (after reading crappy articles like this) proposes accepting a world in which the only reason we love our wives is because our genome thinks she's the best baby machine in town. That is not what love is at all.

This reveals how uninformative and shallow journalists can be in covering a story that they see as nothing more than a cute slice of "science and life," never realizing there is a bigger picture, a far more important context, into which such newsworthy ideas should be placed so the reader can understand how this really affects the human condition. Yet this requires a bit too much thought and education, a bit too much depth and sophistication for the staff at *U.S. News and World Report*. I would never trust reporters to grasp that even though I am a determinist, this does not commit me to thinking that the only reason I love my wife is because I was biologically and environmentally "cursed" by fate to really, really like her a lot. For there is more to it than that. And determinism does not entail fatalism anyway (III.4, "The Fixed Universe and Freedom of the Will").

Though I admit that what I find physically beautiful in my wife is determined by my childhood experiences and my typically-masculine (and thoroughly human) genetic construction, this is not what makes me love her. After all, I am very attracted to Salma Hayek, too, but I don't love her (though I have a hard time explaining this to my jealous wife). Love is not just about attraction. Though I believe at least some attraction is necessary, it is not sufficient to inspire love. For love involves our entire being—our character, knowledge, desires, interests, and ideals. It is what happens when we encounter someone or something, a person or an ideal, an object or an activity, that fits so well with what we want, what we enjoy, what we have adopted as our ideals and dreams, that we are profoundly moved by it. We come to realize that this is one of the most important reasons anyone can ever have to live, in any possible universe.

Love is the realization of an end to every despair—the despair of loneliness, the despair of never knowing comfort or trust, the despair of meaninglessness. And when you find a *mutual* love, you now mean something to someone other than yourself, you can get close to someone and learn more about them than about anyone else, you can enjoy things about them that were otherwise hidden from view. The relationship is itself the height of friendship and partnership, and glorified all the more by a sexual passion far more exciting and enduring than any other. There is more than mere biology and environment at work here: there is an emergent pattern that is unique and precious.

Of course, the natural retort is that all these things—my character, knowledge, desires, ideals—were determined by biology and environment after all. True. But irrelevant. In the first place, it does not matter how I got to be the person I am today. What matters is whether I like who I am. Many people do not like who they are—or would not if they bothered to examine themselves—and my advice to them is to seek every avenue of change, where change is feasible and reasonable, or else realize their dislike of themselves may be groundless. The ultimate test: Could you like a person who was just like you? If the answer is yes, then stop whining. If it is no, get to work.

But if you have lived the self-examined life and are still content with who you are, then it makes no sense to despair that fate has smiled upon you. Fate has given you the one and only thing any conscious being would ever need to be truly happy in any possible world. And what you then fall in love with—the career you love, the mate you love—will simply be an extension of what you already see as the fundamental truths of happiness. And this love will be a hell of a lot more than admiration for a 0.7 ratio between your wife's waist and hips, no matter how determined that love was by fate. It is, after all, the character of love itself that gives it value to us. Where that love came from is moot.

But there is still more to the story than that. For though many of my opinions on physical beauty were decided by biology and upbringing well before my reason could be employed to examine and set myself in order, my opinions on love are far more the product of *me* than is, one might say, my fascination with brunettes. For once I had enough knowledge and awareness (and was given the impetus) to examine myself and correct and mold my being in a more rational fashion, who I was became more *mine* than what I was before, more *chosen* than given. And though even this maturation of character was ultimately determined, the point remains that I—my personality, awareness, reason—was far more causally involved, far more necessary an element, in forming my mature self, than these other outside causes, which merely gave birth to my nascent character and power of reason and consciousness, the things which in turn got to work in building who I am now. And this is no trivial matter.

When I fell in love with my wife, it was not her beauty, and certainly not her potential as a child factory, that truly moved me. It was *who she was*. Her very character and knowledge and manner were what I realized to be most in tune with my own ideals. She represents the sort of human being I wanted the world to be populated with, the sort of person that in my opinion made humankind worth existing at all. To me, the decisive thought was not "Will she be great in bed?" or "Will she be a healthy mother to my children?" but "Will I be happy living with her the rest of my life?" When the answer to this last question is not only "Yes" but a resounding "Damned straight!" then what you have is love.

The *U.S. News* cover story might leave us with the impression that this is all sideshow, that physical attraction is the real thing going on. But anyone who has *really* been in love knows that's silly. Love, after all, is the ultimate sensation we feel when we contemplate anything that makes life not only worth living, but incredibly exciting. An attractive body may thrill me, but it can hardly give my life meaning. But when such meaning is found, the power of that realization is awesome. That is why love is so moving and exciting a sensation—it is telling us that something is moving and exciting us in the profoundest way our brains can calculate. And pardon me if that sounds less like biology than romance. Biology is just the wiring, the chalkboard, the clay. Love is the product, the picture, the sculpture, made possible by this substrate of blood and guts that makes us uniquely human, granting the power to understand, to reason, and to have and cherish ideals.

> For a modern scientific treatment of the emotion of love, with all its faults and glory, see: Helen Fisher, *Why We Love: The Nature and Chemistry of Romantic Love* (2004).

10.4 The Nature of Spirituality

In Part I ("How I Got Here") I described my own spiritual experiences with the Tao, which I later came to realize were entirely natural. Though most people assume being "spiritual" entails being "religious," this isn't a necessary connection. When people talk about a spiritual life, they point to someone who has his mind on higher things, who is not obsessed with property or gain, and who is passionately devoted to a belief about the meaning of life and the path to happiness. But this describes any devoted philosopher. When people talk about a spiritual "experience" they point to the combined sensation of awe, inner peace, and enlightenment, which culminates in a reverence for life and nature, and a sincere self-reflection about these things and oneself. And yet that, too, is the experience of any true philosopher. I live a spiritual life, because I live a self-examined life of the mind, I care deeply about my beliefs, I care more about my ideals and human happiness than about material things, and I experience awe, inner peace, and enlightenment when I fathom human minds and the natural world.

Spiritual experience leads to and reinforces the spiritual lifestyle, and in that respect my path to Metaphysical Naturalism through Taoism was fortuitous. For Metaphysical Naturalism does not have a developed spiritual tradition, even though there have been many spiritual atheists, from Robert Ingersoll to Carl Sagan and Corliss Lamont. Many may be surprised to read the following words of Sagan:

> Science is not only compatible with spirituality; it is a profound source of spirituality. When we recognize our place in an immensity of light years and in the passage of ages, when we grasp the intricacy, beauty, and subtlety of life, then that soaring feeling, that sense of elation and humility combined, is surely spiritual.

This is not an uncommon sentiment among scientists, proving that spirituality is not exclusive to what we normally regard as "religious" life, nor does the term always entail something supernatural.

Meditation as a secular path to spiritual enlightenment is perhaps unequaled, and ought to be mastered by all. It does have beneficial effects on health and self-understanding. But the most common path to godless spirituality is through an appreciation of science: by truly taking in the awe of nature and her complexity, many a scientist has had a spiritual awakening that had nothing to do with God, but everything to do with profound reverence and amazement in the face of tremendous beauty, fearsome power, and the unimaginable depth and complexity of space and time. It sparks the realization of how tiny and insignificant we are, yet how wonderful we are despite this.

If this sounds a lot like my own Taoist experience, I don't think that is a mere coincidence. I suspect that all religious experiences are of the same core nature, being molded and interpreted by this or that theological or metaphysical view. Certainly, I cannot think of a better explanation for the huge diversity of rich, veridical experiences of religious ecstasy reported in all the world's traditions, for this is a reality that cannot be explained by supposing only one of all those religions is true.

We must recognize that experiences we classify as spiritual or religious exist as a subset of ordinary psychological experiences, and not as something separate and different from them. This is not because we are obliged to assume there is nothing supernatural or special that corresponds to spiritual experiences. Rather, it is because before we can reach any such conclusion at all we are obliged to see there is no inherent way to distinguish ordinary psychological events from spiritual events. Because they originate within the same domain (our mental life), the possibility always remains that they are merely different aspects of the same thing. They might not correspond to anything outside of our own, private mental existence.

In particular, spiritual experiences could represent glimpses of ourselves and our understanding of all being, more than of anything beyond us or common to others. Of course, all people can in principle share the same kinds of emotions and arrive at the same kind of understanding—we all share the same biology, after all, and the truth about our relationship to the all is a universal truth. Certainly this and other possibilities must be eliminated before we can assume it is ever the reverse.

I think the evidence is clearly in favor of the natural, internal cause, and not of any transcendent, supernatural cause. Religious claims often seem to be believed in more for their personal worth in answering the human need for an ultimate meaning to life than for their logical or empirical merit. While there are many contradictory yet equally 'real' spiritual traditions, one thing that appears common to every spiritual experience is whether it is (or can be) interpreted in some way that bears on an "ultimate meaning to life." And if an experience, however interpreted, accomplishes well the goal of answering our need for meaning, then this will matter more to someone than whether that belief is consistent, justified, or true.

Yet that leap comes with the false assumption that no other answer is as satisfying, and 'therefore' no other answer should even be considered. This would explain why true believers not only resist attempts to challenge their beliefs, they are often impervious to them. This kind of behavior, which seems inexplicably irrational, appears at least explicable when we recognize that the religiously devout are often interested in things more important to them than the truth—such as an ultimate meaning to life. Since the personal, emotional benefits provided by spiritual beliefs do not depend on those beliefs being true, their truth becomes irrelevant in practice.

So, for example, while Buddhism and Christianity each provide a supernatural explanation for our ills, and an equally supernatural solution, within all this lies a purely practical belief system that not only provides an ultimate meaning to life, but attempts to produce a greater balance of peace and happiness by providing both a moral standard and a reason to live up to it. But none of this has been worked out scientifically, or tested empirically, and all of these benefits are gained merely by the claims being *believed*, and not by their actually being *true*, which is quite unlike scientific claims (much less technological ones), where benefits are usually gained only when we believe in what is true, while definite hazards are often created by believing in falsehoods.

I am not even arguing here that all religious claims are false. Everything I have said so far would apply to all religions regardless of whether their claims were true or false. Since any telltale sign of a spiritual experience can also be found in hallucinations and other mundane phenomena, they are useless *by themselves* for indicating whether an experience can rightly be regarded as granting access to a profound religious truth.

Regardless, it seems clear that spiritual experiences (even godless ones) are at least potentially beneficial, contributing a useful and necessary quality to human life. No matter what religious spin they are given, they often produce emotional harmony and contentment, clarity of thought and perspective, joy and humility. And as this is an emotion, we can assess whether it is an accurate appraisal of its object (see III.10.1, "Emotion as Appraisal"). If its object is our self and our relation to the universe, its appraisal seems quite correct to me. And this benefit can be reliably gained from the experiences *alone*, without any conclusions of an objective nature being drawn from them. But when we *add* to the raw experience, when we use the inner emotion as a "proof" of something about the universe independently of us, the danger arises that we will mislead ourselves down a false path, a path that may be harmful or destructive to us or others, or that might distract us from even more important and wonderful things.

And we know this can happen. Many Islamists use their spiritual experiences to justify mass murder and the inhumane treatment of women, thus corrupting the very joy their spiritual life brings them, translating it into hatred rather than the love and contentment that they would have settled on otherwise. Likewise, many Christians use their experiences to justify slavish obedience to a human tradition or a book (like the Bible or the Book of Mormon), which contains commandments not healthy for humans or their civilization, and lacks a great deal of crucial human wisdom—and what useful wisdom it does have is often buried and obscured in symbol and metaphor, or simply bad prose, and thus all too often missed or misunderstood. Though Buddhism and Taoism are more in touch with human nature and human needs, more easily channeling their spiritual perience into a genuine life of love and contentment, they still promote y false beliefs, such as reincarnation, or the evils of technology—or ssible goals, such as the abandonment of self and desire.

o conclusions should be taken from this:

ot even as a matter of principle, theories about the world outside of *can* arise from spiritual experience. For this is true of all states ness. A mere random whim can produce a theory that could just happen to be right, and this is even more likely when the whim comes from a mind that is thoroughly steeped in the relevant facts and regularly contemplating them. And in a meditative state, your thought can sometimes be clearer than ever, and gain access to information inside you that was not being accessed before, or perceive patterns previously missed. But all such theories are no different than any others: they must still be tested logically and empirically before they can be assigned any knowledge value. We cannot declare the truth of a theory right out of the door, based on a purely subjective insight. That is a bad method, as we well know (see II.3, "Method").

Second, it is vain to appeal to how a spiritual experience transforms someone's life as a 'proof' that a religion is true. Life transformation results whenever anyone pays more attention to an ideal than they do to the details of life, it happens whenever anyone has a natural emotional experience like I describe above. It does not matter whether that ideal is real or not, or how that experience is *interpreted*—hence Kamikaze pilots, Islamic suicide bombers, Marxist fanatics, the Heaven's Gate cult, Jonestown, stories of powerful personal changes through Scientology, all these involve experiencing the transformation of lives every bit as much as any other, such as Born Again Christians or Buddhists describe, and as I described for myself, as a Taoist.

The faith of many socialists in Marxism has indeed transformed some into more loving people, as they come to love the common man and care for his needs and welfare. Does that make Marxism true? The faith of many adherents of Dianetics has transformed them into loving people. Does that make Scientology true? The faith of Jim Jones' followers made all of them, by all accounts, into much more loving people than they had been, even up to the moment they drank the poisoned punch. Was Jim Jones, then, right after all? Hardly. Thus, the transformative power of religion is no indication of its truth, but rather of a universal human longing for a loving society where we can experience happiness and purpose. But we do not need any supernatural dogma to have that. Secular Humanists can fall in love with an ideal, too—with nature and humanity—and their lives are likewise transformed by this just as much as for any religious devotee.

Yet there is more to it still. The starkest cases of transformation that we always read about are those where the converted reached a state of abject misery or confusion, a complete disconnect from any enriching sustaining philosophy of life, often descending into a moral quagmire malaise—and then they hit upon a revelation about how reforming t' lives, and staying in touch with the emotional experience that ins them, will make them happier. This is a generic reality, not one to any religion, or even to religion as such. But the starkness of it can certainly inspire tremendous conviction in any beliefs falsely attached to the event.

We must abandon such false idols and remain within the simple truth (see IV.2.2.4, "Religion as Medicine"). And then we will see, the two qualities often hyped as spiritual, 'hope' and 'faith', are entirely natural. 'Hope' is faith in yourself, in those you love, and in the achievability of your goals and dreams. Even after you've gone, what you do today contributes to humanity's collective dream (see VII.6, "The Secular Humanist's Heaven"). And if you live a life of wisdom and self-reflection, of action and creation, this faith will be justified, anchored in a firm grasp of meaning in your life. I have already shown that meaning comes from us, not from something 'out there' (see III.7, "The Meaning of Life"). As for 'faith', by that word I mean only a faith justified by evidence and reason (see IV.1.1.3, "The Science of Faith"). And so we have our recipe: natural meaning, justified faith, humble self-reflection, and the awe-struck pursuit of wisdom. Put these together, and true spirituality is just around the corner. No god needed.

Quote from: Carl Sagan, "Does Truth Matter? Science, Pseudoscience, and Civilization," *Skeptical Inquirer* 20:6 (1996), p. 29.

Still important is William James, *The Varieties of Religious Experience: A Study in Human Nature* (1902), along with Charles Taylor, *Varieties of Religion Today: William James Revisited* (2002).

We now know religious and mystical experience has an identifiable physical cause in the human brain, and is therefore a natural, biological phenomenon: see Eugene D'Aquili & Andrew Newberg, *The Mystical Mind: Probing the Biology of Religious Experience* (1999) and *Why God Won't Go Away: Brain Science and the Biology of Belief* (2001); Joseph Giovannoli, *The Biology of Belief: How Our Biology Biases Our Beliefs and Perceptions* (2000); John Horgan, *Rational Mysticism: Dispatches from the Border between Science and Spirituality* (2003). See also the bibliography to section IV.2.2.4, "Religion as Medicine."

Important examples of secular spirituality, with discussion of its very nature, include: Robert Solomon, *Spirituality for the Skeptic: The Thoughtful Love of Life* (2002); David Cortesi, *Secular Wholeness: A Skeptic's Paths to a Richer Life* (2002); Matt Berry, *A Human Strategy: Toward a Genuine Spirituality* (1999); Carl Sagan, *The Demon-Haunted World: Science As a Candle in the Dark* (1997) and Richard Dawkins, *Unweaving the Rainbow: Science, Delusion and the Appetite for Wonder* (2000). The issue is also discussed by Martin Seligman in *Authentic Happiness: Using the New Positive Psychology to Realize Your Potential for Lasting Fulfillment* (2002), and secular spirituality is often exemplified in the writings and speeches of Robert Ingersoll (cf. Roger Greeley, ed., *Best of Robert Ingersoll: Selections from His Writings and Speeches*, 1983).

IV. What There Isn't

1. Not Much Place for the Paranormal

So far I have explained how almost everything *does* exist in Metaphysical Naturalism. My worldview requires you to deny very little, and involves believing a lot—whatever evidence and reason tell you to. Most of what is special about this worldview is *how* things are understood, what their *nature* is. But it should be apparent by now that Metaphysical Naturalism holds little place for the paranormal, things unnatural or supernatural: gods, ghosts, psychic powers, faith healing, you name it. Anything involving sentient beings and powers beyond nature (as in not grounded or formed naturally, or existing prior to or independently of nature), and anything purely mental (like true mind-over-matter or purely mental attributes, like a disembodied desire), anything like that excludes naturalism.

We have already discussed reasons to disbelieve in a disembodied soul separable from the brain (III.6, "The Nature of Mind") and reasons to give abstract objects and the experience of spirituality a secular interpretation (III.5.4 and III.10.4, "Abstract Objects" and "The Nature of Spirituality"). And throughout we have touched on the question of God or a Creator many times, always with negative results.

But you can already see from those examples that this rejection of the supernatural is not *a priori*, it is not declared "before examining the facts." It comes only from a scientific investigation of the evidence. Of course, if there are such things as paranormal phenomena, the metaphysical naturalist has good reason to expect they will be fully explicable as natural patterns of matter-energy moving in space-time according to physical laws, just like everything else. But if evidence were to prove otherwise (and scientists could find that evidence if it existed), Metaphysical Naturalism would be proved untrue. But there is no reliable evidence that these beings or phenomena even exist in the first place, much less the incredible mechanisms that would make them truly supernatural. Combine that fact with the ample scientific evidence for other causes of belief in such things, and there remains no reason to believe in the paranormal.

I will not belabor the point by rebutting every popular paranormal claim—there are thousands, from astrology (indeed, at least three completely different kinds) to Big Foot (who seems to have other strange relatives in forests and swamps all over the U.S. and beyond). What I said earlier of Spiritualism and Past Life Regression (III.6.7, "Evidence Against Mind-Body Physicalism?") applies to all other paranormal claims I and others have investigated: they have ready explanations in fraud, self-deceit, sensory illusion, and "myth-making." For in every case where they have been scientifically investigated in the past hundred years, if anything was discovered at all, one of these four causes was proved to be at work.

Instead, I will leave the following bibliography for those who want the real story on the purported evidence for the paranormal. Then I will go on to explain why scientific and historical methods have produced good reasons to disbelieve. I will close this part of the book with a survey of some reasons to doubt there is god.

For further reading, see the Secular Web's sections on "Mysticism and the Paranormal" (www.infidels.org/library/modern/paranormal/) and my essay "Defending Naturalism as a Worldview: A Rebuttal to Michael Rea's *World Without Design*" (2003, at www.infidels.org/library/modern/richard_carrier/rea.shtml), which discusses in more detail my idea of a natural-supernatural distinction.

There are many valuable books on the lack of evidence for the supernatural: Massimo Polidoro, *Secrets of the Psychics: Investigating Paranormal Claims* (2003); David Marks, *The Psychology of the Psychic*, 2nd ed. (2000); Wendy Kaminer, *Sleeping With Extra-Terrestrials: The Rise of Irrationalism and Perils of Piety* (1999); Stuart Vyse, *Believing in Magic* (1997); Michael Shermer, *Why People Believe Weird Things: Pseudoscience, Superstition, and Other Confusions of Our Time* (1997); Susan Blackmore, *In Search of the Light: The Adventures of a Parapsychologist* (1996); Joe Nickell, *Entities: Angels, Spirits, Demons, and Other Alien Beings* (1995) and *Looking for a Miracle: Weeping Icons, Relics, Stigmata, Visions & Healing Cures* (1993); John Cornwell, *The Hiding Places of God: A Personal Journey into the World of Religious Visions, Holy Objects, and Miracles* (1991); John Schumaker, *Wings of Illusion: The Origin, Nature and Future of Paranormal Belief* (1990); and Terence Hines, *Pseudoscience and the Paranormal: A Critical Examination of the Evidence* (1989).

Best of all are the works of James Randi, who has had a one million dollar prize sitting in a bank waiting for anyone to prove anything paranormal—there have been no takers in over twenty years. See his books: *Flim-Flam! Psychics, ESP, Unicorns, and Other Delusions* (1988), *An Encyclopedia of Claims, Frauds, and Hoaxes of the Occult and Supernatural: James Randi's Decidedly Skeptical Definitions of Alternate Realities* (1997), and (with Carl Sagan), *The Faith Healers* (1989).

1.1 Science and the Supernatural

There are generally two approaches to defending belief in the supernatural. The first is to claim there are facts that support it, and this is the point behind the frauds and legends and exaggerated or fabricated stories, the role played by illusions and mistakes and hasty conclusions. But this exposes such claims to scientific and historical study, and when scientists and historians come knocking, either of two things always seem to happen: those making the claim retreat or refuse to submit their evidence to scrutiny, or their evidence turns up bogus or groundless. It would be extraordinarily improbable for this to have happened so widely, so consistently, so routinely, for hundreds of years running, unless there was nothing supernatural to be found.

But there is a second approach, bypassing science altogether, claiming to be a kind of superior "metaphysics." It has other names, parading under the rubrics of theology or "first philosophy" and so on, but it all amounts to the same thing: claiming to get at the truth by sitting in a chair and just reasoning to it. Many of the arguments for the existence of a god or disembodied souls are such things: they claim to prove a point by pure logic from general principles, without even needing evidence, beyond personal armchair observations. Indeed, metaphysicians often claim these conclusions *precede* the evidence in weight and certainty. After all, they say, if the evidence contradicts right reason, so much the worse for the evidence.

Yet you may have noticed that of the six methods of arriving at any credible truth that I accept, I place metaphysics dead *last*. Though in my scheme reason does come first, what I mean by reason is simply the analysis of our own ideas about things, which can only teach us truths about those ideas, *not* about whether those ideas apply anywhere in the world outside our imagination. And when it comes to that question, I place science in first place, for no enterprise is so capable of discovering facts about the world. Why this is so I must now explain.

1.1.1 The "Scientific Method"

The seed from which the success of science was born is a simple three-step process: adduction, deduction, induction. In general, we identify a problem, gather relevant data, formulate a hypothesis (usually an explanatory model of what is really going on), and test the predictions entailed by that hypothesis—looking for whatever would *have* to be the case, and whatever could *not* be the case, if our model were correct. In other words, we creatively "adduce" an hypothesis from some collection of data and questions about that data, then we logically "deduce" what new facts that hypothesis must entail if it is true, and then employ any of a variety of empirical ("inductive") methods to test that hypothesis by seeing if these new predictions hold up.

The virtue of this method is threefold. First, it accommodates creativity. Looking at nature and coming up with guesses about how things work, or what else we might find if we look in the right place, is driven from the start by curiosity and imagination, two things humans have in spades. Second, it doesn't dare let you get down to the business of proving anything without first analyzing the meaning of our guesses, applying rigorous logic to our ideas, forcing us to fully confront just what our theories would entail if they were true or false—in other words, it unites our creativity with our most powerful tool of all: reason. Third, it doesn't let you get away with claiming anything without proof—even better, it especially likes an unusual sort of proof: not mere evidence, but evidence *no one expected.*

Until Einstein's Theory of Relativity, no one would have even thought that matter could be transformed into energy or that gravity bends light, and so it was a surprise to find that they did, just as Einstein's theories predicted. Our ability to predict things we could never have otherwise guessed, from when volcanoes will erupt, to the outcome of mixing two new chemicals, is a daily vindication of the correctness of our scientific knowledge. And this holds even when the phenomena are well-known, but their relationship is a surprise: like Newton's discovery that the same force that makes apples fall on earth also makes the planets orbit the sun.

When a scientific theory predicts a fact no one has ever seen before, or no one even believed before, then it is an increasingly good bet it's right. You can't get a better proof than that. No one can claim scientists are just seeing what they want to see, or making up any old explanations that just fit, if their explanations of existing facts turn up, and explain, totally new facts. When you see that, repeated again and again, only the universe itself could be at work (the criterion of "surprise," per II.3.1, "Finding the Good Method").

However, this procedure does not automatically produce "science." For the devil is in the details. The specific ways one can carry out these three steps are endless, and some particular methods are good and others bad. Science seeks those methods that work well, and progressively abandons those that don't, especially those producing misleading results. But the methods that work well will vary according to what is being studied, and so "the scientific method" really denotes a multifarious hodgepodge of "methods," some shared across fields, some specific to certain subjects of study, and these methods are themselves always subject to scrutiny and change over time, as their efficacy or lack thereof is discovered, and as new, more effective methods are found. The same question can even be explored using several different methods. And scientists often seek to use several, as there is even greater certainty in the results if those same results are arrived at by completely different means (the criterion of "convergence," per II.3.1, "Finding the Good Method").

Of course, this is one reason why it is often difficult for a scientist to cross fields. A biochemist's methods do not resemble a psychologist's anywhere near enough for one to guess just how the other should best pursue his research, unless they each had some prior acquaintance with the other's field. Even some common concepts, like the use of control groups or double-blind experiments, cannot be applied to all scientific problems. So we cannot define the scientific as that which is discovered through the use of control groups or double-blind experiments, or any other subset of particulars. We would have to toss astronomy out as unscientific if we did.

But this is another strength of science: science is not only about testing facts for truth, but testing methods for accuracy. And thus science is the only endeavor we have that is constantly devoted to finding the best means of ascertaining the truth. This is one of the reasons why science is so successful, and its results so authoritative. Yet metaphysics has no room or means for testing different methods for accuracy, and if it ever started producing surprising predictive successes it would become *science*!

However, there are some methodological principles in science that have been proven to be universally effective, which together make 'the scientific method' queen of all. The very fact that some of these principles are themselves counter-intuitive confirms their merit—because the method *itself* meets the criterion of surprise.

1.1.2 The Advantage of Doubt

The truths of science are tentative, not absolute. Popular notions tell us that we need absolute truths independent of individual judgment, because tentative truths are risky, not worth the effort, somehow "not really true," and individuals are biased, fallible and limited in their abilities. Indeed, many conclude that if there are no absolute truths, then there is no knowledge, that if we can't know something for certain, we don't know it at all. Some critics of science even say that science is to be outright discarded because it is always changing, or that since all truth comes from individuals, all truth is subjective and therefore equivalent in merit. But these are all absurd conclusions. History has shown that tentative knowledge is extremely useful if it routinely turns out correct. Our certainty does not have to be 100% for our bid at knowledge to be trustworthy. A gambler doesn't need a sure thing to see a really good bet.

The tentativeness of science is counter-intuitively one of its greatest virtues, because progress toward truth and away from error would never be possible without change, and change would never be possible without perpetual doubt and skepticism. And science does not simply undergo any arbitrary change, as religious ideology or clothing fashions do, nor does it hold out long against contrary evidence, asserting that the facts must surely be wrong if they do not fit the going dogma. When science changes its conclusions, it always does so as a result of further inquiry, and not just any inquiry, but rational, observational, or experimental investigations. So science changes in response to those kinds of discoveries that are more likely to advance us toward truths about the world. And in the end, evidence is king. The dogmas of science will ultimately bend to its will.

The most surprising thing is how we can have knowledge that is both tentative yet still worthy of being called true. We commonly hear a scientific truth being dismissed by critics because it is "just a theory," as if scientific theories are improvable or unproven concepts. But this confuses science with philosophy. A theory of philosophy or speculation is just that: a merely plausible inference, an unproven hypothesis, or a 'best guess'. But a confirmed theory in science must be far from any of those things. It must be extensively tested or confirmed—and it must continue to survive attempts to disconfirm its predictions—in order to win and keep such a title. Otherwise, it is more properly considered a hypothesis or conjecture.

Every currently-accepted theory in science, such as the theories of Evolution or Relativity, is well-supported, and not "just a theory." It is true that some hypotheses are also called theories, but they are still carefully identified as a proposed rather than an accepted theory. And, of course, even among accepted theories, all will vary in how well-supported they are at any given time, while some theories have been refuted and are no longer 'scientific' theories at all. All this is a symptom of the very tentativeness we are talking about as science's virtue: we have confidence *only* when the evidence is and remains overwhelming. Our beliefs can be abandoned—all you have to do is prove them wrong.

As a result, science's dependence on individuals, and encouragement of skepticism, is not the handicap it might be in other endeavors, since these principles require individuals to persuade others with sound evidence and reasoning, and only when an overwhelming majority of those expert in a field are convinced does science declare something "true." This entails another three-step process that defines science over other methods: peer review, corroboration, and consensus. Scientists subject their claims to extensive formal review by their peers, and if a claim pasts that test, other scientists seek to replicate the results to confirm them, and if a claim pasts that test enough times, it will start to convince an overwhelming majority of peers, eventually producing consensus, when nearly every scientist agrees it is correct. That is when a mere claim becomes scientific truth.

This truth is not definitive or final, nor is the process perfect, but it is the most reliable bet around, *precisely because* it must pass these tests of doubt and proof. As history has shown that no other means of inquiry is as successful or as trustworthy, it follows that the mental culture of science is on to something—in other words, skepticism is a virtue. Science sets the highest bar, requiring the highest standards of verification, employing the most experienced and well-trained judges, who are encouraged to be as self-critical as they can be. And that is what makes science scientific.

Moreover, science does not conceal its evidence or rest its case on mysteries or private revelations, but makes everything public, so that all experts, and even the serious layman, can examine and weigh the claims and arguments of scientists to an extent not possible in any other field, creating the most effective check against individual bias that humans can devise. Is it any surprise that metaphysics cannot claim this? With no evidence to sway the skeptic, how could it? Metaphysics only sets the lowest bar credible.

1.1.3 The Science of Faith

Science and faith are not in conflict—unless you want them to be. If someone places faith *before* truth, then they are stepping out of bounds. But if faith is what one has *because* something is true, and not the other way around, science becomes the One True Faith. This is another counter-intuitive feature of science: it is thoroughly empirical, built on observation and evidence, yet "empiricism" is not observation and evidence alone, but a view of things that is constructed from observed facts, and the whole enterprise of science requires at least the provisional belief that those constructions are true. On the one hand science requires faith, a faith that certain principles and models and bodies of knowledge are true. On the other hand, this faith is not blind, but based on evidence. It is *justified* faith.

Scientists build up faith in science's concepts, principles, and conclusions through repeated practice or testing, and when this faith is challenged, they return again to examine the facts, to see if their faith is justified by them. This is what makes science an empirical enterprise, the fact that it ultimately grounds and justifies its faith by appeal to observable evidence.

The idea of an empirically-based faith is hard for many people to grasp, especially if they are used to thinking that "faith" is only a reason for believing something when you *don't* have evidence. The term "faith" does have both connotations, meaning "belief" in some cases, but also "reason to believe" in others. Science has no use for that second kind of faith. But it is not true that a scientist "has no faith" in science: he has faith in it, but a faith that is grounded in empirical evidence and reasoning. By confusing the two notions of faith, common sense creates a false dichotomy between faith and empirical justification. Science unites them.

1.1.4 The Power of Artifice

Another counter-intuitive feature particular to science is the fact that an artificial set of circumstances can teach us about the natural world. This seems inherently illogical, yet we have discovered it true. Many scientific facts can be established, and theories validated, without using experiments. Astronomy, geology, zoology, all include many advances that had nothing to do with experimenting *per se*—but just looking. Geologists cannot exactly create mountains or continental drift in a laboratory. Yet they can confirm such things with extensive and overwhelmingly convincing evidence, simply by observing nature, and documenting what they find. But this is the child of necessity.

The experiment, in contrast, is recognized by all scientists as the most powerful tool of all for explaining nature's mysteries. The secret to this power is that an experiment allows the observer to control the outcome of a process by manipulating the variables that come into play, including the environment itself. And contrary to what our common sense would tell us, this does not mean the scientist proves himself right by manipulating the data. The data will confirm or disconfirm the theory being tested whether the scientist likes it or not. Nor does it mean the manipulation of the experiment invalidates the results. To the contrary, it is the manipulation of the experiment that *validates* the results, by eliminating uncertain causal factors and limiting the observation to a controlled set of causes that the observer can clearly identify and trace. In this way, careful experimentation distinguishes science from 'metaphysics'.

1.1.5 Distinguishing Fact from Theory

Another surprise is revealed in popular confusion between facts and theories. In scientific jargon, facts are what have been carefully observed to be the case. Theories are explanations of those facts. This differs from colloquial and philosophical jargon, where facts are 'whatever is established' and theories 'whatever is speculated'. But to scientists fact is observation, theory is explanation.

Evolution, for instance, is a fact—animal and plant life on earth *has* evolved from simpler to more complex, over billions of years, with current species sharing a clear chain of identifiable ancestors descending back in time, in lineages converging with each other. This is *observed* to be the case, not theorized, and it is observed in several convergent ways: in the fossil record, in the living record, and in genetic relationships. To explain this fact, Darwin proposed the *theory* of natural selection, a theory that has since been experimentally and observationally tested, and confirmed in many different ways.

The theory of natural selection might turn out to be false, but it would take some really impressive findings to overturn the massive body of evidence that stands behind it, and even then we would not have overturned the fact of evolution—that fact would remain, begging for an explanation (see III.8.2, "Evolution by Natural Selection"). Likewise, popular lore has it that Einstein's 'laws' of gravity refuted and overthrew Newton's, but the truth is quite the opposite: Newton's *laws* were and still are an observable fact, while Einstein's *theories* explained Newton's laws. Relativity did lead to some modifications of those laws to fit contexts Newton never observed, but it incorporated and explained his observations rather than casting them out. Again, "laws" are facts: repeatedly observed patterns and relationships in nature. *Theories* are then developed and tested to *explain* these laws.

It is possible for a theory to be so well confirmed that it is nearly as certain to be true as the facts it explains, but that would still not make it a fact in the special scientific sense: theories are not observed, they are only implied by what is observed. This suggests a paradox to our common sense, since here we have something that can be empirically tested yet never observed, a counter-intuitive notion that has led to many an ignorant polemic against science. The correct understanding lies in the role a theory plays in explaining and thus *predicting* facts (in our past or future). Such an explanation must always include one or more unobservable features (the instance of causation, for example) that account for what is observed, since humans cannot observe everything.

In the end, a theory's success in explaining the past and predicting the future is the basis for believing it is a correct account of things, and that is more than sufficient as a justification. After all, only *theories* lead us to those surprise predictions that validate and advance the entire enterprise of science.

1.1.6 The Marriage of Creativity with Truth

Finally, we must not forget how fundamentally imaginative and creative science is. The popular image of the scientist as cynical, unimaginative, and blindly conservative does not reflect the reality that science only progresses when humans are at their most creative, plumbing their imaginations for ideas and ways to understand different facts and theories. Science is the perfect marriage between the eliminating force of doubt, of demands for evidence to back every claim and every element of every theory, and the creative force of invention and imagination.

This also belies the false notion that science is destructive, making us hopelessly skeptical and overly critical and hostile to the feeling of wonder. Science is full of wonder and beauty. It is a source of belief as well as doubt. And it is fundamentally *constructive*. This creative aspect of science permeates all scientific activity. Brilliant and original ways of collecting and analyzing data are developed, clever theories proposed, experiments carried out imaginatively. But this does not undermine the value of scientific results: to the contrary, science needs this inventiveness to make progress. It is another virtue of science that it is so versatile and can find and make use of so many different ways to explore and study people and their natural world. Science is not shackled to one set of rituals or procedures. It is hungry for the new, the innovative.

This does mean that scientific explanations, models, or theoretical entities are human constructions, created by human beings, drawn from their imagination. But the proof is in the pudding. What makes science so successful is that it marries this tireless resource of human fantasy to a dedication to test everything thoroughly against observed facts and the rules of reason and logic. Thus science gains the best of both worlds: the benefits of human creativity without the shortfalls of superstition or fancy, plus the benefits of a demand for empirical proof without the limitations inherent in the data taken by themselves. This is the most essential difference between science and metaphysics.

1.1.7 The Lessons of History and the Burden of Proof

Now we end with the three most definitive tools of science—which truly distinguish it from metaphysics: the Burden of Proof, the Balance of Proof, and Occam's Razor. First: the Burden.

In the 1st century we had no idea what caused lightning, but we had a cogent supernatural explanation: divine or demonic agency, ranging from the anger of Zeus to the combats of evil spirits in the clouds. Yet at the same time *scientists* of the 1st century proposed that lightning was caused by friction between colliding clouds, by analogy with colliding flint stones. This explanation, though still speculative, was much closer to the truth. It only lacked the yet-to-be-discovered sciences of electricity and pressure.

In such a way, scientists throughout history have found that what we call "natural" explanations keep working, unlike supernatural explanations: the stars and planets actually followed predictable routines that had nothing to do with human events, tested drugs cured the sick more often than spells, agriculture flourished under scientific care but floundered under prayers and magic. Then scientists found that atomic and other "naturalistic" explanations for every phenomenon had a much wider explanatory power than supernatural theories, predicting more things, more successfully. Thus, they correctly guessed they were on to something, and stopped accepting "supernatural" explanations because they constantly failed and never had any evidence to sustain them.

This is why scientists even in antiquity pursued only "natural" theories. And in doing so they got very close to the truth, articulating explanations for sound, light, evolution, weather, poison, disease, and getting far closer than any theologian ever came to what actually turned out to be true. Unfortunately, this brilliant discovery was thwarted by the rise of Christianity, which all but put science on hold for a thousand years, relying instead on "supernatural explanations" and the adage that it was vain to study nature when we ought to be saving our souls instead. It was not until the Renaissance, when pagan science was rediscovered, that the bias in favor of what we call the "natural" was taken up again, and lo and behold, every century since has seen unprecedented progress in our knowledge and mastery of the world.

So there is some merit to the presumption of naturalism in scientific practice, what we call "methodological naturalism." It has proven valid as a rule of thumb, one that is more likely to produce success, saving us from wasting time and resources on blind alleys. Still, it is not an absolute law. There may yet be evidence of the supernatural just around the corner. But the evidence against this possibility is so vast, and the evidence for it so feeble, you may as well count on catching Big Foot or finding a Leprechaun's gold (for more on the Inference to Naturalism, see III.2, "A General Outline of Metaphysical Naturalism").

This is merely a global application of a principle essential to any successful quest for the truth: the burden of proof for any claim is on anyone who makes a claim contrary to established facts, or grounded in fewer established facts than any other hypothesis. For when facts are established, with widespread and multi-faceted corroboration, the odds are clearly on the side of their being correct when pitted against anything contrary. This is why the burden of proof weighs heavily on those who advocate belief in anything supernatural—a burden they have so far never even come close to meeting. To the contrary, whenever closely scrutinized, they always come up with nothing.

This is one important reason why metaphysics is the slave, not the master, of science. A metaphysical claim contrary to science fails the burden of proof, for metaphysics is pure speculation, but science is a highly confirmed claim to truth. Even a metaphysical claim that merely lacks support in or agreement with science has little to commend it, certainly nothing so convincing as the evidence grounding the sciences.

1.1.8 The Balance of Proof and Proof of the Extraordinary

Young Earth Creationism holds that God is a great miracle worker, who for some reason once flooded worlds, stopped suns, parted seas and raised the dead and who, above all, created the whole universe in one fell swoop just a few thousand years ago—yet no longer does anything remarkable at all. This belief fails to be credible twice over: it contradicts the actual facts and entails facts that are *not* observed.

For example, it is beyond reasonable doubt that the earth alone is billions of years old, and that no flood could have covered it. Likewise, such a creator, so active in the world and so ready with miracles, so driven by a mission to save man, should be granting us all manner of supernatural events and messages in our lives, too—but guns are not suddenly turned into flowers, churches are not protected by mysterious energy fields, preachers cannot regenerate lost limbs, suns do not stop, seas are not parted, the dead are not raised (see IV.2.3, "The Universe is a Moron" for more on this point). Why, then, should we believe there is such a being, who just changed his mind and his behavior one day and decided to fold up his tent and do nothing magical anymore? That a god worked miracles then, but no longer, is an extraordinary claim. It requires extraordinary evidence.

This is the Balance of Proof. In a public debate with Edward Tabash, Evangelical Christian William Lane Craig challenged this, declaring that the principle "extraordinary claims require extraordinary evidence" is "demonstrably false," and to 'prove' it he used the example of a lottery winner, whose win is extraordinary but we believe it on ordinary evidence. But this is sophistry. In actual fact, that someone should win a lottery is *not* extraordinary, for it happens frequently—the odds are very good at least someone will win. So obviously we don't need extraordinary evidence to believe it.

The logical confusion here lies between general and particular propositions. When a skeptic says "extraordinary claims require extraordinary evidence" she means, or ought to mean, claims that entail one or more general propositions that are extraordinary, not particular propositions. For particular propositions ("I own a car") are ordinarily supported by general propositions ("many people like me own cars") and so only require ordinary evidence. But particular propositions that are not supported by general propositions ("I own a nuclear missile") require evidence well beyond the ordinary: for they require evidence that the necessary *general* proposition is also believable, and far more evidence is needed to establish belief in a generalization than in a particular fact.

Science is built on this very principle: it builds up tons of particular facts that support a generalization, which then can be employed to support or lend credibility to other particular facts or claims. History, in contrast, is dependent upon science as a source for well-confirmed generalizations about the world and human nature, for it otherwise deals almost entirely with unique events, and thus must often be more tentative than the sciences.

Likewise, when scientists make extraordinary claims, they are expected to fork over evidence well beyond ordinary demands—evidence of extraordinary weight. When they fail to do so, they are not believed (witness the fiasco over "cold fusion"), but when they succeed, their extraordinary claim becomes ordinary science: witness the extraordinary documentation collected independently by Darwin and Wallace, and then subsequently multiplied a thousandfold by other researchers in ensuing decades, a pace of accumulation of evidence, in both quantity and quality, that has yet to cease.

Metaphysicians, in contrast, grant themselves the luxury of disregarding any demand for evidence, all too often making extraordinary claims their hobby. One particular example we shall examine a little is the Resurrection of Jesus (cf. IV.1.2, "Miracles and Historical Method"), which hangs on the metaphysical assumption, grounded in no good evidence, that there is a god who does such things. Yet some people absurdly try to use that single, particular case as evidence to support the extraordinary generalization that there is a god who does such things, violating the Balance of Proof.

1.1.9 Simplicity and Occam's Razor

Earlier (III.3.6, "Time and the Multiverse") I described how the universe is patterned in such a way that complex things are the accumulation of simpler things across a span of time, and this is the case even in space, where complex structures are at once comprised of the integration of simpler ones (see III.5.5, "Reductionism"). As a result, the quest for scientific truth has often benefited from a preference for simplicity, even though this is not always what we find. Science's quest for the simpler is, of course, an obvious product of the fact of reduction. Because complex things are made of simple ones, to explain the complex scientists must look for its simpler causal and structural roots. Hence, since physics is a science at the most reductive level, simpler causes and components are naturally the very thing physicists are seeking. But this is far less so in, say, biology or geology.

Even in the reductive quest, the hunting for the simpler origin or basis of complex things, the truth often frustrates the assumptions of scientists. The Standard Model of particle physics is far more complicated than anyone expected, finding over a hundred sub-atomic particles altogether, far more than the simple few that physicists once thought there were. And while scientists had tried for centuries to simplify the number of basic elements to four or five, it turns out there are over ninety!

Thus, though scientists seek the simple, nature does not always give them what they want. This is so much so that we now even have a branch of physics called Complexity Theory. And the case is even clearer in the sciences that study more complicated phenomena. In psychology, for instance, neurophysiology is leading us to realize that the truth lies in far more complicated theories of consciousness than the original promises of

Aristotle or Freud. And in biology, the operation of something so simple as a heartbeat involves a vastly complex system that cannot be captured by any single, much less simple theory, as had long been hoped ever since ancient doctors developed theories of the pulse.

On the other hand, there is also a methodological preference for simpler theories, not because nature always works in simple ways, but because simple theories are more within a human's grasp to test and therefore trust. And since the available and relevant facts are often few, it is usually the case that they can only confirm simple features of what are otherwise much more complicated realities.

The story of solar theory presents a perfect example of this. Solar theory began in Classical Greece with attempts to describe the motion of sun and planets as perfect circles moving at constant velocities around the earth, for nature "had" to be simple, and this was the simplest system that could be. But this didn't work. The planets neither have a constant velocity nor do they move in circles. So theories describing and predicting planetary motion became increasingly complex, culminating in the master work of Ptolemy in the 2nd century A.D., which even he admitted just didn't quite fit, but it was the best anyone would do for a thousand years.

Then Copernicus came along and found a way to make Ptolemy's math much, much easier (and since doing that math was essential for managing the Christian calendar, clearing up such a headache was a welcome improvement). All he had to do was "shift" the planets so they were orbiting in circles around the *sun*. And astronomers were glad to employ this model because it made their lives so much easier. Yet it still did not work quite right. It eventually required Kepler and Newton to make things more complicated than even Ptolemy imagined: for the planets actually move in elliptical orbits at inconstant speeds—no simple matter at all. And as later discovered, Relativity and Chaos theory were required as well, to explain still more complicated deviations from this model.

But at each stage of solar theory's development, the truth could only be known so well as the facts available allowed. And this was one reason Copernicus had for giving up the Ptolemaic model: there simply was no evidence for its rather bizarre arrangements and motions, other than that it predicted where the planets would be seen in the night sky. But that was simply mathematics, and any mathematical system could do that, so Copernicus came up with another, which was preferred because it was easier on the hard lives and cramped hands of human calculators.

But until the telescope started producing data about the planets' motions never before available (such as precise measures of apparent diameter and velocity), the more complicated system developed by Kepler would hardly have been justified. When Newton succeeded in what no one else had ever done—explaining *why* the planets moved like that—he found that the Keplerian system was supported by all kinds of unexpected evidence, such as how an apple falls from a tree. As a result, this system won over all others because the evidence supported it above all others, by leaps and bounds. In contrast, the model of Copernicus "won" at first only because it was easier for humans to work with, even though there was no evidence yet to tell whether it was truer than Ptolemy's (for more on the example of solar theory, see III.3.5.2, "The Origin of Order").

The "Copernican Revolution" is often used to prove that simplicity is an indicator of truth. But it isn't, as the current solar model proves, for it is nowhere near as simple as that of Copernicus. Rather, what the Copernican Revolution exemplifies is Occam's Razor: do not multiply the elements of a theory unless you have to. This rule is correct not because nature is simple, but because humans are ignorant. Occam's Razor aims at preventing you from claiming more than you know, of going *beyond* what the evidence proves. We have a bad tendency to get hung up on marvelously complex systems of thought that distract us from making real empirical progress, so we need to curb that fancy, look for evidence, and keep our claims modest.

This is one of the keys to the success of a scientific method: it only allows the seal of "scientifically proven" to be placed on a fact or theory when every detail is supported by evidence. Ptolemy never had that—nor did Copernicus. But, eventually, and for a time, Kepler *did*. And it took *new evidence* to require expanding it. In short, science succeeds because it requires you to admit when you are ignorant, and to have confidence only in what you *know*. Metaphysics or theology, even at their best, cannot contend with that. Even at best they can only stand in with what is "most plausible" until science can finally step up and answer the question. That is yet another reason why metaphysics can never precede or supercede the findings of science.

> For a good book explaining the difference between science and pseudoscience in lay terms, with entertaining examples, see Michael Shermer, *The Borderlands of Science: Where Sense Meets Nonsense* (2001).
>
> On defenses of science and reason against various postmodern critics: Susan Haack, *Manifesto of a Passionate Moderate: Unfashionable Essays* (2000) and *Deviant Logic, Fuzzy Logic: Beyond the Formalism* (1996); and Paul Gross & Norman Levitt, *Higher Superstition: The Academic Left and Its Quarrels With Science* (1997). For more on scientific methods, see bibliography for section II.3.3, "The Method of Science."

1.2 Miracles and Historical Method

There are natural miracles. Any unexpected, fortunate event can be called a miracle. But nothing supernatural is at work there. For there are just as many unexpected, *un*fortunate events, so the world's luck balances out in the end—or at least it would, without the interference of human design. For in a society where more people are more compassionate and community-conscious, "miracles" will actually happen more often than tragedies, as there will be more Good Samaritans than Joe Criminals. Likewise, in a society that has embraced humanist values and followed them in the pursuit of technological and political ends, miracles will be so common people will start taking them for granted—so common, in fact, that they are actually *expected*, and thus no longer called "miraculous." Here in the U.S. miraculous medical cures are effected every day. Calls to 911 often produce miraculous rescues. People are routinely resurrected from the dead by CPR, medicines, and electric defibrillators. Never mind all the other amazing things we do, which we now take for granted but would have been wonders beyond wonders a few centuries ago, from sending men to the moon to lighting a city without flame.

But there is another kind of thing people sometimes mean by "miracle": a supernatural miracle. I do not mean the mere linguistic superstition whereby people take a patently natural miracle and "assume" it must have a supernatural cause ("A hundred people died in that crash, but I was spared. Clearly a miracle from God!" What? A God who liked you more than the hundred others he let die?). No, I mean here what people say is the *bona fide* act of a supernatural being. This applies equally to all paranormal beings, from angels to ghosts to gods. Are there any of those sorts of miracles, events that clearly could not have been natural but *had* to have been caused by supernatural agents?

Even from the start things add up toward doubt: the lack of evidence for these beings, the lack of plausible mechanisms that explain their existence or activity, the lack of evidence for any such mechanisms, and finally the general Inference to Metaphysical Naturalism (see III.2, "A General Outline of Metaphysical Naturalism" and IV.1.1.7, "The Lessons of History and the Burden of Proof"). That's enough to be skeptical of miracles. This doubt is even further justified in those cases where there is positive evidence *against* a particular entity or phenomenon, as there often is. Of all events claimed to be clear cases of a supernatural miracle, those open to adequate investigation were all found to have natural explanations (amazing but normal coincidences or fortunate turns of events), or to be the exaggerated or erroneous accounts of natural events, or false stories altogether.

First and foremost, however, the reason to doubt miracles is *historical*, not philosophical: we argue *from* the lack of miracles *to* Metaphysical Naturalism. When we look for evidence of any event that could not have happened by accident, human design, or the inevitable course of nature, but instead only by some intelligent, superhuman power, we come up with too little to trust. So there is no reason as of yet to believe any supernatural miracles occur. Such miracles would be both logically and physically possible, if there actually were any supernatural beings around to carry them out, and who actually did so. But as an historian I do not believe any miracle has ever been adequately documented, and thus we have no evidence, and no reason, to believe there are such things. I think you can only begin to understand my conclusion by learning a little of what I know about the historical period I study, and how historians in general deal with evidence.

1.2.1 The Rain Miracle of Marcus Aurelius

In 172 A.D. one of the legions of Marcus Aurelius got itself in a bit of a pickle. Surrounded on all sides by barbarian hoards somewhere in Eastern Europe, in the middle of a hot summer, dying of thirst, the army begged and pleaded with the gods for salvation. The gods replied. Clouds quickly gathered in a previously-clear sky, and a torrential rain fell. As the Romans desperately filled their shields with water to drink, balls of lightning thundered down into the enemy ranks, destroying and routing them utterly.

Just a few years later, this miracle was celebrated in stone, the scene carved into the Column of Marcus Aurelius, still visible today. Carved in the stone, you can see a winged rain god pouring life into Roman shields and raining death upon the enemy hoard before them. About the same time, Aurelius dedicated a statue of Jupiter Thunderbolter, and minted a coin praising the religion of the Emperor. Only eight years after this astonishing event, the Christian apologist Apollinarius claimed that the legion that was there that day was entirely comprised of Christians who prayed to their god for help, and therefore the miracle proved the true power of belief in Christ. Indeed, he says, that legion was then dubbed the Thundering Legion to honor this. Only twenty five years after the event, Tertullian, another Christian apologist, echoed the same story. But fifty years after the event, the pagan historian Cassius Dio tells an entirely different tale. He mentions no Christians. Instead, he says the Egyptian wizard Harnouphis brought it about, summoning the god Hermes with magical spells.

Who is telling the truth here? The first question is whether it ever happened at all. We have no other evidence that Christian prayers or Egyptian magic can call down rain and lightning, although one can perhaps keep an open mind. But in contrast, we have ample evidence that a fortuitous but entirely natural accident could produce the same result. After all, rapidly-appearing and violent thunderstorms are a frequent occurrence in nature, especially in the summer months, while the human propensity to exaggerate is both infamous and ubiquitous. So we have much more reason to believe this was a natural event than a miraculous one. And the fact that the event was commemorated in stone, just a few years later, for all to see, by the officers and soldiers who were there, suggests that something really did happen. It's believable—amazing, but not impossible. We could call it a natural miracle. But not a supernatural one.

The second question, of course, is whether the storm followed the prayers of a Christian legion or the spells of an Egyptian sorcerer. It is obvious that any army in such dire straits would be trying every possible solution to their plight, whether practical or superstitious, and so the coincidence of prayers or magic, or both, with the seemingly-miraculous weather would be natural. But when we start examining the stories, certain doubts arise.

First, it is incredible that a pagan emperor who showed little mercy for Christians would allow any into his legions, much less fill an *entire legion* with them. Despite the fact that some Christians later believed Aurelius to be a "good" emperor, his reign saw the martyrdom of Polycarp, Justin, and a multitude of others at Lyons, and two lengthy apologetic letters were written to Aurelius, begging him to treat Christians with more respect. Besides that, all legionaries had to offer daily prayers to the emperor's guardian spirit and routinely praise Jupiter Optimus Maximus, "Jupiter Best and Greatest," protector of the legions. But since these were things no Christian could do, it is impossible for Christians to have been in any legion at the time. And we have no evidence they were. Second, contrary to the tale told by the Christian authors that the legion responsible won its new name for the miracle, that legion had already been called the Thundering Legion for almost two centuries.

When we look at the pagan's story, our results are decidedly different. An inscription found in Eastern Europe attests that an Egyptian sorcerer by the very name Harnouphis was traveling with the imperial legions at the time, and the coin minted by Aurelius praising his religion specifically depicts the god Hermes standing in an Egyptian temple. This provides support in material evidence for Dio's account. The erection of a statue honoring Jupiter Thunderbolter also shows the emperor understood his army's savior to be pagan, not Christian.

What we have discovered is one among many examples of Christian authors making up stories in the context of deliberately trying to persuade readers to convert. It does not matter that the stories are thoroughly implausible and fairly easy to refute for anyone who took the trouble to check the facts. It is clear no one did. And so, Christian chronicles a century later, and throughout the Middle Ages, celebrated the miracle of the Christian Legion. The Christians won the propaganda war. Though it is clear that Dio's account, written by a sober historian simply trying to convey the facts as best he could, is probably the correct one, hardly anyone paid any attention to it. Instead we have a legend springing up just eight years after the fact, when thousands of eye witnesses were surely still alive, when the government was already promoting its alternative account, when all the necessary records were available. And despite these seemingly unfavorable conditions, this legend beat out the truth.

This is not an isolated case. Historians see this happening again and again, in all ages of history. We recognize that almost any story can be an invention. So the First Rule of Historical Method is: *don't believe everything you read.* A believable history has to be constructed from several converging lines of evidence that have been critically and skillfully examined, and not every piece of evidence is equally trustworthy. Humans are notorious liars, eager exaggerators, and happy to believe almost anything they agree with. Skepticism is a virtue—but unfortunately a rare one, even rarer than honesty. This is the problem anyone faces, right from the start, who wants to find evidence of genuine miracles in historical records.

Evidence that Marcus Aurelius was intolerant of Christians is presented in T.D. Barnes, "Legislation against the Christians," *Journal of Roman Studies* 58 (1968) pp. 39-40.

Ancient sources for the rain miracle: Eusebius, *Ecclesiastical History* 5.5.1-4, *Chronicon* 1.206-7, 2.619-21; Tertullian, *Apologeticus* 5.6, *Ad Scapulam* 4; Cassius Dio 71.8.10 (= epitome of book 72). For depictions of the miracle in stone, see Scene XVI of the Column of Marcus Aurelius in, e.g., Giovanni Becati, *Colonna Di Marco Aurelio* (1957). For photograph of Harnouphis Inscription: Marie-Christine Budischovsky, *La Diffusion des Cultes Isiaques Autour de la Mer Adriatique* (1977), pp. 124-25 (no. 25). For photographs and reconstructions of the Statue of Jupiter Thunderbolter: Werner Jobst, *11. Juni 172 n. Chr.: der Tag des Blitz- und Regenwunders im Quadenlande* (1978), cf. Abb. 29 - 30 (No.

24 / Bd. 335). For images of the Hermes Coin: Mattingly & Sydenham, *Roman Imperial Coinage* (1930), vol. 3, pl. 12, no. 247.

For modern scholarship: Garth Fowden, "Pagan Versions of the Rain Miracle of AD 172" and Michael Sage, "Eusebius and the Rain Miracle: Some Observations," both in *Historia: Zeitschrift für Alte Geschichte* 36 (1987), pp. 83-113; Michael Sage, "Marcus Aurelius and 'Zeus Kasios' at Carnuntum," *Ancient Society* 18 (1987), pp. 151-72. See also: Julien Guey, "La Date de la «Pluie Miraculeuse» (172 Aprés J.-C.) et la Colonne Aurélienne," *Mélanges d'Archéologie et d'Histoire de l'École Française de Rome* 60-61 (1948-49), pp. 105-27, 94-118 and "Encore la «Pluie Miraculeuse»: Mage et Dieu," (1948), *Revue de Philologie* (3e ser.) 22 (1948), pp. 16-62.

1.2.2 Understanding the Ancient Milieu

The first task of any historian worth his salt is to understand as thoroughly as possible the era and culture he is studying. And the first thing that strikes any student of antiquity is how the miraculous was so strangely common then, yet became less and less common as literacy, education and science spread, until today when it is nowhere to be found but in urban legends, obscure backwaters, and other suspicious circumstances.

As early as the 18th century, renowned philosopher David Hume observed how strange it was "that such prodigious events never happen in our days," yet were once routine. And his contemporary, historian Edward Gibbon detailed the frequent "fraud and sophistry in the defense of revelation" seen in antiquity, and wondered why even then miracles never seemed to be noticed by anyone except believers. They were so common, he says, that "a miracle, in that age of superstition and credulity, lost its name and its merit." Modern historians agree. Gibbon and Hume were not making a merely philosophical argument, but stating a simple case demonstrated by historical evidence: we observe so much fraud and credulity in those days, miracles were so widely believed on so little evidence, that it would be irrational to believe anything that smelled of the same character, without the best of evidence.

Hume did expand on that reasoning with a more thorough philosophical argument, but his case has a strong historical component that commentators generally ignore. There were *so many* miracles all over the Roman Empire in those times that they became as common as natural events. What can explain their sudden disappearance in the past few centuries? Observing the first fact, Gibbon and Hume, like all sound historians, saw in it an ideal explanation of the second: fraud and credulity were far more common, and far more successful, then than they are now, though even now they are routine features of human behavior. And this is a very reasonable explanation, especially when we comprehend the time and place at hand.

The early Roman Empire in particular was replete with kooks and quacks of all varieties, from sincere lunatics to ingenious frauds, even innocent men mistaken for divine, and there was no end to the fools and loons who would follow and praise them. Skeptics and informed or critical minds were a small minority. Although the gullible, the credulous, and those ready to believe or exaggerate stories of the supernatural are still abundant today, they were vastly more common in antiquity, and taken far more seriously.

This was an age of fables and wonder. Magic, miracles, ghosts were everywhere, and almost never doubted. Yet this is not because ancient peoples were all ignorant or uncritical. Some among the well-educated elite had enough background in science and skepticism not to be duped—they saw a world much more natural and normal, and saw through most lies, exaggerations and tall tales. But these men were a rarity even among their peers. The vast majority had none of these advantages. And this can be credited to the complete lack of any mass public education—much less in science and critical thinking—and the lack of any mass media, or any organizations dedicated to investigating and getting at the truth and publishing the results.

By the estimates of William Harris, author of *Ancient Literacy* (1989), only 20% of the population could read anything at all, fewer than 10% could read well, and far fewer still had reasonable access to books. He found that in comparative terms, even a single page of blank papyrus cost the equivalent of thirty dollars—ink, and the labor to hand copy every word, cost many times more (p. 195). As a result, books could run to the thousands or even tens of thousands of dollars in value each. Consequently, only the rich had books, and only elite scholars had access to libraries, of which there were few. Education was likewise expensive and rare, but even when had was not much steeped in science or critical thought.

To give you an idea of just how much different things were then from now, I will start with a prominent example I studied for my Master's Thesis at Columbia University, "The Cultural History of the Lunar and Solar Eclipse in the Early Roman Empire" (1998). We have several accounts of what the common people thought about lunar eclipses. They apparently had no doubt that this horrible event was the result of monsters trying to devour the moon, or witches calling her down with diabolical spells. So when an eclipse occurred, everyone would frantically start banging pots and blowing brass horns furiously, to scare away the monsters or confuse the witches' spells. So tremendous was this din that many better-educated authors complain of how the racket filled entire cities and countrysides. This was a superstitious people.

As noted, only a small class of elite well-educated men adopted more skeptical points of view, but because they belonged to the upper class, both them and their arrogant skepticism were scorned by the common people, rather than respected. The Greek priest and prolific author Plutarch laments how doctors were willing to attend to the sick among the poor for little or no fee, but they were usually sent away in preference for the local wizard ("On Superstition," *Moralia* 167e). Plutarch also engages in a lengthy digression to debunk a popular belief that a statue of Lady Luck actually spoke in the early Roman Republic (*Life of Coriolanus* 38.4). He claims it must have been an hallucination inspired by the deep religious faith of the onlookers, since there were, he says, too many reliable witnesses to dismiss the story as an invention (although a careful reader can see how easily pious trickery could have been afoot). But he digresses further to explain why other miracles such as weeping or bleeding—even moaning—statues could be explained as natural phenomena, showing a modest but refreshing degree of skeptical reasoning that would make the Amazing Randi proud. It is notable that we have thousands of believers flocking to weeping and bleeding statues even today. Certainly these people should concede that pagan gods must also exist if they could make their statues weep and bleed as well.

Miraculous healings were also commonplace. Even the emperor Vespasian once cured the blind and lame, and statues with healing powers were common attractions for sick people of this era. Pilgrimages to these wonders were a regular event. These magical statues included not just gods, but the images of wise men, military generals, even famous athletes, the ancient equivalent of Michael Jordan in bronze. But, above all, the pagans had their god Asclepius. Surviving testimonies to his influence and healing power throughout the classical age are common enough to fill a two-volume book. Of greatest interest are the inscriptions set up for those healed at his temples. These give us almost first hand testimony of the blind, the lame, the mute, even the victims of kidney stones, paralytics, and one fellow with a spearhead stuck in his jaw, all being cured by this pagan savior. And this testimony goes on for centuries. Inscriptions span from the 4th century B.C. to the 3rd century A.D. and later, all over the Mediterranean.

Clearly, the people of this time were quite ready to believe such tales. Miracles were not remarkable at all. One can add to these examples all manner of things people claimed to have seen: temples where water was turned to wine every year, talking dogs, flying wizards, animated statues, and monsters springing from trees. There were wandering wise-cracking sages like Apollonius of Tyana who could teleport or resurrect the dead, or Proteus Peregrinus who burned himself alive in the Colosseum to prove his faith in reincarnation (and was seen resurrected shortly thereafter). Relics associated with such men became prized talismans hailed for their magical powers for centuries.

But it wasn't just honest wonder and gullibility that was at work. There were hucksters, too. The best-known example is Alexander of Abonuteichos. The literary wit Lucian dedicates a lengthy and biting account, both detailed and entertaining, of his personal contacts with this man, whom he calls "the quack prophet." His story illustrates how easy it was to invent a god. Alexander's scam began around 150 A.D. and the cult he created, which drew countless adherents from every class, lasted well into the 4th century.

The official story was that a god named Glycon, an incarnation of Asclepius, the Son of the God Apollo, was born in the form of a giant snake with a human head, and Alexander was his keeper and intermediary. With this arrangement Alexander gave oracles, offered intercessory prayers, and began his own religion. Lucian tells us the inside story. Glycon was in fact a trained snake with a puppet head, and all the miracles surrounding him were either tall tales or the ingenious tricks of Alexander himself. Yet so credulous was the public as well as the government, that a petition was heeded to change the name of the town where the god lived, and to strike a special coin in his honor, and we have direct confirmation of both facts: such coins have been found, dating from the reign of Antoninus Pius and continuing up into the 3rd century, bearing the unique image of a human-headed snake, and the town's name was indeed changed—from Abonuteichos to Ionopolis (City of Apollo's Son), and its present Turkish name is a derivation of that (Ineboli). Even statues, inscriptions, and other carvings survive, attesting to this Alexander, his god Glycon, and their cult (see bibliography below). As for his influence, Lucian tells us that his devotees included Roman governors. People would travel a thousand miles to see him. Even the emperor Marcus Aurelius sought his prophecy.

It is most notable that Lucian's skeptical debunking never persuaded the believers, showing that even the rare skeptic, no matter how convincing his arguments and evidence, could have no practical effect on the credulous. The vast majority would never read or hear anything he wrote, and most of those who did would dismiss it. Believers were often hostile to critical thought and would shout the skeptics down, driving off suspected doubters in their midst. In the case of Alexander, before every ceremony his congregation would cry "Away with the Epicureans! Away with the Christians!" since these were the two groups with a reputation for trying to debunk popular religion. On other occasions this hostility could come to slander and violence. Challenging a popular legend might start a riot, even get you killed

From all of this one thing should be apparent: the ancient world was not an age of enlightenment. It was an era filled with con artists, gullible believers, martyrs without a cause, and reputed miracles of every variety. Legends were taken as history, and exaggerations could snowball into fabulous wonders practically overnight. The world was filled with people lacking in education or the skills of critical thought. They had no newspapers, telephones, photographs, or access to public documents to consult to check a story. There were no reporters, coroners, forensic scientists, or even police detectives.

If someone was not a witness, all they had was a man's word. They would base their judgment not on evidence but on the display of sincerity by the storyteller, by his ability to persuade, and to impress them with a show, and by the potential rewards his story had to offer. And even if someone *was* a witness, their skills of critical reflection would often be lacking, and their readiness to believe in wonders great. Certainly, this age did not lack keen and educated skeptics. Rather, the shouts of the credulous crowds overpowered their voice and seized the world from them, boldly leading them all into a thousand years of darkness. Perhaps we should not repeat the same mistake. After all, the wise learn from history. The fool ignores it.

> Quotations of Edward Gibbon's *The History of the Decline and Fall of the Roman Empire* (1776-1788) are from the Womersley ed. (1994), vol. 1, ch. 15, pp. 511-512 and vol. 2, ch. 28, pp. 92-3. Quote from Hume: "Of Miracles," *An Enquiry Concerning Human Understanding* 10.94 (1777). On the value of historical records in the ancient period, see Michael Grant's exposition on the nature of historical unreliability even in the most reliable sources: *Greek and Roman Historians: Information and Misinformation* (1995).
>
> On popular gullibility of the time, the Syrian satirist Lucian is a vital reference, writing in the 2nd century. Of his works, the most important here are the "Death of Peregrinus" and "Alexander the Quack Prophet," which are available in a very readable and engaging English translation in Lionel Casson, *Selected Satires of Lucian* (1962), while Lucian's equally important work "Lover of Lies" (*Philopseudes*) is accessible in the Loeb Classics Series published by Harvard. For scholarship, see: Ramsay MacMullen, *Christianity & Paganism in the Fourth to Eighth Centuries* (1997); Graham Anderson, *Sage, Saint, and Sophist: Holy Men and their*

Associates in the Early Roman Empire (1994); Robin Lane Fox, *Pagans and Christians* (1987); C. P. Jones, *Culture and Society in Lucian* (1986); E. R. Dodds, *The Greeks and the Irrational* (1951). See also Ramsay MacMullen, *Paganism in the Roman Empire* (1981), which includes a photograph of a Glycon statue on the cover (and on pp. 120-21).

On Vespasian and healing statues: Suetonius, *Life of Vespasian* 7.13; Tacitus, *Histories* 4.81; Lucian, *Council of the Gods* 12, *Lover of Lies* 18-20; Pausanias 6.5.4-9, 11.2-9; Athenagoras, *Legatio Pro Christianis* 26. The "two volume" work I mention is Edelstein & Edelstein's *Asclepius: A Collection and Interpretation of the Testimonies* (1945), esp. §423-450. For photos of a Glycon statue and coin: Francis Greenleaf Allinson, *Lucian, Satirist and Artist* (1926), pl. opp. p. 108.

On ancient miracles generally: Wendy Cotter, *Miracles in Geaeco-Roman Antiquity: A Sourcebook for the Study of New Testament Miracle Stories* (1999); Harold Remus, *Pagan-Christian Conflict over Miracle in the Second Century* (1981); Robert Grant, *Miracle and Natural Law in Graeco-Roman and Early Christian Thought* (1952). Also relevant is Bengt Ankarloo and Stuart Clark, eds., *Witchraft and Magic in Europe: Ancient Greece and Rome* (1999); and Thomas Matthews, *The Clash of the Gods: A Reinterpretation of Christian Art* (1993)

On Apollonius: Maria Dzielska, *Apollonius of Tyana in Legend and History* (1986), translated by Piotr Pienkowski.

1.2.3 Historical Method Saves the Day

But wait, you might ask: if history is so plastered with lies, exaggerations and errors of judgment, how can we trust any history at all? What if it's all bunk? How do we tell the difference? How do we identify a well-supported fact of history anyway? Well, I'll tell you. It takes experience, hard work and a good dose of critical thought.

A reasonable belief in historical facts is usually established by two kinds of arguments: the argument to the best explanation, and the argument from evidence. It does *not* begin with the 'assumption' that all stories are to be trusted unless you have a specific reason otherwise. Rather, you must have specific reasons to trust a story before you can regard it as true. Without such reasons, any particular claim is as likely false as true (indeed, even more likely, given the overabundance of false stories on record).

So the first thing the historian must do, and that anyone must do if they want to get to the bottom of any historical claim, is analysis. The historian has to ascertain as best he can where all the evidence comes from, the date of its composition or creation, the background and objectives of its author or creator, and what the claim or object actually means in the context of its time, society and culture. He must try to find out what sources were used by an author. How did the author come to know what he reports? How could he have come to know it? When did he write it down? Why did he write it down? Has anyone tampered with it? Is the author already known from past studies to be trustworthy with this sort of material?

In all, there are four stages of analysis one must engage in to thoroughly examine a historical claim. First is textual analysis. We must use the methods of textual criticism and paleography to find out whether the document we presently have is authentic and accurately reflects its original (since usually only copies of copies exist today). Second is literary analysis. We must ascertain what the author meant, which requires a thorough understanding of the language as it was spoken and written in that time and place, as well as a thorough grasp of the historical, cultural, political, social, and religious context in which it was written, since all of this would be on the mind of both author and reader, and would illuminate, motivate, or affect what was written. Third is source analysis. We must try to identify and assess the author's sources of information. Finally, only then can we begin historical analysis proper, and thus assess the reliability of the claim as now understood. As you can see, history is a difficult business.

The historian needs to know everything he can about where certain stories or artifacts came from, and why they exist, in order to assess their relevance and weight. Conversely, if the historian cannot ascertain these things, then the evidence becomes less compelling or useful. The less you know about the sources of your evidence, the less reliable it is in establishing any truth. This entails one particular lesson above all, which we shall call Rule Number Two: *always ask for the primary sources of a claim you find incredible.* Many modern scholars will still get details wrong or omit important context or simply lie. Since proper history gets done by citing the actual evidence, so claims can be checked, always challenge people who make amazing claims without backing them up.

The historian must also endeavor to gather *all* the evidence. He can't just count on one item. He has to go and look around for everything that could relate to the case, and bring it all together without omission or abject distortion. And what is relevant is not always immediately obvious or easy to find, which is one reason why expertise in a particular historical period and culture is often indispensable to doing history well. Another reason is that the historical context often changes our perspective, as I've shown regarding the state of ignorance and gullibility in the ancient world. Indeed, this makes for the Third Rule of Historical Method: *make sure you understand all aspects of the historical context.* So again, beware of scholars who make amazing claims about some historical event but who are not experts, or even experienced historians at all.

Finally, the historian must evaluate and analyze all the evidence and see if anything believable can be justified by the argument to the best explanation or the argument from evidence.

1.2.4 The Argument to the Best Explanation

The argument to the best explanation works something like this: any statement about the past that we are justified in believing true to any degree must be tested against five criteria, and if no other competing statement about the same event comes close in meeting the same criteria, then we are more than justified in believing it. I will list and illustrate these criteria. In general, the more one explanation exceeds all others on each criterion, the more confident we can be it's true.

(1) *Explanatory Scope*

The best theory of events will explain more facts than any competing explanation. Consider the rain miracle: the theory that the storm happened explains all the evidence we have, including the coin and statue, as well as the monument and the many stories. The theory that Christians were credited, however, does not explain this evidence very well, since, apart from the reports of Christian authors, it all pays tribute to pagan deities. In contrast, the theory that an Egyptian wizard named Harnouphis gave a display of working his magic explains both the pagan report that he did so, as well as the inscription indicating he was with the army in the general period of that war.

(2) *Explanatory Power*

The best explanation will render the facts more probable than any other. The facts of this case are *very* probable if the event happened and Harnouphis gave a magic show—indeed, if this happened, then all the evidence we have is almost certain to have come about in some form, even if we could never predict, for example, what of such evidence would survive the two thousand years of intervening time. The challenging fact of a Christian alternative is also made probable by this theory, since some Christians would surely not allow proof of demonic magic to triumph over the Gospel, and would likely try to spin the story to their advantage, especially in works directly aimed at persuasion.

In contrast, the Christian account does not make the pagan account very probable, since Dio's version is not polemical or defensive in any way, but simply reports a story, in the context of an otherwise sober history, and the story does not appear from its content to have been concocted to praise pagan magic. The other evidence, all soon after the event, praising traditional pagan religion without any indication of Christian involvement, does not become very probable either.

(3) *Plausibility*

The best explanation will be more plausible than any other theory. I mean this here in a narrower sense than usual (see III.3.1, "Plausibility and the God Hypothesis"). To have plausibility a claim must be *historically reasonable*, fitting true generalizations about the time, place, and circumstances in question. That is, we have to be able to say with good reason that such a thing could have happened in that general time and place, even if it was improbable. As long as it was reasonably possible, the sort of thing that did or could happen in that time and place, no matter how rarely, then we can say it is *plausible*. Of course, that does not mean it is true, or even the most plausible account.

I have already noted the immense *im*plausibility of even having Christians in the legions of Marcus Aurelius, much less an entire Christian legion. In contrast, a freak summer storm, though rare, is not implausible, nor is the presence of a pagan wizard in a pagan army. We can say that the *exact description* of the storm is implausible, but it is not implausible that the details would be *exaggerated*.

(4) *Ad Hocness*

The best explanation will also rely on fewer undemonstrated assumptions than any competitors. We call these *ad hoc* features of a theory, things simply made up out of whole cloth in order to explain what is otherwise inexplicable. This is not unjustified, since we can't have evidence for everything that happens, especially things going on behind the scenes, so some features of any theory will be more or less *ad hoc*. The believable theory, however, will usually be less *ad hoc* than any other.

In the rain miracle case, the Christian explanation requires inventing some reason why Christians would be in a pagan army and this fact never recorded elsewhere, whereas the pagan explanation requires inventing some reason why Christian authors would lie about this so soon afterward, and their version win out. When we compare these features, however, we see that any excuse one might invent to explain Christians being in a pagan army would be based on no precedents and thus would be completely *ad hoc*. But we have plenty of examples of Christian apologists telling lies to sell their faith—and when specifically trying to sway people to believe one's religion is more powerful, we know full well that human nature is inclined toward fibbing: lawyers and politicians are notorious for it, and preachers are just as human as they are. Thus, the pagan explanation is less *ad hoc*.

Likewise, that the event actually happened requires positing a lucky coincidence, which is slightly *ad hoc*, but the claim that it didn't happen requires a much more implausible sequence of events to explain the production of all the ensuing evidence, and would therefore be more *ad hoc*.

(5) *Fit to Evidence*

The best explanation will least contradict well-known or well-established facts. Of course, apparently contradictory or inexplicable facts can always be explained away somehow, and sometimes with good reason. But how one defends a claim is very pertinent to its believability, and this is where the other criteria come in, honing in on the best possible explanation of the evidence, which must be some historical fact or other. But if the claim is not even credible on its face (it is based on no evidence whatever, or flatly contradicts abundant reliable evidence), that is a big strike against it.

So the historian has a choice. You can leave problematic evidence unexplained and thus lower your claim's "fit to evidence" or you can increase its fit by making it more *ad hoc*. But the more you have to explain away to make your claim fit the evidence, the more unbelievable it gets. Take the rain miracle: the claim that a Christian legion was at the event requires explaining away the lack of evidence for a Christian legion at the time, and the false story about the legion's name. But more importantly one would have to explain away all the evidence against the plausibility of a Christian legion at the time, and all the evidence for *pagan* involvement.

The claim that the event never happened at all, however, would require explaining away the early erection of a dedicatory monument to the event, as well as the existence of several different stories about it arising within ten to fifty years. In contrast, the claim that it happened, and an Egyptian magician was present, has a lot of evidence behind it and meets with no contrary evidence at all, except the omission of that detail by Christian authors.

Thus, when we deploy the argument to the best explanation, we find that it is most believable that this freak storm occurred, and that an Egyptian wizard gave the impression at the time of causing it, and that no Christians were likely involved at all. But we *also* discover that we cannot justify believing this was a miracle. We cannot use this as evidence of the efficacy of Egyptian magic, or the real intervention of the god Hermes. For this explanation fails on all the criteria but one.

For example, it lacks scope. The efficacy of Harnouphis's magic on this occasion does not explain the complete absence of evidence of other amazing spells cast on the enemy throughout the same war: it does not explain why he was so successful apparently only once in his life, whereas the theory that it was just a lucky coincidence explains this fact perfectly. It is also seriously challenged by the fact that we have no evidence that such magic works at all, much less at causing storms—otherwise, why aren't Egyptian armies marching invincibly across the earth calling lightning on their enemies and supplying their men with water from the sky? And this requires something exceedingly *ad hoc*. To propose that Egyptian magic works, but that for various reasons it isn't evident now, and only worked occasionally then, entails inventing a whole matrix of *ad hoc* assumptions for which there is no evidence. Finally, it is implausible that the Roman legions began to be defeated more and more after this year, eventually being overrun and the Western Empire sacked altogether, despite the fact that they continued to employ wizards of all varieties.

We must grant that this theory does win the day on explanatory power, since if Egyptian magic works, and Hermes really can be persuaded to play with the weather, this would make the occurrence of the storm very probable indeed. But the failure of this theory on every other point, and the fact that the alternative, of a lucky natural coincidence, also makes the event probable enough to be believed, leads us to conclude that no miracle happened that day.

1.2.5 The Argument from Evidence

Another way of approaching historical questions is the argument from evidence, which precedes the argument to the best explanation: for before we can apply that tool, we must first prove that we have a believable fact to explain in the first place. And the degree to which we can trust a claim about the past is directly related to the scope and variety of the evidence, and how much we can trust it. There are five specific classes of evidence to account for: the more a claim fits these features, the more believable it is (leaving out here indirect evidence that increases the *plausibility* of a claim, by supporting the underlying generalization). These categories of evidence are:

First, what I call "physical-historical necessity."
Second, direct physical evidence.
Third, unbiased or counterbiased corroboration.
Fourth, credible critical accounts by known scholars from the period.
Fifth, an eyewitness account.

To illustrate the application of these criteria in assessing evidence, I will draw upon a comparison made by modern Christian apologist Douglas Geivett, who declares that the evidence for the physical resurrection of Jesus meets, and I quote, "the highest standards of historical inquiry" and "if one takes the historian's own criteria for assessing the historicity of ancient events, the resurrection passes muster as a historically well-attested event of the ancient world," as well-attested, he says, as Julius Caesar's crossing of the Rubicon in 49 B.C.

Well, you heard me mention earlier the known tendency of Christian writers to exaggerate or outright lie, and here is a good example. Let's take Geivett up on his claim and apply the historian's own criteria to these two events and see how they come out. Let's start with Caesar's crossing of the Rubicon river.

First, the physical-historical necessity of this event is exceedingly great. The history of Rome could not have proceeded as it did had Caesar not physically moved an army into Italy. Even if Caesar could have somehow cultivated the mere belief that he had done this, he could not have captured Rome or conscripted Italian men against Pompey's forces in Greece. On the other hand, all that is needed to explain the rise of Christianity is a belief—a belief that the resurrection happened. There is nothing that an actual resurrection would have caused that could not have been caused by a mere belief in that resurrection. Thus, an actual resurrection is not necessary to explain all subsequent history, unlike Caesar's crossing of the Rubicon.

Second, we have lots of direct physical evidence. We have a number of inscriptions and coins produced soon after the Republican Civil War related to the Rubicon crossing, including mentions of battles and conscriptions and judgments. On the other hand, we have absolutely no physical evidence of any kind in the case of the resurrection. No documents exist, and no inscriptions were commissioned by the resurrected Jesus, or by witnesses like Peter or Joseph of Arimathea—and the Shroud of Turin is a proven Medieval forgery (and even if authentic, it would not prove Jesus was resurrected any more than, say, vaporized or flash-frozen).

Third, we have unbiased or counterbiased corroboration. An unbiased source is someone who certainly would know if the story was true or false (such as an eye-witness or contemporary), but for whom there is no identifiable or plausible reason to be credulous, or to lie or distort the account in the ways (or with respect to the details) that concern us. Though they could still be wrong, at least we can rule out some very common causes of falsehood, and this is what makes sources like this so weighty in assessing the historicity of an event. On the other hand, a counterbiased source is someone who is actually notably biased *against* the event being reported, so that if even *they* admit it happened, there is good chance it did. And so, we find that many of Caesar's enemies, including his nemesis Cicero, refer to the crossing of the Rubicon, as did friends and neutral observers, whereas we have no hostile or even neutral records of a physical resurrection of Jesus by anyone until over a hundred years after the event, fifty years after the Christians had already been spreading their stories far and wide, and well after any facts could be checked.

Fourth, we have credible critical accounts by known scholars in antiquity. In fact, the story of the "Rubicon Crossing" appears in almost every history of the age, by the most prominent scholars, including Suetonius, Appian, Cassius Dio, and Plutarch. Moreover, these scholars have a measure of proven reliability, since a great many of their reports on other matters have been confirmed in various ways, and they are especially trustworthy on public political events like this. In addition, in their books they all quote and name many different sources, showing a wide reading of the witnesses and documents, and they often show a desire to critically examine claims for which there is serious dispute. If that wasn't enough, some cite or quote texts written by witnesses and contemporaries, hostile *and* friendly, of the Rubicon crossing or its repercussions.

In contrast, we have not even a single historian mentioning the resurrection until two or three centuries later, and then only Christian historians, who show little in the way of critical skill. Others simply repeated what the Christians told them. And of those Christians who describe the resurrection within a century of the event, none of them show any wide reading, never cite any sources, show no sign of a skilled or critical examination of conflicting claims, have no other literature or scholarship to their credit that we can test for their skill and accuracy, are mostly unknown, and have an overtly declared bias towards persuasion and conversion. No one of these facts renders a source useless, but each diminishes its weight, and their cumulative force greatly reduces credibility.

And fifth, there is an eyewitness account of the Rubicon crossing. For we have Caesar's own word on the subject. Indeed, *The Civil War* has been a Latin classic for two thousand years, written by Caesar himself (and completed by one of his generals, a close friend). In contrast, we do not have anything written by Jesus—and we do not know for certain the name of the author of any of the accounts of his physical resurrection. Contrary to popular belief, the names of the four Evangelists were assigned to their respective Gospels decades after they were written, and on questionable grounds. And Paul, of course, did not actually *see* the resurrection, since he only encountered Jesus years later in a vision, and he mentions no other kind of evidence than that.

It should be clear that we have many reasons to believe Caesar crossed the Rubicon, all of which are lacking in the case of the resurrection. In fact, when we compare all five points, we see that in four of the five proofs of an event's historicity, the resurrection has no evidence at all, and in the one proof that it does have, it has not the best, but the very worst kind of evidence—a handful of biased, uncritical, unscholarly, unknown, second-hand witnesses.

You really have to look hard to find an event in a worse condition than this as far as evidence goes. So if we are charitable, Geivett *at best* is guilty of a rather extreme exaggeration. This is not a historically well-attested event, and it does not meet the highest standards of evidence. If Caesar's crossing of the Rubicon were miraculous—if, say, the Rubicon had been a sea of lava, and his army flew over it at the behest of an Egyptian spellcaster—then we would have pretty strong evidence that a miracle happened in history, and I think it really would be a believable miracle. It would certainly be mysterious enough to wonder at.

Instead, when we look at the evidence for actual miracles, we always come up with very poor evidence indeed, a trend that cannot be an accident. Christ's resurrection is one of the best attested and most widely believed and celebrated miracles in history, and yet here we have seen that it is one of the worst supported historical claims we have. The readiest explanation for this lack of evidence is that it isn't really true, given all we know about the time and place in question, about historical sources and human nature and the natural world in general.

There are plausible natural explanations of an apparent resurrection. The four most prominent are the survival theory, the mistake theory, the conspiracy theory, and the bodysnatching theory. I present the best case for the possibility of a mistake or a theft in "The Burial of Jesus in Light of Jewish Law" and "The Plausibility of Theft." Both in Jeffrey Jay Lowder and Robert Price, eds., *The Empty Tomb: Jesus Beyond The Grave,* to be published by Prometheus Books in 2005.

But this is not the most plausible natural explanation of the origin of Christianity. *That* would be the theory that there was no physical resurrection, but that Christ was seen risen in visions, as Paul reports, while the stories in the Gospels are the accumulated result of exaggeration, symbolism, and doctrinal and legendary development, over two or more generations. I present the case for this in "The Spiritual Body of Christ and the Legend of the Empty Tomb," also in the same anthology cited above.

The quotation of Douglas Geivett is from "The Evidential Value of Miracles," *In Defense of Miracles: A Comprehensive Case for God's Action in History* (1997), pp. 185-6. Many others have made similar claims that the resurrection of Jesus is "supremely" well attested historically and thus 'believable': e.g. Josh McDowell, *The New Evidence That Demands a Verdict* (1999), esp. §9.5A, 9.8A; and Lee Strobel, *The Case for Christ: A Journalist's Personal Investigation of the Evidence for Jesus* (1998). A good, neutral summary of extra-biblical mentions of Jesus (and his Resurrection) is Robert Van Voorst, *Jesus Outside the New Testament: An Introduction to the Ancient Evidence* (2000). Important reading is also all the material in the Secular Web on the "Resurrection" specifically (www.infidels.org/library/modern/theism/christianity/resurrection.html), and on "Christianity" in general (www.infidels.org/library/modern/theism/christianity/).

On Caesar's crossing the Rubicon, ancient sources that survive include Julius Caesar, *The Civil War*; Appian, *The Civil Wars*; Cassius Dio, *History*; Plutarch, *Life of Caesar*; Suetonius, *Life of the Divine Julius*. Many others are cited or mentioned by ancient authors but do not survive. Modern scholarship on all this (including the material evidence) include Ronald Syme, *The Roman Revolution* (1939); M. Gelzer, *Caesar: Politician and Statesman*, 6th ed. (1968); L. Kreppie, *Colonization and Veteran Settlement in Italy: 47-14 BC* (1983); P.A. Brunt, *The Fall of the Roman Republic and Related Essays* (1988).

1.2.6 The Criteria of the Good Historian

Now it is important to digress a little further on the fourth type of evidence, and explain the criteria that establish a historian as skilled and therefore credible—the criteria that distinguish a good historian from a bad one. These criteria admit of degrees: historians will fall on a spectrum from most to least reliable, with many finding a place somewhere in between to the degree that they meet, or fail to meet, the following criteria. A good, and therefore relatively trustworthy historical writer, will exhibit these qualities:

(1) He will show in his writing a critical awareness of problems with his sources or with the intrinsic believability of an event. In other words, he will admit when something sounds amazing, or when his sources contradict each other or are not completely reliable. A bad historian will show no awareness of his sources at all, will report the incredible as if it were ordinary, and will never even mention alternative accounts of the same event even when he knows they exist.

(2) He will engage in logical historical argument addressing various forms of evidence and assessing their merit. A bad historian will never mention what evidence he has for what he is claiming and thus will never discuss its merits. He simply tells stories as if not a single event or fact is doubtful or uncertain. Yet in even the simplest of true stories, there are always details that are uncertain, like the exact name of a person or place, or just when or where something happened, or just what was said or done there. Competent historians will admit this, incompetent ones will not. Likewise, good historians will tell readers who their sources are and what good they are. Bad historians will name no one.

(3) Lastly, because he is an historian, and a good one, if he writes enough, eventually we will find him correct not only on many matters of fact (for even bad historians get some things right), but on notable or difficult historical questions. In contrast, a bad historian will be caught in overt falsehood or errors that a competent scholar of the same period would not make. In the one case, we have evidence of good scholarship; in the other, of bad scholarship.

The quality and therefore trustworthiness of an historian will stand in direct proportion to how often and how thoroughly he fulfills these criteria in any given work. It is easy to see how modern historians do far, far better here than even the best of ancient historians, but even the average historian of antiquity meets all three criteria to some degree, whereas, for instance, the authors of the Gospels meet none of them in any appreciable degree. In fact, and this is the salient point, no record supporting a supernatural miracle in all of human history meets all three of these criteria, and most meet none of them. This cannot be a coincidence. Since we are left with the scientific observation that no supernatural miracles actually happen today, we are quite reasonable to conclude that they never did.

(For sources on the historical method, see bibliography concluding section II.3.5, "The Method of History")

1.2.7 Prophecy and History

I have surveyed all the tools you need to apply critical historical methods to miracle claims. But there is another kind of miracle in history that is an animal all its own: prophecy. Unlike mere marvels, a prophecy is itself an ordinary statement, and refers to otherwise normal historical events. What makes the prophecy miraculous is its ability to predict those events. Thus, at first glance you don't need to rely on anyone's claims that a miraculous prophecy happened: you can check for yourself. If the prophetic statement truly predates the event foreseen, then you have all the evidence you need: proof of a miracle right before your eyes—that is, if such a thing could not happen by accident or through natural means, and only some superhuman intelligence could be at work. Of course, some "natural" psychic power might be to blame (or some other supernatural method like astrology). But we have no evidence of this, much less for the power of prophecy, as we shall see.

Robert Newman presents five criteria by which a prophecy could be proved a supernatural miracle, and they are as applicable to psychics and astrologers as to prophets. His criteria are quite good and I will repeat them here, quoting him verbatim:

(1) "the [prophetic] text clearly envisions the sort of event alleged to be the fulfillment"

(2) "the prophecy was made well in advance of the event predicted"

(3) "the event actually came true"

(4) "the event predicted could not have been staged by anyone but God"

(5) the "evidence is enhanced" if "the event itself is so unusual that the apparent fulfillment cannot be plausibly explained as a good guess"

This is a structurally sound approach. Three of his criteria merely need slight expansion.

For Point (5), not only can't the prophecy be an easy guess, it can't be inevitable either. In other words, if a prophet was likely to say some particular thing for cultural or traditional reasons, then any "hits" (that is, outcomes that match what was said) cannot be distinguished from accidents. This is not a "good guess," because this is not a guess *per se*, but an accidental correspondence. For example, if I was liable to say over and over again that a bird will land on my head, because it was a chant I inherited from my father, and then a bird landed on my head, it cannot be said that I predicted this. It cannot even be said that I guessed it. My reason for issuing this prediction had an entirely separate cause, and because it is always repeated, its success can have nothing to do with prophetic powers. We see this phenomenon in Judaic prophetic tradition, when the doom of the Jews or their enemies—the "punishment" of Jehova—is constantly proclaimed for cultural reasons, often as a threat aimed at encouraging piety and morality. When prophecies like these come true they cannot usually be counted as a prediction, because they would've been made anyway, and the predicted outcome is not all that improbable to begin with, given hundreds of years.

Also for Point (5) we have to rule out selection bias. How do we know that the compilers of the Bible didn't just pick the books or passages that got lucky and throw away the ones that didn't? This would artificially bloat the apparent success rate of the Bible, by disguising the more numerous failures, hence making mere guesses seem miraculous. This relates to Point (2): to meet *that* criterion, we also have to be sure no one tampered with the documents later to make them successful *after the fact*. As for the Bible, how can we really be certain no one did that?

For Point (1) we must be especially wary of retrofitting. In other words, if a prediction is suitably vague, then any future event that can be made to fit the prediction can be claimed a success. This is a common trick well known to those who investigate psychics. In effect, if any event that will fulfill the prediction is likely to happen anyway, then it cannot be called a miraculous prediction—even if it was neither staged nor a guess. For example, if I said that Zimbabwe would suffer under an evil ruler who would start a war with a great nation, that would never become a miraculous prediction, since that might happen anyway. Since I have not specified a time, even if Zimbabwe's history meets with such an event in the next two or three thousand years you can claim I predicted it—but given thousands of years, who would be surprised? And even if the event were to happen tomorrow my prediction would not be miraculous, since I allowed that it might happen *any* time, meaning that my prediction was deliberately loose enough that I could claim any such event at any time a success, and there are many ways just such a thing could naturally happen in Zimbabwe at almost any time. Other details are ambiguous, too. What counts as a 'great nation', for example?

Assuming we can get past these issues, I will close with an example of how to apply historical method to examining claims of miraculous prophecy. One of the most popular cases, frequently cited as one of the best examples of a miraculous prophecy in the Old Testament, is a pair of predictions made by the Prophet Ezekiel about the fate of the two principal cities of Phoenicia—Tyre and Sidon—who were among the most loathed neighbors of Israel.

First, in Ezekiel 28:22 the prophet issues the most vague of predictions about Sidon that could never be counted as miraculous. That a city should be sacked and its people slaughtered in antiquity is already a highly likely event, and that a prophet should declare such a doom upon an enemy city is likewise commonplace, so that the one is certain to follow the other by chance alone. This is an example of an inevitable, and perfectly natural correspondence between a prophecy and an event.

On the other hand, Ezekiel 26:3-14 predicts that the city of Tyre will be attacked by many nations, its walls torn down and its rubble cleared away, and it will be a bare rock. Then "out in the sea she will become a place to spread fishnets" and will never be rebuilt. The passage specifically predicts that Nebuchadnezzar of Babylon will do this, and his army will throw the stones, timber and rubble into the sea. Is this a miraculous prediction?

The historian's first task (once we have established to a reasonable extent that the document is authentic and properly understood) is to examine his source: who was Ezekiel and where and when did he record this prophecy? Ezekiel was a Jew held captive by Nebuchadnezzar since the sack of Jerusalem in 597 B.C. Already we see some important problems. It is all too likely that Ezekiel is issuing propaganda flattering his captor to get on his good side, while wishing ill on an old enemy of the Jews. Moreover, Ezekiel could easily have intelligence about the king's plans since he would see the preparations. His prophecy about Tyre was issued in 586 B.C. Nebuchadnezzar began the siege of Tyre only a year later. So this fails Point (5).

Tyre resisted his every attack for over a decade, and finally came to terms with Nebuchadnezzar in 573. He did not sack the city after all—

forcing Ezekiel to *retract* his prediction in verse 29:18, and instead predict a victory against Egypt, after Nebuchadnezzar had already turned against *that* country. So what do we have here? We have a man who sees the world's most powerful army preparing to besiege a city and then predicts it will be taken and destroyed—hardly something he could not guess would happen. Yet even his guess failed, and so did the prediction! A failed prediction can hardly be a miracle. So this even fails Point (3).

To try and save this failure, apologists like Newman spread false claims about the history of this event that make the prophecy seem miraculous. For instance, he claims that Nebuchadnezzar "took" the city in 573 B.C., but we have no evidence of that. The city submitted to Babylonian rule without being sacked, as Ezekiel's retraction of the prophecy confirms. Indeed, since it was a trade powerhouse with two outstanding all-weather naval ports, a conqueror would be a fool to destroy it. Moreover, Tyre has a long history of working out just such favorable agreements to get besiegers off its back. This was the key to its survival.

Newman's answer is to offer an absurd *ad hoc* thesis: he makes up the story, told in no ancient source, that the inhabitants of Tyre resettled inland during the siege, forcing Nebuchadnezzar to "settle for very little plunder." Not only is this false—for there had been a mainland suburb since before 800 B.C., and it was not called Tyre but Ushu—it is extremely implausible: people are supposed to defeat Nebuchadnezzar by leaving a nearly-invincible island city, nestled behind a monstrous wall, to resettle, with no fortifications, on the mainland, in open ground and with no port, while the Babylonian army apparently sits by and twiddles their thumbs? The absurdity of this is palpable, never mind that it has no basis in any evidence.

Newman also tries to rescue the claim by proposing another *ad ho*c thesis: that Alexander the Great, centuries later, fulfills the prophecy, since he famously used the rubble of the mainland city of Ushu to connect the island of Tyre to the mainland. But Ezekiel leaves no doubt that he means Nebuchadnezzar's men will do this, not some other guy centuries later, and the prophecy incorporates charioteers in the victorious forces, yet chariots were no longer used in warfare by the time of Alexander. So this is a classic case of retrofitting—indeed, it is worse than that, since the prophecy as stated actually forbids attributing this event to anyone else but Nebuchadnezzar. We might as well call the Israeli rocket attack on the ruins of Tyre in the Arab-Israeli war of 1981 a "fulfillment." So the prophecy also fails Point (1).

Newman then says the mainland site was "scraped clean" by Alexander and has "never been restored" while "parts of the former island are used even today for spreading fishnets." But this is a bit deceptive. First, Ushu is not Tyre. Alexander's successors built up Tyre as the powerful naval and merchant port it had always been and it remained an influential city for over a thousand years. It has never been a bare rock. It still stands even to this day. The modern city of Tyre sits beside and atop the ancient ruins (many of which still stand), and has a population of over 70,000, twice what it was in Alexander's day. It is now a major Lebanese financial center. And what about the fishnets? Authors like Newman are fond of citing a 19th century tourist who saw fishnets stretched over the rocks of Tyre as proof of the fulfilled prophecy, ignoring the fact that fishnets have always been stretched over bare rocks in every city with a fishing industry since the invention of the net, and they were no doubt stretched across the rocks of Tyre long before Ezekiel was even born. So this prophecy again fails Point (5).

Apologists do try to argue, of course, that Ezekiel really meant that Ushu would be leveled, not Tyre. But Ezekiel specifically says the nets and scraping will happen not on the mainland but "in the midst of the sea," so he clearly did not mean Ushu, but Tyre. In fact, Ezekiel always refers to the mainland site as among the "daughters" of Tyre, never as Tyre herself. And that is to be expected. Ushu was neither rich nor powerful, since it had no ports—unlike Tyre, which had two ports situated to allow year-round sailing, making it one of the most powerful military and trade cities in the world. It would be silly to make elaborate claims about the fall of "mighty" Ushu. So this prophecy again fails Point (1).

I offer this as a paradigm case. When we examine other claims of miraculous prophecy in the Old Testament in the light of history, we quickly find that they don't hold up. In fact, Newman never tells his readers that Ezekiel actually went on in verse 26:19 to predict that Tyre would be covered by the sea, and in 26:21 he says it would never be found, two clearly failed predictions. There is definitely no miracle here. And yet this is supposed to be the best case in the Bible. I can say it is certainly a typical case. This is yet another reason why I don't believe in miracles.

I have provided the reader with an entire toolbox for analyzing historical claims: the argument to the best explanation, the argument from evidence, the criteria of a good historian (which make up part of an 'argument to authority', cf. II.3.6, "The Method of Expert Testimony") and the criteria for miraculous prophecy. I have also given examples of how historical context (like the credulity of the ancient world) must be considered alongside any particular claims made from that period. It is important to emphasize that these analytical tools *could* be used to prove that a miracle genuinely occurred in history—if there were any genuine miracles supported by evidence adequately enough to survive all the relevant tests and criteria. To this date, I have never encountered any such claim.

> The prophetic criteria come from a Christian apologist who actually ignores or abuses them: Robert Newman, "Fulfilled Prophecy as Miracle," *In Defense of Miracles: A Comprehensive Case for God's Action in History* (1997), p. 215. In contrast, on other blatant failures of Biblical prophecy, see Tim Callahan, *Bible Prophecy: Failure or Fulfillment?* (1997), and the Secular Web library on "Prophecy" (www.infidels.org/library/modern/theism/christianity/prophecy.html).

2. Atheism: Seven Reasons to be Godless

The grandest paranormal claim, made by the most people, most vehemently—so vehemently they routinely kill each other over it—is "Our God Exists." But he doesn't. This I have concluded after years of study and investigation, considering all the evidence and every argument ever presented in its defense.

Certainly, Metaphysical Naturalism is a worldview without a god in it. But I can imagine forms of Deism, for instance, or even Taoism, which would be almost entirely consistent with the Metaphysical Naturalism I defend in this book, so much so that the combination could be made entirely consistent with the evidence we now have. There is just no reason to believe in such things. Yet they would fit the evidence far better than any actual God believed in today, with far less resort to *ad hoc* explanations. That right away should clue us in: if the gods people actually believe in are even more unbelievable than the most believable god we can come up with, and there is no reason to believe in even *that* god, how on earth can those popular gods be credible? This is a serious question every serious person should ask.

For instance, imagine this:

> In the beginning was God. And He was alone. Lamenting that there were no others to enjoy Being, no others to love and be loved, no others to think and create, He resolved to give His life so that beings like Him might be, and love, and enjoy life. So He exploded His body, and He was no more, but out of His body came the physical Universe, and out of His blood came the realm and possibility of mind. There was no other way, such were the limits even upon the

all-powerful and all-knowing. And He so arranged His death that the embers would one day generate mindful beings such as us, who may be, and love, and know, and do.

This is pure fiction. I just made it up. Yet it explains so many things that traditional or popular theism doesn't, such as the silence and hiddenness of God, and the imperfection and brutality of the created world, as well as religious diversity and the physical nature of our minds and existence generally.

It can also claim in its defense almost all of the arguments for the existence of God, such as the so-called cosmological and ontological arguments, the argument from moral objects, the transcendental argument, the argument to design, even the argument to divine purpose. It could even claim the argument from religious experience, as all forms of religious contact with the divine would be contact with the cosmic memory of God's choice of a selfless death for us, and all the diversity and disagreement among religions would be the result of variously perceiving this in imperfect ways and coloring it with various human assumptions and presumptions. After all, the one thing all religions seem to have in common is a belief that love has something to do with the meaning of life. On virtually everything else they disagree—so virtually everything else is probably false.

I could even expand my fiction in various ways. For instance, if I wanted to add a comfort factor, I could say that God did not "die" but by giving up his body for us was merely rendered powerless because he was now immaterial and not causally connected to the world, but that in death our disembodied souls would ascend to be with him in the realm of pure mind, after being "grown" in our physical bodies, which would be a temporary shell or seed. Thus, I could invent an eternal paradise for all people. Even the evil among us would see the light after death and become good and repent their sins in sorrow, and their victims, also transformed by paradise, would understand and forgive them. Or maybe the universe is just a computer, a way for God to work out a set of souls worth saving, so that at the end, when the universe collapses and his body reunites, he will be able to physically "recreate" those people who were good, in a new physical world made of his body in a second sacrifice, just for them.

And so on. I could make up religions all day long. But for all their merits, they are still fiction, as I believe are all other religions, which have even fewer merits than the ones I just invented. This will take some explaining. So I will present seven reasons why you shouldn't believe in any god. But first, of course, I must say a little something about this word "atheism." In its most common meaning, as it is most often used and as it was originally conceived, *atheism* is the compound *a*-theism, which means the absence of theism, the absence of belief in a god. It is parallel to *anoxia*, the absence of oxygen in the blood, or *agnosticism*, the absence of knowledge, or *anarchy*, the absence of government. Atheism does not necessarily mean the outright denial of god, since the mere lack of belief is sufficient to make you an atheist. Many an atheist will echo the creed "I don't believe in what I don't know to be true." This is called "negative atheism," "weak atheism" or "soft atheism" to distinguish it from positive, strong or hard atheism, which is the certainty or positive belief that there is no god. The negative atheist says, "There is no reason to believe." The positive atheist says, "Besides that, there are reasons to *dis*believe."

When there is no trustworthy evidence of something and no valid reason for it, you should not believe it. That is a good rule to go by, for there are a million things we have no reason to believe that still *could* be true, and to pick some and reject others is arbitrary and thus a hypocrisy. So if no one can show you any good reason to believe in God, you should be an atheist. Many people in this position, to avoid the ignominy falsely placed on the word "atheist," prefer to call themselves "agnostics," people who do not "know" whether God exists and therefore are undecided whether to believe, or simply do not believe because they lack the requisite knowledge to justify belief. But either way, agnostics lack belief, and thus are atheists in the common sense—unless they *believe*, despite not knowing (believing solely on faith, for instance), in which case they are agnostic *theists*, believers in God.

But if the idea of a god is inherently illogical (if the very idea is self-contradictory or meaningless), or if it is contradicted by the evidence, then there are strong positive reasons to take a harder stance as an atheist—with respect to that particular god. For in this sense, even believers are strong atheists—they deny the existence of hundreds of gods. Atheists like me merely deny one more god than everyone else already does—in fact, I deny the existence of the same god already denied by believers in other gods, so I am not doing anything that billions of people don't already do.

Still, I am an atheist in the softer sense with regard to gods like the one I invented above. I have no evidence against such a god. There may well be one, or hundreds of them. Heck, for all I know, God is a super-intelligent time-traveling tabby cat named Boo. But since I have found all the arguments for belief in God to be logically invalid, and as there is no *evidence* for any god, there is no reason to believe in any of them, certainly no more reason than there is to believe in Boo. But I am a harder atheist with regard to some gods, ones that simply make no sense or fail to fit the evidence. And I have found that the most popular gods are gods of such a sort. The following seven reasons to be an atheist pertain to them, and are presented in order from what may be the least to the most compelling.

These seven reasons consist of positive *contrary* evidence, which falls mainly into two categories: the lack of evidence that *ought* to be present if God exists, and the presence of evidence that *shouldn't* exist if God exists. Attempts to explain away such evidence are all *ad hoc*, because they have no evidence whatever in support of them, except the sole fact that they "rescue" the theory that a particular god exists. But since *any* god's existence can be established by such a tactic, this leads either to hypocrisy or absurdity: the moment it is admitted as a valid argument, belief in *every* god becomes *equally valid*, which is impossible.

2.1 Metaphysical Naturalism is True

The first reason to be an atheist is what we have shown throughout this book already: all the evidence we have so far, about us and the world, is best explained by Metaphysical Naturalism. Naturalism is therefore probably true, and that means there is no god. For there are no miracles, no infallible scriptures, but instead only a blind, mechanical universe.

Likewise, the truths of Metaphysical Naturalism, as supported by the facts, render invalid or unpersuasive every logical argument for belief in a god: what theologians call the Transcendental Argument is effectively refuted by the facts and arguments presented in III.9, "The Nature of Reason"; the Argument from Religious Experience, in III.10.4, "The Nature of Spirituality"; the Ontological and Cosmological Arguments, in III.3, "The Nature and Origin of the Universe"; the Argument to Design, in III.8, "How Did We Get Here?"; the Arguments from Miracles and Holy Scripture, in IV.2, "Miracles and Historical Method," including much of IV.1; and the Moral Argument will be refuted next, in Part V, "Natural Morality." Pascal's Wager ("it's safer to believe") is collectively invalidated by material presented in I.2, "How I Got Here," II.3, "Method," III.7, "The Meaning of Life," and III.10.4, "The Nature of Spirituality." And so on.

Beyond that, the same conclusion can be reached by several other independent lines of observation. For instance, as we have shown (III.6, "The Nature of Mind"), there are no minds without complex physical brains, therefore there can be no divine mind, since there is clearly no gigantic brain for it. On the other hand, if God can have a mind without a physical brain, it is inexplicable why *we* need them. It is far more probable that such a god would create beings with minds like His, minds that could not be damaged or destroyed, rather than minds needlessly dependent on something so fragile as a brain.

Or since natural explanations of the origin of the universe, and of life and ourselves, are altogether more plausible than theistic ones (III.3, "The Nature and Origin of the Universe" and III.8, "How Did We Get Here?"), it is likely there is just nature, and no god at all. There is no plausible reason why an Almighty would need billions of years and trillions of galaxies to accomplish his ends through long, deterministic causal processes. But that is exactly what we should expect if there is no god, but only nature.

Thus, since the facts strongly fit Metaphysical Naturalism more than any popular kind of theism, we should be Naturalists, and therefore atheists. Certainly, some new evidence could turn up one day to change all this. But until then, the best bet is Naturalism. In various ways, all the remaining arguments expand on this one. But they also stand on their own, on independent evidence, without being part of an entirely naturalistic worldview. You will recognize from II.3.1, "Finding the Good Method" that these represent seven independent lines of observation that converge on the same conclusion: there is no god. This means they converge from the same evidence to another conclusion: Metaphysical Naturalism is true.

2.2 The Religious Landscape is Confused and Mundane

Another argument for atheism is that god-beliefs have not been produced by the actual existence of a god, but by interests and circumstances unrelated to solid evidence and sound argument. And that implies a god does not exist except as a false belief, for two reasons. First, if a compassionate god exists, he would not allow people to damn or kill or deceive themselves in this way, but would nurture and guide them and steer them to the truth. Like any good person, he would correct us when we err, mediate between groups when our disagreements get out of hand, and make sure to supply the reasoning power he gave us with clear and convincing evidence.

I would do this, and surely I cannot be better intentioned and more compassionate than God. On the other hand, if God is not compassionate, he is unworthy of worship anyway. But even then his complete disinterest in setting his subjects straight still remains inexplicable. Meanwhile, most god-beliefs are of a compassionate god, which certainly entails observations contrary to what we actually see, so those beliefs are probably false. Religions are endlessly fragmented in disagreement about the most important things about our salvation and our welfare, things no caring god would let us be without.

Second, if half the world falsely believes in Allah while the other half correctly believes in Christ, how do we know it isn't the other way around? How can we even be sure any of the Big Religions are right? Maybe the Jews are right. Or the Zoroastrians for that matter. Or the Taoists or Deists...or the atheists! We should be able to tell by observing which religion has shown clear divine aid and inspiration—that is, which religion's success is *not* explicable due to normal human cultural diffusion, but *only* due to divine inspiration or intervention, or the unassailable truth of the evidence for the divine. But when we look for this evidence, we are sorely disappointed. *No* religion shows any special heavenly support (see, for example, III.10.4, "The Nature of Spirituality" and IV.1, "Not Much Place for the Paranormal," as well as I.2, "How I Got Here"). To the contrary, all the largest theistic religions exhibit signs of a memetic virus rather than a healthy conquest of truth, as we shall see. And smaller religions certainly can't claim anything special over their vastly more successful competitors. So if the Big Religions are unaided by God, so are all others. This supports atheism, which holds that there is no divine aid to be had. This claim requires a much deeper examination of the facts before moving on.

2.2.1 Religion Didn't Win by Playing Fair

No religion, in the history of its development and success, shows any sort of divine backing. Rather, every single one shows a fallible, human, and often brutal history. As a virus impairs or kills its host, a viral meme impairs or kills your mind, your power of reason. Even those whose minds survive it act as carriers. So, drawing on what I discussed earlier (III.8.4, "Memetic Evolution"), it is easy to see how the ultimate memetic equivalent of a *virus* would have nothing to do with things as harsh and difficult as the truth, but everything to do with silencing competing memes, preying upon our fallible intuitions, our ignorance, and our intellectual laziness, and stirring purely emotional attachment to the viral memes, which, sometimes, play to one's selfish ego—like memes that tell a man he has the authority of Truth behind his every desire, and is the center of the universe, the purpose for which the whole cosmos was made, and that he will get everything he wants later if he plays the sheep now—or, other times, play upon one's fears—like memes that tell a man he will never *really* die, or that there really is no senseless chaos, but every boon and every misfortune is the intended outcome of Other Beings who can be blamed, thanked, or bribed.

    This is often what people get out of the popular religions, painting a far less humble picture of man than Metaphysical Naturalism does. It is ironic that, as if engaging in a classic Freudian "projection defense," theists attack atheists for escaping fears and arrogantly trying to make themselves the center of the universe, when in fact atheists, at least Naturalists, do the exact opposite. In our worldview, we are just another tiny byproduct of nature, special in no sense to anyone but among ourselves, subject to a plethora of random accidents and forces, and there is no perfect or supreme being at all, least of all us. In contrast, it is *theism* that often encourages arrogance, making man the center of the universe, exaggerating his importance in the grand scheme of things, asserting his immortality and divine backing, making him more (and more important) than he really is, granting him the dangerous feeling that he has the Authority of the Almighty behind him, and that everything that happens is somehow deserved. In many ways like this, religion dazzles you with sweet talk, making it easy to forget (or you unwilling to admit) that it isn't true.

If major religions have the attributes not of a memetically-propagated truth but of a memetic virus, this is itself a good reason to dismiss them. And once we've done that, we are back to zero, with no reason to believe in any alternatives but what the evidence alone presents. And as I have shown in the rest of the book so far, that is Metaphysical Naturalism. Now, the first and strongest clue that religions are viruses rather than healthy truths is in their selfishness, a selfishness that makes no reference to humble or objective standards or the pursuit of truth, but solely to their own preservation as beliefs, and the destruction or abandonment of all contrary beliefs.

Compare:

> If your very own brother, or your son or daughter, or the wife you love, or your closest friend secretly entices you, saying, "Let us go and worship other gods." ... Show them no pity. Do not spare them or protect them. You must surely put them to death. ... Stone him to death who tries to turn you away from the Lord your God.
> Jehova (Deuteronomy 13:8-10)

> The fool says in his heart, 'There is no God'. These people are corrupt, they have done vile deeds. None of them do good.
> King David (Psalms 14:1)

> If anyone comes to me and does not hate his father and mother, his wife and children, his brothers and sisters—yes, even his own life—he cannot be my disciple. ... Whoever does not believe stands condemned already because he has not believed in the

name of God's one and only Son. ... He who has believed and has been baptized will be saved, but he who has disbelieved shall be condemned ... [And] this is how it will be in the end: the angels will come and separate the wicked from the righteous and throw them into the fiery furnace, where there will be weeping and gnashing of teeth.
Jesus Christ the Lord
(Luke 14:26, John 3:18, Mark 16:16, Matthew 13:49-50)

He will punish those who do not know God and do not obey the gospel of our Lord Jesus. They will be punished with everlasting destruction.
Saint Paul (2 Thessalonians 1:8-9)

Those who misbelieve and die while still in misbelief, on them is the curse of God, and of the angels, and of mankind altogether, and they shall dwell in Hell forever. Indeed, the torment shall not be lightened for them, nor shall they find mercy. ... for them are cut out garments of fire, boiling water shall be poured over their heads, which shall melt what is in their bellies and their skins as well, and for them are whips of iron, and whenever they want to leave, from grief, they shall be turned back, and taste the chastisement of burning.
The Holy Koran (2.85, 22.19-22)

O the wise, and the learned, and the rich, that are puffed up in the pride of their hearts, and all those who preach false doctrines, and all those who commit whoredoms, and pervert the right way of the Lord, wo, wo, wo be unto them, saith the Lord God Almighty, for they shall be thrust down to hell!
The Book of Mormon (2nd Nephi 28:15)

Fix reason firmly in her seat, and call to her tribunal every fact, every opinion. Question with boldness even the existence of a God; because if there be one, he must more approve of the homage of reason than that of blindfolded fear.
Thomas Jefferson
(Letter to Peter Carr, 10 August 1787)

What do we see right away? There is a radical difference between Jefferson's religion and all these others. His was an Enlightenment Deism that rejected the divinity of Jesus, the existence of miracles, and the authority of the Bible, and accepted in their place secular ideals of science and humanism. His religion is best defended and explained by another Founding Father of the United States, Thomas Paine, in his three-volume work *The Age of Reason*.

We see no selfish meme in Jefferson's religion, but the exact opposite: a meme encouraging doubt and inquiry, defending it as morally good, even approved by God. His religion is nonviral. So are many other minority religious views today, like contemporary Deism, or Philosophical Taoism, or the kinder, gentler Christian Methodism of my childhood (cf. I.2, "How I Got Here"), or Christian Universalism, which is undogmatic, humble in its assertions, tolerant of alternative views, and always encouraging doubt, independent thought, and investigation into other religions.

In contrast, all the other religions whose holy books are cited above outright condemn and slander not only atheists and doubters, but anyone of a different religious creed. That is a sign of a virus, not anything true or good. For threats are the hallmark of a wicked creed. These religions must be wicked indeed, for they all make terrifying threats against doubters and unbelievers. The reasons why many of the major religions persist today reinforce this conclusion.

2.2.2 To the Victor Goes the Spoil

It is obvious that religion is attractive because it provides easy and comfortable answers to the Big Questions of Life. It is a handy explanation for every mystery, an easy way to bear every evil, deny our mortality, and condemn anyone who doesn't conform to our ideals. And it doesn't require much independent thought. This is, after all, what most people use their religion for, no matter how much religious reformers try to get them to stop.

But the most fundamental reason for the persistence of religious beliefs is the very simple fact that we believe as we are taught. Tradition and popular opinion can take decades to change, and usually do so only when they jeopardize success or survival, or when people slowly come to their senses and their sensibilities rub off on their children. As Thomas Paine once said, "time makes more converts than reason." It was only six human lifespans ago that we were still hanging witches, and barely four lifespans ago when natives of America were widely condemned as worshipers of Satan and classified as animals rather than human beings—in spite of their often superior moral character and their rather suspicious resemblance to ordinary men and women. It was just two *generations* ago when people with black skin were widely regarded as biologically inferior to people with white skin, and this convention was so thoroughly and widely held that even some blacks believed it was true. More relevant to the subject of this book, it was barely twenty years ago when people thought communism was the evil result of godlessness, and yet the Soviet Union had barely dissolved before religious tensions broke out there into waves of ethnic cleansing. Clearly God had never left many of these peoples' minds, nor was his presence on their minds any guarantee of good behavior.

All of these ideas and more were perpetuated because of the simple rule of social learning. We are literally surrounded every day by the cultural references and effects of Christianity in America. Many of us grow up being told that God exists and that Jesus was his son and loves us, that there is a heaven that we go to when we die or, just as conveniently, a hell for those who are bad or who don't believe in the rest of it. We even grow up believing that the Bible is a holy and good book without ever having read it. Christians take all of this for granted, and use these widespread beliefs and notions as proof they have it right, yet they conveniently put out of their minds the other half of the world's population that lives with very different religious convictions. Muslims grow up just as honestly believing that the Koran is the inerrant message of God, and that everyone who dies defending the faith is guaranteed a place in paradise, while those who do not recognize Allah as the true God are infidels destined to hellish wrath. They use the same evidence to prove their case, and the things they believe, the superstitions they hold, are just as real and profound to them as Christ's love is to those growing up in Christian cultures.

The science of psychology has taught us how much of what we take for granted is actually learned, taught to us by other people or the environment from the day we open our eyes. In fact, almost everything we take for granted is entirely learned, even things as rudimentary as how to eat or have sex or care for our young. Genetics may influence how we perceive, how we feel, or how we physically develop or react to our environment, but most of what we do and think (in many cases, even how we think) is taught to us, either through trial and error or by our parents or teachers or society in general, through test, example, and education.

Even our concepts of God and an immortal soul are taught to us. This is driven home by the fact that many cultures have existed that did not believe in personal immortality, and the vast majority of cultures throughout history have held a belief in many gods, not merely one—and the gods' appearance, character, temperament, and abilities are as endless in diversity as the cultures themselves, reflecting the ideals of each culture more than any real cosmic truth.

Since the designs and moral expectations of every god and religion differ, and they cannot be reconciled, it would be quite impossible to attempt to find any common thread among all religions that is not already common to human beings in general. Love, for instance, is not a religious virtue, but a human one, and no religion is needed to define or explain it. Likewise, justice is a gift of necessity in any society, and religion is merely its sometimes-cheerleader, not its origin. Hammurabi invented The Law a long time before Moses copied it.

Christianity is no different. As society has changed from a feudal and agricultural existence to a democratic and industrial one, our view of God—his character, temperament, and physical description—has changed accordingly. Once upon a time the Christian God was a god of wrath who actually approved of killing and torturing people who didn't believe in him, a God who condemned women to inferiority through the folly of Eve, a God who chose our kings for us and was quite unmistakably male (for only Adam was created in His image).

Now, popular opinion holds God to be indescribable, conveniently androgynous, and even a staunch advocate of freedom and democracy—an attitude suspiciously absent from God's plan until the philosophers of the Enlightenment, not religious leaders, brought it up less than four centuries ago. Indeed, it took nearly two thousand years since Saint Paul wrote "do not permit a woman to teach or have authority over a man" (1 Timothy 2:12) before women achieved the right merely to *vote* in the United States. For centuries, the right to participate in government was denied to women because it was truly believed that God did not approve of such a thing. Did God change his mind? Or were all the Apostles and Priests wrong about what God approved of? If they were wrong about what God thought then, they are just as likely wrong about anything else now.

Likewise, reincarnation is as much a certainty to a Hindu as the salvation of Christ is to a Christian. A child growing up in a Hindu culture truly believes in the love of Krishna or the force of spiritual destiny. To them, the power of karma and the fate it engenders is simply accepted as a fact. Likewise, a child growing up in a Buddhist culture truly believes that seeking refuge in the Compassionate One's teaching will save them from unfavorable reincarnations.

All this shows that the religion you believe to be true is rarely decided by you, but by where you are born. Not many Christians are to be found in Iran, nor are there many Muslims in Mexico. Most Buddhists are found in the Far East, and Hindus are mostly in India. Religion throughout the entire world is largely tied to culture, and the culture you are born into is ultimately determined by chance. Christianity is as much a cultural circumstance to an American as Buddhism is to a Tibetan.

But that is not the sum of it. Christianity, as with all religions, started in one tiny place, when the rest of the earth was populated by a wildly marvelous diversity of religious beliefs—and yet, curiously enough, the concept of warfare over religious differences was virtually nonexistent. Most people in ancient times believed it was proper to respect the gods of other peoples. This changed on a global scale when Christianity was spread, quite literally, by the sword. Those who attempted to assert their religious differences were harassed, tortured, robbed of their land and belongings, even killed. Before it achieved political power, Christianity was a small sect, a heresy against the Jewish faith, that had to accept equality among all the other religions of the Roman Empire. Yet it was the first religion to openly attack the religions of other people as false (the Jews, at least, were a little more tactful). Needless to say, Christianity only truly flourished when it had the ability to eliminate the competition—when it had the full support of Rome's Emperors after 313 A.D., and when, in 395 A.D., every religion other than Christianity was actually *outlawed*. Through force and decree Christianity was immersed in the cultural surroundings of lands near and far, and in an environment where it was widely accepted, if not the only thing accepted, it spread and planted itself among subjugated peoples. As kids grew up taking Christian ideas for granted, they often did not realize that only a few generations ago those ideas were entirely alien.

Colonization of the world, more often than not by robbery and warfare, spread Christianity into the Americas and other corners of the earth, just as Islam was spread throughout Asia and Africa. It is not a coincidence that the two most widespread religions in the world today are the most warlike and intolerant religions in history. Before the rise of Christianity, religious tolerance, including a large degree of religious freedom, was not only custom but in many ways law under the Roman and Persian empires. They conquered for greed and power, rarely for any declared religious reasons, and actually sought to integrate foreign religions into their civilization, rather than seeking to destroy them. People were generally not killed because they practiced a different religion. Indeed, the Christians were persecuted for denying that the popular gods existed—not for following a different religion. In other words, Christians were persecuted for being *intolerant*.

Such absolute religious intolerance is an idea that found its earliest expression in the Old Testament, where the Hebrew tribe depicts itself waging a campaign of genocide on the Palestinian peoples to steal their land. They justified this heinous behavior on the grounds that people not chosen by their god were wicked and therefore did not deserve to live or keep their land. In effect, the wholesale slaughter of the Palestinian peoples, eradicating their race with the Jew's own Final Solution, was the direct result of a policy of religious superiority and divine right. Joshua 6-11 tells the sad tale, and one need only read it and consider the point of view of the Palestinians who were simply defending their wives and children and the homes they had built and the fields they had labored for. The actions of the Hebrews can easily be compared with the American genocide of its native peoples—or even, ironically, the Nazi Holocaust.

With the radical advent of Christianity, this self-righteous intolerance was borrowed from the Jews, and a new twist was added. The conversion of infidels by any means possible became the newfound calling card of religious fervor, and this new experiment in human culture spread like wildfire. By its very nature, how could it not have? Islam followed suit, conquering half the world in brutal warfare and, much like its Christian counterpart, it developed a new and convenient survival characteristic: the destruction of all images and practices attributed to other religions. Muslims destroyed millions of statues and paintings in India and Africa, and forced conversion under pain of death (or by more subtle tricks: like taxing only non-Muslims), while the Catholic Church busily burned books along with pagans, shattering statues and defacing or destroying pagan art—or converting it to Christian use. Laws against pagan practices and heretics were in full force throughout Europe by the sixth century, and as long as those laws were in place it was impossible for anyone to refuse the tenets of Christianity and expect to keep their property or their life. Similar persecution and harassment continues in Islamic countries even to this day, officially and unofficially.

Many cultures were won merely by converting their kings or chieftains, who, in return, required their subjects to adopt the new faith of their ruler. Still others mistook the numerical and technological superiority of their conquerors as evidence that they had the better god. Thus, the spread of Christianity was not due to its truth, God's grace, or its unique attractiveness to foreign people. Simply imagine two competing religious points of view, one holding the idea that other religions are to be respected and that war is justified only in defense, the other holding that war is justified in converting infidels to the only true faith, and that this faith must by its very calling be spread across the world. Which religion will survive and grow, and which will be stamped out and forgotten? The answer is self-evident—and yet it has nothing to do with which religion is actually *true*.

So that is how Christianity got in our backyard. It is a logical result of cultural evolution. The ultimate memetic virus, it developed unique characteristics ideal for its survival and domination. It had by its very nature a calling to spread itself throughout the world, and it rejected, even to the point of physical obliteration, alternative religious views. This demeanor still resides deep within Christian thought today. Christians view their faith and ideology as "right" and all other religions as just superstitions, whose followers are misguided—or misled by Satan, as many Christians still seriously believe—even though they know virtually nothing about those other religions, and hardly much more about their *own*. Ultimately, most Christians generally accept the "fact" that non-Christians will not be saved, since by definition "salvation" belongs only to those who have faith in Christ (this is the very heart of the New Testament teaching, as stated in Matthew 10:28-40, 12:30-32, Mark 16:16, John 14:6, etc.).

The attitude of Muslims is identical. Those who reject Mohammed and fail to follow the Koran are doomed. They do not permit any discussion of the matter, for it is a certainty. Indeed, atheism is still punishable by death in many Muslim countries. Even where it is technically legal it is still a death sentence, from gun-toting Muslim fanatics who have no qualms about taking justice into their own hands. Most people think this has always been an attitude common to religion in general, but that is not true. Most religions in history had plenty of room to accept other views as valid. Even Taoism and Buddhism, to their eternal credit, have rarely opposed or attacked competing religions, and have co-existed happily and constructively with each other, and with countless folk beliefs and festivals, for over two thousand years. Sadly, attacked and infected by Islamic-Christian doctrines of brutal righteousness, violent fundamentalist movements are now rising among Indian Hindus and Sri Lankan Buddhists, but these remain unusual exceptions in the history of religion.

In contrast, the intolerance among the two fiercest Western religions, Christianity and Islam, very often led to the permanent elimination of competing faiths. Because of their intrinsic exclusiveness and intolerance, countless other religious ideas—ideas that might have been nobler or even truer—never had a chance to grow or survive. Thus, today it seems that other religions are small or rare or obsolete, when in reality this is only because they have never been tolerated or given a chance. However, though this intolerance has paved the way, it is the missionary aspect of these religions that is the most significant reason for their pervasiveness in the world today. For the three most widely practiced religions—Christianity, Islam, and Buddhism—root themselves in the idea that the "faith" must be spread to all people. It is probable that without this missionary calling

Americans would almost certainly be polytheistic, having accepted and adopted many of the religious and cultural ideas of the Native American peoples, perhaps having imported many of the old Greco-Roman deities as well. Instead, we wiped them all out in our 'god-sanctioned' zeal. This was the fault of a new breed of intolerant, missionary religion called Christianity.

So the new idea that only one religion is true and all others are evil or false, and the idea that this true faith must be carried across the globe in order to save everyone from doom, are the very attributes that guaranteed the survival of Christianity and Islam, and the elimination of nearly all other religions in the world. Both these characteristics are much more plausible explanations for the widespread acceptance of Christianity and Islam than the claim that "they are widespread because they are true" or "this is evidence of God's design." How could both Christianity *and* Islam credit their spread to their unique truth? Clearly, at least one of them has to be false, proving that such vast success does not need truth behind it. And they can't both be the result of God's design, unless God is confused.

So the claims of theologians don't hold water in light of plain observation. Even the equally widespread presence of the "godless" view of Buddhism puts such claims into serious question. The fact is that we believe in God and an immortal soul because of the missionary zeal and religious intolerance intrinsic to the Christian religion. We owe our superstitious ideas to sword and gun and flame. In this corner of the globe, the Christian church was the victor, and our minds were the spoil.

> This is all standard, established history. Any mainstream college textbook will reveal the details, and one can follow its bibliography for more. I recommend John McKay, Bennett Hill, and John Buckler, *A History of Western Society*, 6th ed. (1999), in two volumes (*I: From Antiquity to the Enlightenment* and *II: From Absolutism to the Present*).
>
> But for more specific detail on Christianity's abuse of power and use of violence, in contrast with greater pagan tolerance, see: Helen Ellerbe, *The Dark Side of Christian History* (1995); Jim Hill and Rand Cheadle, *The Bible Tells Me So: Uses and Abuses of Holy Scripture* (1996); Ramsay MacMullen, *Christianity and Paganism in the Fourth to Eighth Centuries* (1997), *Christianizing the Roman Empire: A.D. 100-400* (1986), and *Paganism in the Roman Empire* (1981); and combine the very-readable C. Warren Hollister, *Medieval Europe: A Short History*, 7th ed. (1994) with some actual primary sources, like: *The Carolingian Chronicles: Royal Frankish Annals and Nithard's Histories*, tr. by Bernhard Scholz (1972), and Bede, *Ecclesiastical History of the English People* (Penguin

Classics ed., 1955). Various histories of slavery and of tribal peoples in America add even more to the horrible story.

On world religions, a good place to start is the *Oxford Dictionary of World Religions* (1997). Just follow its bibliographies for further reading. Also, see H. L. Mencken, *Treatise on the Gods* (1946). Finally, an elegant example of how mundane social forces, not divine aid or even truth, decide the success of religions even today see Roger Finke, *The Churching of America, 1776-1990: Winners and Losers in Our Religious Economy* (1993), and Stephen Prothero, *American Jesus: How the Son of God Became a National Icon* (2003). And applying sociology to Christianity's early days: Rodney Stark, *The Rise of Christianity* (1996).

2.2.3 Dissent is Checked at the Door

We can see why a particular religion and its superstitious beliefs can surround us and permeate our cultural heritage, but why haven't we outgrown it? Of course, we are still a very immature, superstitious species, as the popularity of psychic hotlines, astrology, and alien conspiracy theories proves. But certainly, we have made advances in common sense. Most people generally regard blacks as equal to everyone else in rights and abilities, Native Americans are widely accepted as moral human beings, and the more sensible folk know that communism failed because it was a bad economic theory, not because it abandoned religion (which it did not—the Russian Orthodox Church remained alive and well throughout the communist legacy, as Catholicism has in Cuba, and other religions in every other communist state).

So why are convictions still so strong when it comes to the certainty of such amazing claims as God and heaven or the miracles of Jesus? Though theologians still answer this question with the claim that it is evidence of their truth and God's grace in the world, the real answer is that many are essentially afraid *not* to believe, a fear whose encouragement is built into each religion itself—as we saw, for example, in the quotations from scripture above. Besides threats of hell, we face an almost insurmountable degree of peer pressure when it comes to our religious convictions. Some still say to themselves that if everyone else believes it, it must be true. After all, how could so many people be that crazy or misguided? Some might jokingly call this the "lemming" effect, but it is no small force in society. Ask yourself how often you have held a serious discussion about the reasons and evidence supporting your beliefs. If you think about it carefully you might realize that the support for some of your beliefs is so vague that the same reasons and experiences could just as easily support any religion on earth, including voodoo and alien mind control.

In contrast, as atheists know better than anyone else on the planet, if you say you don't believe you often become a social outcast. You might already know this, having seen people in movies or read about them in books, characters who display doubt, who don't believe in God or immortality or claims of the paranormal, are often portrayed in a negative light. Magic, ghosts, psychic powers, or answered prayers always seem to catch them off guard, proving they are wrong. Or they are presented as feeling empty or amoral, despairing at the cold realities they attempt to grasp. Of course, this is fiction far apart from reality. Atheists are generally quite happy and asserted in their principles, while magic, ghosts, psychic powers, or answered prayers are continually proven fictitious. Nevertheless, preachers and other devout individuals continually bombard us with the message that our misfortune, confusion, or despair is to be blamed on our lack of faith—rather than our lack of effort and attention to life, which is a much more sensible explanation. This is merely evidence of a fact present in all cultures: dissent is feared, and must often be checked at the door.

In contrast, if we accept belief, especially if we go to church, we are surrounded by a group of like-minded people, rather than being labeled an outcast or looked at strangely or harassed with attempts to return us to the faith. We gain an invaluable sense of community, a sense of joy and belonging. Thus, the human desire to be accepted, to be a part of something, contributes to religion. And for that reason churches have long been masters of advertising: just like selling soup or diet pills, if you associate your product with Mom and Apple Pie and Everything Good, and claim it will solve all your problems and really pay off in the end for just $19.99, a lot of people will buy it.

Finally, there is one thing people naturally fear. It is so frightening to us that when we face it we often attempt to ignore it, distracting ourselves endlessly in every possible way. That feared fate is the loss of a basis for our values. Since we have rarely been given any chance or encouragement to explore alternative worldviews, if we reject our superstitious beliefs, hence our religion, we lose the basis of the only philosophy we know. We lose our cultural identity and our moral roots. This is no trivial matter. This is, in my opinion, the single greatest reason that religion is perpetuated in every culture, and why it endures in any individual—it is even more important than habit, peer pressure, or fear of death.

Sometimes if we lose our faith, circumstances allow a swift replacement. We seek out another religion, or another finds us. More often than not, however, we simply end up confused and neglect to bother with the issue, living by some undefined standard of conduct that we borrow from our surroundings, becoming superficial or apathetic by default. This is then blamed on a "lack of faith" by the religious, and unless we think about it for a minute we might make the mistake of thinking they're right. But the truth is, it is not a "lack of faith" that drives people to act selfishly or to otherwise become deficient in compassion and integrity, but the lack of a philosophy—or the lack of a sound and constructive one. Even the religiously devout are often wanting in character and moral integrity, and there is no lack of faith among them. They simply lack a philosophy firmly based in reality, and its flaws become expressed in their behavior and attitudes.

It is absolutely crucial to have a sound and successful philosophy of life, and yet almost no one studies philosophy. It is rarely taught in public schools—in fact, it is barely even taught in universities unless a student specializes in the field. The books available in stores or libraries are often dry, and almost as often polluted by unsound thinking, or obscured behind high-brow jargon and symbols, or divorced from any relevance to everyday life. Rarely are complete and understandable philosophies of life available by the book, clearly written, to be studied, compared, selected. Of course, so few people even bother to look. Religion does not benefit by teaching its pupils to investigate other philosophies, nor does religion prosper by giving people the tools to think, but only by giving them the tools to believe. And that requires suppressing freethought. After all, if everyone found and embraced reasons to be good and enjoy life without the religious superstitions claimed to be necessary, religion would become obsolete—a fate, I imagine, that believers and their churches cannot emotionally or economically afford.

2.2.4 Religion as Medicine

Religion is also seen as a medicine, curing grief and fear of death, for example—and in principle (though rarely in practice) healing divisions and hatreds in society and thus bringing peace. But there are secular ways to all these same ends, which lack the singular fault of Big Religion. This fault is the selfish and arrogant condemnation of doubt and difference inherent in Christianity and Islam, a feature that has in practice fragmented rather than united society, fostering war and violence of every kind, healing grief and fear only through false comforts whose side-effects are often a detriment to society in the end. Religions are full of bad ideas, and yet they encourage a fanaticism and a bigotry in favor of such ideas that is altogether unhealthy.

Now, whether the Koran, or the Bible, or any other religious text contains noble ideas is not the point here. Noble ideas stand on their own.

They do not need "holy" texts to support them, they do not need miracles, or religious systems or supernatural entities, in order to possess their nobility. Wisdom is wisdom, from wherever it comes, and for all practical utility we should seek it where it is most carefully, correctly and usefully described and explained. I rarely find this to be the case in any religious text. Philosophy has been far more successful at this, with a better grasp of the concept of explanation, definition, and logical analysis and argument. And all Great Literature captures the essence of things in beautiful prose or verse. Thus, there is nothing that religious texts have to offer that is not better said in philosophical texts—especially those written well for a popular readership—and in profound and moving fiction.

If, then, religious texts like the Koran are all just ancient human works of mortal and fallible piety, we should not be obsessed with them, or revere them as anything other than they are—the cultural dogmas of ancient peoples. Instead, we should seek wisdom in art, reason, logical argument, and scientific investigation. Thus, I reject the Koran, and all holy scriptures like it, because it is a human, fallible, primitive work of a bygone age, ineloquent and inappropriate for our times. It may have other salvageable virtues, in some of its moral teachings, or its literary quality or historical interest, but these do not justify making a religion out of it.

Ultimately, the failure of the "religion as solution" argument is analogous to the failure of Traditional Medicine ('TM'), wherein natural roots and plant products are held to be "better" for you than real medications. The fact is that the real meds often contain exactly the same efficacious chemicals as their natural sources, while lacking the other chemicals present in those plants that do not effect the cure but instead contribute to unneeded side effects. And the real meds are measured in precise doses, whereas natural sources have variable and thus unknown doses and are as a result more dangerous, or more inconsistent in their effect. The mistake at work here is the point I want to bring out: TM proponents fail to notice that what we *ought* to do is look for what works, which is held in common among all the noted effective agents, focusing more precisely and effectively on *that*, and then eliminate the superfluous agents that actually have no real effect on the problem but often have unneeded and even dangerous or uncomfortable side effects.

We should indeed treat religion just like medicine: a thousand chemicals go into a root, but only one cures the disease, and it needs proper use and measure to succeed. Simply producing that one chemical in a factory may be a thousand times cheaper and more accurate than harvesting the root. A thousand ideas go into a religion, but only one (or at best a few) actually helps people, and it needs proper use and measure to succeed. Simply focusing on that one thing may save you a lot of time and money—time that could be better spent on acquiring useful knowledge and enjoying the one life you actually have, and money that could be better spent on real problems or achievements.

When we see the same benefits claimed by a Catholic being enjoyed by Taoists and Secular Humanists, like the scientist who has identified that the real medicine is chemical *A* and not root *B*, we see at once that religion is a red herring, the superfluous contents of a root that only by chance includes a useful agent. So we can dispense with the rest of the root, and stick with the one thing that really works, and by focusing on it and analyzing it, we can use it better, more effectively, more profitably, and more safely.

The advocates of religion are like the TM proponent who says only Bonobo root cures headache, and that Anga root will actually make it worse, when in fact both roots contain an aspirin-like compound, and both demonstrably alleviate headache, but neither as well, or as consistently, or as inexpensively, or as side-effect free as the actual chemical by itself, well-measured and competently used. This is the difference between thinking scientifically and thinking traditionally: the scientific mind analyses, tests, examines, and compares, to get rid of the obfuscation and land on the real fundamental causes of this and that. The traditionalist just repeats dogma passed on to him by others. As it happens, when we extract all the effective elements from all the world's religions, and hone them to completion, what we end up with is Secular Humanism (see esp. Parts V and VII, "Natural Morality" and "Natural Politics").

> For more detailed argument that all belief in gods is caused by various cognitive illusions with cultural and behavioral explanations, see: Pascal Boyer, *Religion Explained: The Evolutionary Origins of Religious Thought* (2002); Scott Atran, *In Gods We Trust: The Evolutionary Landscape of Religion* (2002); Robert Buckman, *Can We Be Good Without God? Biology, Behavior, and the Need to Believe* (2002); Michael Shermer, *How We Believe: Science, Skepticism, and the Search for God*, 2nd ed. (2003); Stewart Guthrie, *Faces in the Clouds* (1993).

2.3 The Universe is a Moron

So confusion and mundanity is the second reason to be an atheist. Now for the third. The nature of the world is manifestly dispassionate and blind, exhibiting no value-laden behavior or message of any kind. It is like an autistic idiot savant, a marvelous machine wholly incomprehending of itself or others. This is exactly what we should expect if it was *not* created and governed by a benevolent deity, while it is hardly explicable on the theory that there is such a being. Since there is no observable divine hand in nature as a causal process, it is reasonable to conclude that there is no divine hand. After all, that there are no blue monkeys flying out my butt is sufficient reason to believe there are no such creatures, and so it is with anything else.

But the point is even stronger. All the causes whose existence we have confirmed are unintelligent, immutable forces and objects. Never once have we confirmed the existence of any other kind of cause. And that is most strange if there is a god, but not strange at all if there isn't. This is the basis for what I call a *teleological argument for atheism*. "Teleology" is the study of goals, of designs with intended ends. A teleological process is something goal-oriented, aiming at a final purpose, ever-correcting itself toward it. The Teleological Argument for God is that the universe exhibits teleology, teleology entails a mind (since only minds have desires or intentions), therefore a mind must lie behind the universe, and that would be God. But as we've seen throughout this book already, this argument fails twice: teleology *doesn't* entail a mind, and the universe *doesn't* exhibit any sort of teleology distinctive of a mind (see III.9, "The Nature of Reason," III.3, "The Nature and Origin of the Universe," and III.8, "How Did We Get Here?").

One particular way to put this is that there is no *high teleology* anywhere in the organization of the universe. By that I mean the sort of intention or goal one can only expect from a conscious being like us, as opposed to the sort of goals exhibited by, say, a flat worm or a computer game or an ant colony (or a machine like the solar system: see section III.3.5.2, "The Origin of Order"). The most teleological force we observe in nature, apart from the goals and intentions of animals—whose cause we already understand to be evolution by natural selection—is that of natural selection itself, which shows no more intelligence than an ant hill or, more to the point, a desktop computer, which we know even today can model the entire process, and from a simple set of rules can spontaneously produce organisms just as surprising and complex as humans, especially given the same ridiculous lengths of time nature has clearly needed to get so far.

Any dumb process can exhibit a blind teleology, winnowing behaviors or outcomes away, leaving only those few that satisfy the particular criteria of survival.

In contrast, even a coldhearted superintelligence would not be so stupid as to take billions of years of meandering and disastrously catastrophic trial and error to figure out how to make a human. It would just make humans. But the evidence does not pan out that way—that is not what happened. Instead, a moronic teleological process did the work, sloppy and slow. That is incredible if God exists. But it makes perfect sense if he doesn't.

On the other hand, a superintelligence would certainly design a universe with very abstract goals built in. For example, if I were to make a universe, and cared how the people in it felt—whether they suffered or were happy—I would make it a law of the universe that the more good a person really was the more invulnerable they would be to harm or illness; and the more evil, the weaker and more ill. Obviously, such a law would not be possible unless the universe "knew" what good and evil was, and cared about the one flourishing rather than the other. And unlike mere survival (which does its own choosing through the accidents of nature), to have something other than mere survival as an end, but a highly abstract good instead, would be inconceivable without a higher mind capable of grasping all these deep abstract principles, and *caring* about them (as we know humans do and the universe does not). So this sort of law would indeed have been a *very* strong proof that the universe was created by an intelligence—and a benevolent one at that!

That's just one example. I can think of dozens of possible teleological laws that could have been built into the very fabric of the universe, which would not be possible unless a mind with highly abstract goals put them there. But we have found no such laws. None at all. That is exceedingly strange for a universe supposedly designed by a caring superintelligence. And if what we would expect on the assumption of a superintelligent creator is not found, there probably is no such creator. And since what we would expect on the assumption of a mindless, uncreated natural universe *is* what we find, and nothing else, the most reasonable conclusion is that such a universe *is* all there is.

This need not be limited to the fabric of the universe itself. Any obvious divine activity would be a very strong proof. In IV.1.1.8, "The Balance of Proof and Proof of the Extraordinary," I noted how according to old Creationism, God is a great miracle worker, who flooded worlds, stopped suns, parted seas and raised the dead—in short, someone who left no one mistaken that he existed. How could any Hebrew reasonably doubt there was a God, and a God on his side no less, after watching events like these?

Even I can see that if the earth flooded entire in just forty days after a voice warned me to build a boat to survive it, the sun stopped moving for a few hours without harm to anything on earth, the sea parted so I could cross it to escape an oppressor, and dry bones filled with flesh and the person whose bones they were came alive again before my eyes, there could not possibly be any sane or rational doubt that God exists, or something like him. If millions of Jews got to see such things all the time, but I don't get to see anything of the sort, God can hardly blame me for being an unbeliever. He would have a good case against unbelievers among the witnesses of those events, but I am not one of them. All I see is a universe that is nothing but a great big moron, stupidly doing its business with no awareness or care for anyone.

The existence of a divine creator driven by a mission to save humankind entails certain things *should* be the case about his Creation. For instance, if I were omnipotent, whenever I got fed up with all the killing I would just snap my fingers and turn all guns into flowers. Such a worldwide miracle would be undeniable proof not only that I existed, but that I am a very compassionate being who is merciful and kind. Likewise, if I wanted people to know which church was teaching the right way to salvation, I would protect all such churches with mysterious energy fields so they would be invulnerable to harm, and its preachers would be able to work great acts of mercy, such as regenerating lost limbs, and its bibles would glow in the dark so they could always be read, and they would be indestructible, immune to any attempt to mark or burn or tear them, or change what they said. Indeed, I would regard it as my moral obligation to do things like this, so my children would not be in the dark about who I was and what I was about, and what was truly good for their happiness (see also V.2.2, "How Naturalism Accounts for Value" and III.3.1, "Plausibility and the God Hypothesis").

But since the universe exhibits no miraculous activities or value-laden design features, much less doing so in any way indicative of the values of a compassionate being, and this is improbable if God exists, but inevitable if he doesn't, atheism is the only reasonable view to take.

2.4 The Idea of God Doesn't Make Any Sense

The fourth reason to be an atheist is that most god-concepts are illogical. To be fair, most arguments from Incoherence, as they are called, are often frivolous. A typical example is the taunt "If God is all-powerful, can he make a rock so big even he can't lift it?" This is supposed to prove that omnipotence is illogical and therefore God (who is supposed to be omnipotent) doesn't exist. There are many arguments like that. But I don't buy them. These are generally not valid, since any definition of god (or his properties) that is illogical can just be revised to be logical. So in effect, Arguments from Incoherence aren't really arguments for atheism, but for the reform of theology. For instance, if we define omnipotence (like many do) as "having all the power that exists" or "being able to do everything that can ever be done" then we avoid silly objections like the impossible rock. Likewise, we can define omniscience as "knowing everything there is to know or that can be known" and omnibenevolence as "loving all things with the most profound compassion possible."

But in a few cases, the theological reforms that would be required to avoid defeat at the hands of an Argument from Incoherence are reforms that fly in the face of all popular beliefs about God. For example, it is obvious that a perfect being, by any definition, could not and would not create an imperfect universe, yet the universe is imperfect, therefore God cannot be perfect. This does not prove there is no God, but it does prove that, given the way the universe plainly is, if any God exists, he is imperfect. We can know this with almost absolute certainty—the evidence is *that* overwhelming, far more overwhelming than any evidence to the contrary.

After all, the design of man, and nature generally, is wasteful and messy, inefficient and full of needless vulnerabilities and imperfections, pitfalls and limitations (see III.8.2, "Evolution by Natural Selection"). There is no logical ground, much less any need, for an omnipotent, omniscient, omnibenevolent being to create us and this universe as we are. Rather, such a being would have made a universe where there was no need of rape and murder and all other forms of natural suffering among animals, where death and killing was not required for any creature's survival, where injury could never be permanent or traumatizing, where there is no disease, where human minds were all very intelligent and reliable, and education available to all, and where there were no natural disasters. Even a heartless deity would at least make his creation less messy and chaotic, unless mess and chaos were the very things he wanted. And no matter what, a perfect god *by definition* could only ever produce perfect worlds—for even the capacity, much less the tendency, to make what is imperfect would itself be an imperfection in the creator. So whether he's kind or cruel, we can be quite sure that if there is a God, he isn't perfect.

This also entails something more, refuting all popular conceptions of God: it entails that any God who promises to take us to heaven cannot exist. For if such a God existed, we would already be in heaven. For there would never be any need for this world—nor could anyone say our present world was the best God could make, unless heaven is no better (and if it wasn't, then heaven would be pointless). We cannot say the present world is a test, for example, since God already knows everything, so he has no need of tests. And he made us, so we can't be any better than we already are anyway. And if someone wanted to lamely argue that this world exists so we can choose between good or evil even to God's *surprise*, we would *still* have that freedom to turn to evil in heaven, so God would gain nothing by putting us here that he would not automatically and unavoidably have by just putting us in heaven in the first place. So putting us down here serves no purpose at all, even on *that* ridiculous argument. So if God exists, there cannot be a heaven—or if heaven exists, there is no God who wants us there—unless God is enfeebled, like the being I invented at the beginning of the chapter, or God didn't create the universe, or we accept some other astonishing rejection of popular notions.

This is just one type of incoherence argument that is *not* frivolous. It leaves us with an inexplicable god-concept, proving that there just isn't any way God can make sense to us, unless we adopt an idea of God wholly alien to anything anyone has ever found believable or comforting. There are other incoherence arguments just as damning, mainly refuting attempts to explain away other arguments for atheism, so I will return to those incoherent notions later, in argument number seven.

2.5 Too Much Needless Cruelty and Misery

The same points made already also combine to make their own argument against the existence of a caring God, a fifth reason to be an atheist. As I've said already, the inherent role of violence, injustice and cruelty in the operation of nature, bringing about the pointless suffering of animals as well as humans, is inexplicable as the creation of a kind being, but entirely to be expected as the outcome of a blind process of natural selection in a mindless universe.

Even from a design perspective, there is no moral justification for disease or predation, for example. In fact, there is no way to deny that it is a truly *bad* design to create a world with limited resources—not enough food or water or space to satisfy every creature's needs, much less its happiness, forcing everything to viciously compete, suffer, and have millions of babies that must die so only a few may live—that is, *if* I wanted to minimize suffering and maximize love and happiness. So, if this universe was made by a god, he must either be really bad at designing universes, or sorely limited in ability, or (even worse) *not care* about minimizing suffering and maximizing love and happiness. None of these is acceptable to anyone holding to a popular conception of God. And if we retreat into an unpopular conception, we are left with no reason to believe such a being exists in the first place.

The list of needless horrors is endless. Consider the abundance of poisonous and diseased food and water sources, as well as infertile or inhospitable land. No considerate civil engineer would include such things in any city he was building. Indeed, if he did so, he would have to be a villain with a very perverted sense of humor. And if he did so by some accident or limitation, he would hardly be worthy of the title "God." Just imagine God telling you, as you walk into heaven asking what the hell was up with all that crap, "Oops! I didn't think of that! Sorry. Couldn't be helped. Did my best! No. No way to fix it now. My budget ran out!" What sort of God is that?

So incompetent or perverse would this engineer have to be that the very acts of compassion he supposedly encourages in us—like selflessly defeating disease and hunger in Third World countries by eating the expense of billions of dollars of free shipments of food and vaccines every year—end up causing profound misery far worse than we were trying to alleviate in the first place. For such an "act of mercy" causes populations to skyrocket out of control, now that disease and starvation are no longer killing enough people off (two population control mechanisms already more worthy of a Mengele than a God), leading to massive poverty, overcrowding, misery, and an ever-spiraling increase in demand for more food and vaccines, not to mention the inevitable outcome of inspiring desperate people to turn to radical, suicidally violent movements like Islamic Jihad, or the support of dictators and thugs.

It is easy to see how this perversion of the way things are *supposed* to be could not possibly happen in a world well-designed and well-managed. For it would be no trouble to simply permit, by divine act of will, only enough conceptions to occur every year as can be sustained. Indeed, since we know that people raised by good, loving parents are always better off, and more morally and mentally stable, God could simply allow only good, loving parents to conceive children. There is certainly no logical need for bad, heartless parents to do so. And this, again, could have been a law of nature, not even requiring God's attention.

We can frame this argument in three distinct ways that all converge on the same conclusion:

First, if any one of us had the power, we would immediately alleviate all needless suffering. For this is exactly what we already do: with drugs and surgery and therapy. No one says "Stop getting better at alleviating suffering! Don't open too many hospitals!" Of course not. Every decent, compassionate person asks, even labors, for exactly the opposite: more hospitals, affordable to more people, with more drugs and techniques, better drugs and techniques. We can't throw enough money at it or complete our research fast enough to satisfy the loving hearts among us humans. So why isn't God joining in? If we are this compassionate, if we can see it is obviously our *moral obligation* to end all needless suffering through the perfection of medical means, it plainly follows that if a god exists, he is not compassionate at all. For he does *nothing*, despite being able to do so much more, while we do *everything*, under the withering strain of limited time, resources, knowledge, and ability. Thus, either God does not exist, or he is a heartless demon.

Second, if any one of us had the power, we would immediately enforce true justice everywhere. For this is exactly what we already do: with the pursuit of forensic techniques and the use of identity cards and police and juries and judges and prisons and on and on. No one says "Stop getting better at catching criminals! Stop improving your ability to reform them! Stop making so many correct judgments upon people!" Of course not. Every decent, compassionate person asks, even labors, for exactly the opposite: less error in the justice system, more ways to catch criminals and thwart crime, more success in turning criminals around. Again, we can't throw enough money at it or complete our trials and investigations fast enough to satisfy the loving hearts among us humans. So why isn't God joining in? If we are this compassionate, if we can see it is obviously our *moral obligation* to enforce a more perfect justice by getting better at solving crimes, and judging and reforming criminals, it plainly follows that if a god exists, he is not compassionate at all. For he does *nothing*, despite being able to do so much more, while we do *everything*, under the withering strain of limited time, resources, knowledge, and ability. Thus, either God does not exist, or he is a heartless demon.

Third, if any one of us had the power, we would immediately intervene to prevent all natural disasters—maybe even inadvertent manmade ones, too. For this is exactly what we already do: with the development and improvement of 911 emergency response, suicide hotlines, building and workplace safety codes, the Red Cross, rescue teams and vehicles and equipment. No one says "Stop saving people in distress! Stop preventing floods! Stop making buildings impervious to earthquakes!" Of course not. Every decent, compassionate person asks, even labors, for exactly the opposite: better, stronger, safer buildings and civil resources, more firemen and ambulances, more rescue boats and helicopters, more, and more skilled, 911 dispatchers. As with medicine and justice, we can't throw enough money at it or be in enough places or complete our research and develop and enforce new standards and materials fast enough to satisfy the loving hearts among us humans. So why isn't God joining in? If we are this compassionate, if we can see it is obviously our *moral obligation* to prevent more tragedy by building safer and getting trained people with the right equipment and support in the right places at the right times, it plainly follows that if a god exists, he is not compassionate at all. For he does *nothing*, despite being able to do so much more, while we do *everything*, under the withering strain of limited time, resources, knowledge, and ability. Thus, either God does not exist, or he is a heartless demon.

It is not even the point that God doesn't do *enough*. It is that he does *nothing whatsoever*. Even if God did too little, if he at least did *something*, we would be justified in believing he existed, leaving us to debate only about his character or limitations or whether we ought to love or worship him. But we don't even get that. Someone is contemplating suicide. Is God there to counsel them? Someone is drowning. Does God throw them a lifejacket? A gunman is about to walk into a church and kill thirty children. Does God warn the children? Does he try to talk to the gunman and help him work out his problems instead? Or persuade him to put down the gun? Does he turn the bullets into popcorn? It goes even deeper, to the fundamental nature of the universe itself. Why design a planet that has earthquakes in the first place? Why a universe that can produce gunpowder? It simply makes no sense at all. Unless there is no god. Then it makes perfect sense.

2.6 Not Enough Good from God

A sixth reason follows: not only should there be less pointless suffering if there is a God, there should be more *benefits* from such a God as well. Yet there are none. God is supposed to be your bud, your pal. He is a shepherd, a father, a mother, a friend. Yet he does none of the things such people do. You cannot deny that God would act like a friend and a parent if he really were one. After all, it is only "by their fruits that ye may know them." You cannot say whether someone is good or evil or indifferent if you never see them doing anything, by which their character can be known. And unless God is tied up and stuck in a box somewhere and unable to chew his way out, he would surely make a regular appearance in our lives, well beyond the vague emotional illusions and contradictory revelations people claim to be from God.

As I've argued before, the mistaken interpretation of spiritual emotions and sensations, and hallucinatory altered states of consciousness, are natural and solely personal phenomena, not communications from outside your own mind (III.10.4, "The Nature of Spirituality"). But even if they were believable, they are not indicative of any real friendship or love, unless God is so powerless and numbed of mind that he can barely even communicate to us with ambiguous impressions and random symbols.

As a friend, I would think it shameful if I didn't give clear, honest advice to my friends when asked, or offer comfort when they are in misery or misfortune. I loan them money when they need it, help them move, keep them company when they are lonely, introduce them to new things I think they'll like, look out for them. God does none of these things for anyone. Thus he is a friend to none. A man who calls himself a friend but who never speaks plainly to you and is never around when you need him is no friend at all. And it won't do to say God's with only some people, speaking to and comforting and helping them out, because this means he doesn't really love all beings, and is therefore not all-loving, which would make him less decent than even some humans I know. And it's sickeningly patronizing to say in the midst of misery or loneliness or need that "God's with you in spirit," that he pats you on the head and says "There! There!" (though not even in so many words as *that*). A friend who did so little for us, despite having every resource and ability to do more, and nothing to lose by using them, would be ridiculing us with his disdain. Thus, we cannot rescue the idea of God as Friend to All. The evidence flatly refutes the existence of any such creature.

Likewise, as a loving parent, I would think it a horrible failure on my part if I didn't educate my children well, and supervise them kindly, teaching them how to live safe and well and warning them of unknown or unexpected dangers. If they asked me to butt out I would. But if they didn't, it would be unconscionable to ignore them, to offer them no comfort or protection or advice. Indeed, society would deem me fit for prison if I did. It would be felony criminal neglect. Yet that is God. An absentee mom, who lets kids get kidnapped and murdered or run over by cars, who does nothing to teach them what they need to know, who never sits down like a loving parent to have an honest chat with them, who would let them starve if someone else didn't intervene. As this is unconscionable, almost any idea of a god that fits the actual evidence of the world is unconscionable.

And such a line of reasoning is not limited to good gods. It would be most bizarre to have no evidence of any divine activity or consistent communication even from an evil or disinterested god. I am reminded of the God Torak in the *Belgariad* of David and Leigh Eddings. A rather spiteful, evil being. Yet because of that fact he was ever-present among his chosen people, communicating with them, appearing by their side to fight battles with them. At the time, he left no doubt of his existence. Or think of the insane creator-god conceived by H.P. Lovecraft, the Mad Azathoth who created the universe around him in a fit of mad genius, while later whiling away his time listening to alien flutes at the center of it, with not a care for any of the creatures that arose in the world he made. Even he could be seen: he and his attending flute players occupied a physical swirling mass at the center of the universe, within the range of a strong enough telescope. And even he would talk to people who figured out how to reach him. It would be sad indeed if there was a god and he was like one of these characters—though there is no more reason to deny that than to believe the opposite. But it is just as improbable, since even such gods as these would leave *some* clue, engage the world in *some* observable way.

One is left to ask in the end, "Where's *our* Christmas Carol?" I am often asked, and rightly, what evidence would convert me to some religion, convincing me a god existed. I've already given throughout this book numerous examples of the sort of evidence that would sway me. Any of it would do. But Charles Dickens' *A Christmas Carol* also comes to mind. Here is a perfect example of God doing something good: showing a bad man how his badness is caused by his ignorance, and then enlightening him, so that with this new knowledge he realizes how pointless and wrong being bad was. In a single night the nastiest of fellows is transformed by very real visitations. He is hardly lectured at, nor forced into much, but simply shown the sequence of facts that he needed in order to learn, to understand.

No one would say this was wrong. That the conversion of Mr. Scrooge was an unqualified good for him, for his whole community, for everyone, is undeniable. I dare anyone to come up with some ridiculous, harebrained argument that God shouldn't have done that. Yet this is only fiction. No ghosts have ever sought to educate any evil man. Nothing of the kind happens in this universe. And that is all but impossible for a universe with a god in it. If such ghosts visited me tonight, and took me around, showing me what's right and true, I would believe and be saved. Isn't that what God would want? Since he clearly hasn't done it, he clearly doesn't exist.

2.7 Anything Defended with Such Absurdities Must be False

Last but not least is the *coup de gras*. The surest reason of all to be an atheist is the simple fact that the only things people can come up with to make excuses for God, in their desperate attempt to explain away all six arguments above, are things so absurd (not to mention wholly *ad hoc*), that it is equally absurd to give belief any credit whatever. No fact that people need such ridiculous contrivances to defend is ever likely to be true. If it were true, the facts would speak to it. You would not need to resort to the absurd. But this is just what everyone does. The six most common "I can't believe you're saying that!" excuses for God are as follows. All lead to irreconcilable contradictions and are thus plainly false:

2.7.1 The Argument from Mystery

First is the Argument from Mystery, or the "God Has His Reasons" Argument. It is said, for instance, that God must have a reason to be completely silent, obscure, unfriendly, and unhelpful, and to have built a dangerous, painful, limited, and flawed world for the poorly-designed creatures he is supposed to love. We can't know what this reason is because it must be so complicated or profound that human minds could never grasp it. It's a "mystery."

Of course, there is no reason to believe such a mysterious reason exists, apart from its unwholesome use in preventing us from facing the fact that there is no God. Moreover, I can justify belief in the existence of hundreds of dreadful beings if I am allowed to use this excuse to explain away the complete lack of evidence for them. So clearly such an excuse is unallowable, especially to use it so arbitrarily in just the one case—that is simply hypocrisy.

But that isn't the worst of it. For even if there were such a mysterious reason, it would entail something unacceptable about God: it would entail he was not the Supreme Being. For if there is anything that exists that is so powerful even God cannot overcome it, and so overwhelming it literally *compels* God to remain totally quiet and inactive against his every longing to express his boundless compassion, then God is a miserable prisoner, a slave to a far superior entity. Whatever this dreadnought was, this cosmic bugbear, it would clearly not be created by God, for no compassionate being would create something that would so hog-tie him as to ensure the needless misery of billions of creatures and stymie his every desire to stop it. Nor could it be a part of God, since God is supposed to be perfect and all-compassionate (not to mention omnipotent and omniscient). But, as every theologian would have it, there cannot be anything that is not God and not created by God. And even if for some reason such a thing existed, and was superior to God in power—as this thing would have to be—*it* would be the Supreme Being, uncreated and mightiest of all, reducing God to a quivering, helpless piss-ant.

Needless to say, that is absurd.

2.7.2 The "Free Will Defense" Deployment Number One

It is often said that all the misery and suffering in the world is not God's fault, but is the fault of human sin, which God must allow, lest he violate our free will—that is, our freedom to do good *or* evil. Of course, those who argue this have to come up with yet another *ad hoc* excuse for why God doesn't set right the mistakes of his children, as any responsible parent would, and yet another *ad hoc* excuse for why he doesn't stay their hands before they really hurt somebody with their folly, and yet another *ad hoc* excuse for why he is a silent no-show (since you can't claim *that* is the fault of human sin), and on and on. By now we have added several completely unproven assumptions about God, without any basis in evidence. Never mind that it is wholly irrelevant (and thus wholly invalid as a defense) if you adopt a compatibilist account of free will, as I think one must (III.4, "The Fixed Universe and Freedom of the Will").

But even if you adopt a libertarian view of free will (despite it being incoherent and unintelligible), this defense still fails, and for two obvious reasons. For one thing, most misery in the world cannot even in the wildest of imaginations be blamed on humans—as noted above, much of it is *inherent in the design of the universe*, inherent even in the design of human and animal bodies. Most of it is needlessly endured by animals who have no free will and can hardly be blamed for 'sinning' and thus hardly deserve to suffer. And a whole lot of it comes from "Acts of God," not acts of man: earthquakes, floods, plagues, genetic defects, and on and on.

Moreover, we routinely restrain or prevent people from doing evil. We lock them up, we hire cops to capture, stop, even shoot them if necessary. We regard it as a morally acceptable, even a morally obligatory thing to act in self defense or to come to someone else's aid when they are being victimized by another. We regard it as an unqualified good to set up institutions and procedures that prevent people from doing or wanting to do evil acts. And we know for a fact most evil is done in ignorance, and thus we strive to educate people as well as possible. We also punish and reward by many different means so as to encourage good and discourage bad behavior. If all this is good for us, even morally obligatory, and is not a "violation" of the free will of evil doers, it is absurd to say it is wrong for a god to do it, that it violates free will only when *he* does it but not when *we* do.

I have actually heard someone say "But God would do it too well!" Yes, they actually said that, even though it is absurd to think we can have a society that was *too fair*. One wonders, again, what heaven could possibly be like, or why we would want to go there, if there is such a thing as too nice a place to live. If heaven is better than this place, then God has no excuse not to make this place better, too. And even assuming there is such a thing as too much niceness and justice, who better to give us exactly the amount that is right, but an all-knowing, all-powerful, superintelligent being? Are we to say that we are already there, that more justice than we have now, more good than we have now, would be *too much*, would somehow take away our free will? If you really believe that, then you should oppose with fierce horror any attempt to improve crime control or prevention, or the justice system, or our medical system, or medical abilities and technologies, or any compassionate enterprise whatever, including soup kitchens and Doctors without Borders. For all these things would be unconscionably violating our free will. Yet that is clearly absurd.

So you must face facts: the world *could* be a better place, *should* be a better place, yet it isn't. Therefore, there cannot be any compassionate God to make it so—for if there were, he would have done so already.

2.7.3 The "Free Will Defense" Deployment Number Two

Expanding on the fantastically absurd argument that God can do nothing, not a single thing, because to do so would be doing too much, and would violate our free will (even though *our* doing more somehow doesn't), the "Free Will Defense" is used in another astonishing way. God must remain silent and inactive and leave no good evidence of his existence, because if he gave any more, the evidence would be so convincing he would violate your free will. That is, you could not choose to disbelieve, so your belief would not be freely given. This suffers from all the same objections, and is even more absurd than Deployment Number One. Even from the start, it makes no sense: God should only care about our free choice to *love* him, not our freedom to believe he exists. After all, how can we make a reasoned and informed choice about whom to love if we can't even figure out who *exists*?

This also totally fails to explain the misery-inducing flaws in the very design of nature itself, or the suffering of animals, and is entirely invalidated by my arguments for compatibilist free will (III.4, "The Fixed Universe and Freedom of the Will"). But even if you come up with another excuse to explain away the flaws in nature's design and yet another for the suffering of animals, and illogically adopt a libertarian view of free will, you are *still* stuck in a quagmire of absurdity. For the very argument entails that convincing evidence deprives us of free will. But there are a great many things for which we have convincing evidence. Are we to destroy that evidence, avoid it, so as to stay free? Don't you *want* convincing evidence before committing yourself to something? To assume we don't, to assume it is *ever* a good thing to believe a claim on *unconvincing* evidence, is to take a position on method that is wholly unlivable and inherently absurd, inverting all rationality, and divorcing faith from reason. It would mean that teachers, scientists, lawyers, are routinely committing unspeakable crimes against humanity, depriving everyone of their free will by providing convincing evidence to believe their claims.

There can be no merit in a belief that is held for bad reasons, in a loyalty that is given on uncertain knowledge of to whom you are pledging it, in a trust that is placed in something wholly unproven. And there can be no evil in telling a man what he needs to know to save himself and be happy, and proving to him, like Mr. Scrooge, that it is true. If you believe a man is obliged to prove a claim to you before believing it, then you *cannot* believe it is in any way wrong for a god to do so. If you believe it is a good thing for a preacher, an apologist, a missionary to give me *more* evidence and *better* reasons to believe, then you cannot believe it wrong for a god to do so. Otherwise, missionaries must be villains, and apologetics a wanton violation of man's free will. These are absurd conclusions. Therefore, this defense is absurd.

2.7.4 The "Arrogance Defense" Deployment Number One

It is often charged, "Who are you to question the Supreme Being? You must be really arrogant to think you can make up a better design than the most intelligent and all-knowing being there is!" The absurdity of this charge is that it is far *more* arrogant to assume you know so much about God and his plans and limitations that you can confidently excuse him for being a do-nothing no-show—and not only that, but with clearly absurd excuses even a human can see are ridiculous, much less a god. It is also hypocritical to say it is arrogant to claim to know something, and then arrogantly claim to know *that*.

There is nothing arrogant in using reason, or in believing only what there is evidence and reason to believe. Nor is there anything arrogant in honestly stating what we see to be the case, which is all I have done. If God is a reasonable being, and the world is so strange that *even God* can't explain why he appears not to exist, he could not in good conscience condemn an atheist for her unbelief, since unbelief is then entirely reasonable. And if God is neither reasonable nor compassionate, there can be no comfort in seeking his eternal company, for you would then be condemned to an existence almost as terrible as hell itself, forever wallowing in the misery of seeing billions of good people suffer a needless and cruel torment below, while shivering above in the shadow of a coldhearted, all-powerful lunatic. And that would truly be absurd.

2.7.5 The "Arrogance Defense" Deployment Number Two

"I can't be wrong, you are just sick or confused, and you'll suffer for it!" This is little more than the "Nya-Nya-Nya-Nya!" defense or "I know you are, but what am I!?" One might call it the "Schoolyard Copout." When a man has run out of arguments because his position is indefensible, he will often resort to idle but mean-spirited threats ("You'll burn for not believing!") which are the hallmark of a wicked creed, the faith of a cruel, merciless, and unjust heart. That someone would want another human being to suffer, or would even tolerate the idea, for committing no crime at all but *being reasonable*, is truly frightening. A religion that breeds such people is a genuine plague upon the earth.

On the other hand, someone set on defending an indefensible faith, if he is too kind or timid or embarrassed to employ threats, will often resort to some form of the Monkey Defense, clamping his eyes and ears shut and humming a loud tune so he will not hear the truth. Or *he'll* make the arrogant claim that he just "knows" he is right, despite the fact that he cannot present a single reason for it that wasn't just refuted. There is no point in arguing with such irrational people. If you are that set on believing, then your belief is truly pathological. This book is only for sane, reasonable people. It is not for fools and loons.

2.7.6 The Great Deceiver Defense

Yes, this is argued, again and again: God put all this contrary evidence there, and hid all the evidence for his existence, to deceive you and thus test your faith. After all, for there to be a God *and* for the universe to lack all signs of superintelligent design or concern, *and* to possess abundant evidence instead of slow, blind, natural processes of creation like evolution, God would have to have gone *out of his way* to conceal all evidence of his existence, and to deliberately deceive us by planting a good deal of contrary evidence.

That's undeniably true. But this would make God a Cartesian Demon (cf. II.3, "Method," with III.9.4, "Alternative Accounts Are Not Credible"), which is not at all worthy of worship. Indeed, such a malicious liar could never be trusted. Nothing you thought you knew about such a being would be free of suspicion, for it could easily be a lie, a plant, a trick, the product of divine deception. And that would render all beliefs about God quite absurd indeed.

It is especially strange that people would resort to insulting their God's character this way, since the Being they usually call the Father of Lies is Satan, the personification of Evil. It follows that if God is the Father of Lies, then *he* is Satan, and anyone worshipping such a God is a Satanist, a servant of Evil. Lucky for them there is no Satan. But unlucky for us there are people around in the service of a non-existent demon, who believe grand lies to be righteous.

2.7.7 Facing the Absurd and Calling it Bunk

So here we are. This is the sort of litany of absurdities people are forced to resort to in order to defend their belief in God. It should be clear that no *true* belief would ever need any of these absurd defenses, or anything like them. So we can be fairly certain this is not a true belief. There may be some other god, like the dead or helpless god I invented at the beginning of this chapter, or like the god of Deism or the Eternal Tao, or the Great Cat Boo. But no one will believe in those gods because they already know there is no reason to believe in them. Yet they fail to see that belief in any other god is far *less* reasonable. In fact, it is often so absurd it is unmistakably false.

If you think *that* claim is absurd, consider this observation. I ask the believer, if none of this persuades you—all the evidence and arguments I have compiled from page 1 to here—then what evidence *would* convince you there was no God? That is a serious question. For if you can think of nothing, no evidence of any kind that would ever sway you, then your belief is irrational. On the one hand, such a belief could easily be false, and you would have no way of knowing it. And you know this is true, since you know of many people who believe in false gods—and you can never be sure you aren't one of them. On the other hand, if there is no evidence or argument, even in principle, that would change your mind, then your idea of God is *meaningless*. For if you cannot imagine what would be the case if he didn't exist—or worse, if you actually think nothing *would* be different if he didn't exist—then whenever you say "God" you are uttering an empty word with no content at all. It predicts nothing, conveying no information about the world. And that is hardly different from simply being false (cf. II.2, "Understanding the Meaning in What We Think and Say").

For more on this last point, see Antony Flew, "Theology & Falsification: A Golden Jubilee Celebration," *Philosophy Now* October/November 2000, pp. 28-9, reproduced on the Secular Web at www.infidels.org/library/modern/antony_flew/theologyandfalsification.html. On the many natural, nonrational causes of religious feelings and beliefs, see bibliographies concluding the preface to IV.1, "Not Much Place for the Paranormal," and after III.10.4, "The Nature of Spirituality."

For more on atheism generally, see Nicholas Everitt, *Non-Existence of God* (2003) and Michael Martin & Ricki Monnier, eds., *The Impossibility of God* (2003), as well as the Secular Web's libraries on "Arguments for the Existence of a God" (www.infidels.org/library/modern/theism/arguments.html), and "Theism" generally (www.infidels.org/library/modern/theism/), as well as the rather copious library on "Atheism" (www.infidels.org/library/modern/nontheism/atheism/). There is also a growing section of "Testimonials" written by all sorts of atheists telling the story of their path to nonbelief (www.infidels.org/library/modern/testimonials/), and also a section devoted to "Ex-Christians" telling their own unique stories (www.infidels.org/library/modern/theism/christianity/ex-christian.html).

There is a boundless supply of literature on atheism. And amazingly, much of it is not slanderous. Decent defenses or discussions of atheism include: Michael Martin, *Atheism: A Philosophical Justification* (1992); Theodore Drange, *Nonbelief & Evil: Two Arguments for the Nonexistence of God* (1998); Robin Le Poidevin, *Arguing for Atheism: An Introduction to the Philosophy of Religion* (1996); George Smith, *Atheism: The Case Against God* (1980); Douglas E. Krueger, *What Is Atheism? A Short Introduction* (1998).

There is also a rich literature on leaving or rejecting Christianity for atheism, where many eloquent thinkers have told their story, including: Dan Barker, *Losing Faith in Faith: From Preacher to Atheist* (1992); Michael Martin, *The Case Against Christianity* (1993); Ludovic Kennedy, *All in the Mind: A Farewell to God* (1999); Charles Templeton, *Farewell to God: My Reasons for Rejecting the Christian Faith* (2000); Marlene Winell, *Leaving the Fold* (1993); Bertrand Russell, *Why I Am Not a Christian, and Other Essays on Religion and Related Subjects* (1977). For one man's story of rejecting Islam for atheism, see: Ibn Warraq, *Why I Am Not a Muslim* (1995).

One should also not forget the collected essays on religion by the likes of Robert G. Ingersoll and Mark Twain, which can be found in many editions and anthologies.

V. Natural Morality

1. Secular Humanism vs. Christian Theism

"Secular Humanism" is any philosophy that holds to two basic doctrines: that the progress and welfare of all human beings is the greatest good, and that only secular solutions to achieving this end are credible, not supernatural ones. Christian philosopher J. P. Moreland attacks the ethical foundation of Secular Humanism in "The Ethical Inadequacy of Naturalism," an article published in the May/June 1996 issue of *Promise* (pp. 36-9), and "God and the Meaning of Life," a chapter in his book *Scaling the Secular City: A Defense of Christianity* (1987, pp. 105-32). As Secular Humanism describes the ethical component of my Metaphysical Naturalism, I shall answer his critique, then launch into a full description and defense of my moral philosophy. I will demonstrate, first, that Moreland has no better alternative to offer and, second, that natural morality makes complete sense, and is in no need of anything supernatural or mysterious. Indeed, an examination of Moreland's arguments reveals that the Secular Humanist is in the same position as Moreland's Christian, if not a better one. I will then show his claim that American values have been corrupted by Secular Humanism is also not credible.

1.1 Do Secular Humanists Have No Reason to be Moral?

According to Moreland, "the ultimate values of [secular] humanism are incapable of rational justification" and "Christian Theism is a background theory that makes the existence and knowability of morality more likely than does the background theory of atheism" (*Scaling*, p. 121; J.P. Moreland and Kai Nielsen, *Does God Exist: The Debate between Theists & Atheists*, 1993, p. 119). But while he claims that Secular Humanism can supply no rational reason to be moral, while Christian Theism can, the reasons he offers for Christian theists are effectively the same reasons offered by Secular Humanists. And in actual practice, the secular reasons appear more certain and knowable than the Christian ones (following my survey of epistemology in part II, "How We Know").

Moreland defends the Christian theory of value by declaring that "my motive for being moral should be because I love God, I recognize him as my creator, I want to do what is right for its own sake, and I desire my own welfare in this life and the life to come" (*Scaling*, p. 128). He also says that it is rational to obey a kind person, that it makes sense not to dehumanize or trivialize your own existence, and that we are best protected, satisfied, and fulfilled by living morally.

A Secular Humanist can make a similar declaration of equal merit. A secular version of Moreland's first statement might be, "my motive for being moral should be because I love humankind, I recognize my debt to society, I want to do what is good for its own sake, and I desire my own welfare in this life as well as the welfare of generations to come." These statements provide equally good reasons for being moral, because they appeal to essentially the same values, and a Christian has no better reason to commit to those values than a Secular Humanist. Likewise, Secular Humanists also find, for all the same or equally sound reasons, that it is rational to obey a kind person, that it makes sense not to dehumanize or trivialize your own existence, and that we are best protected, satisfied, and fulfilled by living morally.

This equivalence of the two worldviews will now be demonstrated, revealing how Secular Humanism has certain advantages over Moreland's Christian Theism.

1.1.1 Love as Reason to be Moral

First, Moreland offers "because I love God" as a reason Christians have to be moral. But this only begs the question, "Why love God?" Secular Humanists suffer from the same objection, since it could easily be asked, "Why love humankind?" Both worldviews stand on the same footing. The Christian can offer no better reason to love God than the humanist can offer to love humankind. In both cases, it is ultimately just a commitment to love.

On the one hand, it can be argued that we should love God because he loves us (while humankind doesn't). However, it does not necessarily follow that we should love those who love us, as in the case of the battered wife whose husband loves her but beats her to death anyway. And though one might say that true love is proven by acts of love, and that therefore a wife-beater does not really love his wife, this argument would also go to prove that God does not love us, since he also "beats us to death anyway," as the horrifying treatment that millions of people receive at the hands of Mother Nature adequately demonstrates, no matter how one might justify it. Indeed, the Christian idea that we 'deserve' it or 'brought it on ourselves' because of our own sins, sounds exactly like the wife-beater, who gives exactly the same excuses.

On the other hand, it does not follow that we should *not* love someone simply because they do not love us. Love for us (or acts of love toward us) are not generally the conditions we set for loving someone. Rather, we usually choose whom to love according to certain qualities they possess apart from how they feel or act toward us, and we certainly love many things that are not even people, such as our country or our profession. I love my wife because of who she is and what she is. I love America because of what it represents and what it has accomplished. My reasons for loving humanity are similar (for more on love, see III.10.3, "The Nature of Love").

Naturally, with regard to God, it cannot even be adequately proven that God exists, much less what his qualities are. It is easy to describe a God worth loving, but it is something else to prove that such a thing actually exists. An atheist finds adequate proof that a lovable God does not exist (see IV.2, "Atheism: Seven Reasons to be Godless"). But even if a lovable God *did* exist, we would love him not simply for what he does for us, but for his character and qualities.

The fact remains that Secular Humanists, by the very definition of 'humanist', love humankind, whatever their reasons. This therefore stands as a reason to be moral equally as strong as the Christian's "love for God." One may even say the Secular Humanist is on stronger ground here: the good qualities of humankind are visible, demonstrable facts, whereas we have no facts at all to judge God's character by. Moreover, the love of God can lead to acts of immorality toward humankind, as exemplified by Abraham's willingness to murder his own son because of his love for God, and the Hebrews' willingness to slaughter men, women and children at the command of God (and whether God really said or intended such things is not relevant here), whereas love for humankind would only produce moral acts toward humankind—whether God were good or evil, or real or not.

This is one of the fundamental and irreconcilable differences between the values of Christian Theism and Secular Humanism: as a Secular Humanist, I see Abraham's action as thoroughly immoral (as I explained in Part I.2, "How I Got Here"). A moral response in that situation would be to rebuke God, since the very act of asking Abraham to kill his son merely to prove his own faith would in itself entail that God was malevolent, a villain, and God's standing as the supreme creator would not change the fact that his character was thereby reprehensible, and thus *not* worthy of love. A humane story would depict Abraham being praised by God for having dared to issue such a rebuke, for that would have demonstrated that his character and moral courage (his compassion for his son and respect for human life) was greater than the fear of angering God himself. But this is not the message the Bible conveys, a classic example of how the Bible is in conflict with humanist values.

1.1.2 Debt as Reason to be Moral

Moreland's second reason for Christian theists to be moral is "because...I recognize [God] as my creator." This is no more justifiable than the secular version. On the one hand, simply being our creator is not a sufficient reason to listen to God. For he could be a cruel or indifferent or insane creator, and it would actually be more rational *not* to listen to our creator if that were the case. And given the evidence of science and history, these possibilities cannot be ruled out (cf. IV.2.5 and IV.2.6).

This is identical to the Secular Humanist's situation. A humanist who recognizes his debt to the society that "created" him, and the moral teachers and agents who created his society, has just as much of a reason to be moral for their sake. Even if a God exists and created the universe, that would not change the fact that we all owe an even greater debt to the actions of countless moral agents whose works God can not truly take credit for—if you believe in libertarian free will as Moreland does (see chapter III.4).

Of course, the Christian theist can argue that if a humanist's society were evil or degenerate, then respect for society would cease to be a good reason to follow that society's morals or to return any favors. But this is identical to the Secular Humanist's objection to Christian Theism's reasoning: if God were evil or degenerate, then the theist's debt to him would also cease to be a valid reason to adopt God's morals or to return any favors. And regardless of what Christians believe about God, they must recognize the fact that many Secular Humanists sincerely believe the evidence proves that if a God does exist, he must necessarily be evil or indifferent.

Ultimately, Secular Humanists *do* recognize a certain debt to their society and its heroes and seek to return the favor. After all, Western democratic societies are not as evil or degenerate as Christians sometimes make them out to be (as we shall see in section V.1.2). They have many admirable qualities which humanists recognize. Humanists therefore have the same reason to be moral on this ground as the Christian theist. Indeed, the humanist reason is more epistemically secure, for the merits of society and its praiseworthy heroes are empirically demonstrable and visible to all, whereas the merits of God are wholly theoretical (see, for example, III.10.4, "The Nature of Spirituality" and, again, chapter IV.2).

1.1.3 Goodness as Reason to be Moral

Next, Moreland offers "because...I want to do what is right for its own sake" as another reason to be moral. Of course, this begs the question, "Why do right for its own sake?" Both the Christian theist and the Secular Humanist can offer this as a reason to be moral, and neither would be any more justified than the other. The Secular Humanist would say they do good because they want to be a good person, or they want to create and embody the moral values and ideals they believe to be good, and this is essentially the same reason Moreland implies for Christians.

Indeed, Moreland's justification for moral action here cannot be entailed by, nor dependent on, the truth of anything else, much less Christian Theism. By definition, we either want to do good for its own sake or we do not. That is, to offer a reason to do good for its own sake (such as by appealing to some aspect of Christian Theism) would be to do good for something other than its own sake. This therefore cannot be a peculiarly Christian justification. If it is valid at all, it is valid for everyone. Perhaps Moreland means to say we should be moral because moral behavior produces goodness in the world, and goodness is desirable in and of itself. But the Secular Humanist completely agrees with that, too, and for essentially the same reasons (above and below).

1.1.4 Self Interest as Reason to be Moral

The fourth reason Christian theists have to be moral, according to Moreland, is that we should "desire [our] own welfare in this life and the life to come." Ultimately, this reason boils down to an ordinary appeal to self-interest, and differs from Secular Humanism in only one respect: the inclusion of a "life to come."

On the face of it, this does seem to be the one point on which Christian Theism has a better reason to be moral than Secular Humanism. After all, in the Christian view, appropriate rewards and punishments are guaranteed. Secular Humanism can never offer such a guarantee, since doing the moral thing can sometimes lead to things going badly for oneself, or doing the immoral thing might not lead to any negative consequences at all. Of course, this uncertainty affects everyone, moral or immoral: even if you reject morality completely and seek to live entirely for yourself, you *still* cannot be sure that any given act will actually benefit you in the long run. And in practice, it is the probabilities that get you: a moral life does in fact produce a statistically greater chance of an enduring and genuine happiness, in contrast with an immoral life which does in fact produce a statistically greater chance of chronic misery and disappointment (as explained in the next chapter).

However, if Christian Theism is false, then its guarantee is false, and it then ceases to be a good reason to be moral. Thus, while the rewards and punishments of the mundane world are at least observably real, and reliable enough to study and take into account, there is no reliable evidence of any kind to prove the certainty or even the *nature* of any supernatural rewards and punishments. And this places a dependence on the afterlife on very poor footing. Moreland himself admits that if one "lives one's life in a self-induced delusion...then satisfaction comes from living a lie" (*Scaling*, p. 121). Thus, if I do not honestly see adequate empirical evidence to justify believing in Christian Theism, then I cannot use its claims about the afterlife as a reason to be moral.

Worse, even if this Christian claim were in any way true, it *still* does not offer a better guarantee than Secular Humanism. This is because there is no certainty as to which acts or beliefs God will reward or punish, and to their very death a Christian can never be any more certain than a Secular Humanist that living a moral life (or not doing so) will turn out best for them in the end. God may, for instance, send all Catholics, or all Lutherans, or all Christians to hell (after all, Jesus may have been a fraud, and Mohammed may have been right) regardless of how moral (or immoral) a life they've lived.

The Christian must acknowledge the very real likelihood of this, since his own Bible claims that only those who approach moral perfection shall attain heaven (Matt. 5:20), and that God will only forgive those who forgive all evil men (Matt. 6:15). Both of these conditions effectively exclude nearly every Christian in existence no matter how much "faith" they have. Indeed, how many Christians in human history have actually adhered to the principles defined in Matthew chapters 5 and 6? We cannot rationally conclude that Jesus "didn't really mean" we had to follow those guidelines, since why else would he emphasize them? Though one can theorize one's way around such things, these solutions would be mere human opinions, which may or may not have anything to do with what God really has in mind.

More troubling for his position, as an Evangelical, Moreland himself believes we are saved by faith, not works: thus, even an immoral person goes to heaven as long as he 'believes on Christ', and a moral one goes to hell if he does not (see Gary Habermas and J.P. Moreland, "Hell: the Horrible Choice," *Beyond Death: Exploring the Evidence for Immortality*, 1998, pp. 285-319, and p. 378). Thus, on his own sectarian view, self-interest is *not* served by being moral. Immoral people are saved if they truly believe and accept the atonement of Jesus for their sins before they die, while moral ones are damned. So it is perhaps a bit hypocritical for him to use this argument. In the end, there is still no guarantee that God will judge wisely or compassionately or recognize your religious denomination, or disregard religious denomination altogether. That he will do so can only be taken on faith.

This is the same position a Secular Humanist is in. But the humanist has at least some empirical evidence in his favor, whereas the Christian has absolutely none. For a humanist can point to the awful life of a criminal or reprobate as proof that a life of evil is bad for you, but when the Christian is asked to point to that same villain in Hell to prove that his punishment will extend even past death, the Christian has nothing to point to. The same applies to the rewards of a moral life. Though the Christian has a book full of theories, he is woefully short of actual evidence for them.

Thus, that living morally will lead to the greatest sum of happiness or satisfaction in a person's own life can never be known for certain, nor can it be known for certain that not living morally will lead to a better life, *not even by a Christian*, and that is the central point at hand. The Christian is thus in no better position than the Secular Humanist, and is actually in a worse one, for there is abundant evidence that a moral life is more likely to produce happiness in this world than an immoral life will, whereas for the theist there is no evidence whatsoever as to who has actually made it (or will make it) into heaven and who not.

1.1.5 Trust as Reason to be Moral

Moreland also argues that "it is rational to obey a kind Being," i.e. God (*Scaling*, p. 128). However, if this is true, then it is equally rational for a Secular Humanist to obey a kind *human* being, such as a compassionate moral teacher (though Moreland does not say, I assume 'kind being' also means 'wise'). Indeed, if one can actually prove that it is rational to obey a kind being—and I have no idea what such a proof would look like, but I'll assume for the sake of argument that such a proof is available—then this would remain a valid Christian argument *even if God did not exist*.

How so? If God were known to be a fictional character who was kind, it would still be rational to emulate him or follow his advice, if he and his advice were truly kind. For the actual existence of a hero or moral teacher is irrelevant in this case: if it is wise to emulate a kind being, then even the example of a fictional kind being would be valid. And in this light, the atheist would be no less justified in emulating divine virtues than a Christian, for belief in God is not a precondition for doing so. However it may be, the Christian has no better reason than the Secular Humanist to be moral on this account. It is even worse that the evidence is strongly against God being a kind being, fictional or not.

1.1.6 Self-Image as Reason to be Moral

Moreland also argues that he is rationally justified in living a moral life because he was made in God's image and attends to himself "as an image bearer" in that he does not "dehumanize or trivialize" his own existence. In short, by being moral he affirms that he is "a creature of value who is worthy" of rewards like heaven (*Scaling*, p. 131). He is again begging the question, "Why care about God's image?"

First of all, it is uncertain whether God's image is a good act to follow. If the Old Testament is anything to go by, God is presented as a cruel and irrational being (see Part I.2, "How I Got Here"). The second commandment makes it clear that God's jealousy drives him to proclaim unjust laws, since he says he is "jealous" and "punishes the children for the sin of the fathers to the third and fourth generation" (Ex. 20:5, Deut. 5:9). I imagine even Moreland would agree that jealousy, and punishing the children of criminals, is not an act anyone should follow. Yet God is said to do this with cruel zeal, ordering his followers to "put to death men and women, children and infants, cattle and sheep, camels and donkeys" (I Sam. 15:2-3). Wanton rage and genocidal war crimes are not evidence of a good character. God does not get much nicer in the New Testament either. There he introduces eternal, agonizing torture for unbelievers. I certainly would never want to emulate such unjust cruelty.

Moreland would also agree, I hope, that executing people for using bad language (Lev. 24:13-16) or picking up sticks on Saturday (Num. 15:32-36) is not a good image to follow. Note, also, that Jesus is never said to have laughed, and by all accounts he had no family life, no children or wife. Is that really an image we ought to follow? And contrary to popular belief, he was not a peace-loving man (Mt. 10:33-6). He could not even restrain himself from violence in the marketplace (Jn. 2:13-16), and this makes even Gandhi a better man than Jesus. But even if we could excise all this from the Bible, or explain it away through untestable, subjective human opinions regarding what the Bible "really means" when it says such things, even if we could paint an ideal image of God, we are back to the same place we just were: if it makes sense to emulate an ideal being, it would not matter whether that being existed or not, so this reason to be moral would be no less available to the atheist.

In my opinion one of the best and most important reasons Secular Humanism offers to be moral is that we are indeed, in Moreland's own words, "image bearers." But in humanism, the image we are bearing is not that of some God, whose precise image is unascertainable with any confidence, but of our own ideals of what a good human being is and can be. The Secular Humanist sees this in the same way a Christian does, who thinks about what would be good, and then attributes those qualities to God. For instance, Christians conveniently ignore or try to 'explain away' the bad stuff in the Bible, only some of which I cited in the previous paragraph, because it does not fit their pre-conceived notion of the ideal. They praise the Ten Commandments, but ignore the "corruption of blood" clause, and they also ignore the other commandments beyond the first ten. For example, "the Lord said to Moses" that "anyone who blasphemes the name of the Lord must be put to death" (Lev. 24:13-16). Almost any decent Christian today would say that this was an unjust law. But it is the law of their God nonetheless. Nowhere in the Bible does it say "just follow the first ten." Thus, Christians pick and choose not according to the example of their God, but according to their *own* notion of what is just. And then they add other things to the character of their God, like the love of democracy and the abhorrence of slavery, despite the fact that no such notions can be found in the Bible, not even in the words of Jesus. Thus, humanists and theists are simply doing the same thing: creating an image of the ideal and choosing to follow it.

Ultimately, Secular humanists, just like Moreland, choose to be moral because they do not wish to "dehumanize or trivialize" their own life, because then life would lose its meaning and quality. We all want to affirm that we are "a creature of value who is worthy" of the gift of life itself. For gifts need not come from persons: accidents produce gifts, too, just as they can produce injuries. And thus the humanist has exactly the same reason to be moral as the Christian. Self-respect would not be possible otherwise, nor would any true happiness be possible without self-respect, something we will examine later.

This belies a principal fact which Christians often fail to understand about Secular Humanists: the meaning in our lives is derived directly, and in numerous ways, from the significance and beauty of our own humanity and conscious existence. It is not necessary to be someone's creation for our lives to have value, as I have explained already (III.7, "The Meaning of Life"). The inability to see how an accidental existence would be no less valuable than a planned one is a great stumbling block before many Christian intellectuals.

1.1.7 Worldly Self-Interest as Reason to be Moral

Finally, Moreland appeals to worldly self interest. He argues that "we are protected, satisfied, and fulfilled best" in this life "by doing what is right." He means that God "will harmonize happiness and duty" (*Scaling*, pp. 129, 128). But the Secular Humanist says the same thing about friends, family, and the good society that Moreland says about God.

According to the secular view, we cannot count on there being a God to ensure our protection, satisfaction, or fulfillment. And we believe all evidence supports us here. We are protected only by fellow human beings whom we can trust and who trust us. And trust can best be won by committing ourselves to a moral life and living by that commitment. We are truly satisfied only by living in a society that is safe and prosperous, and we cannot rationally expect such a society to exist for us if we are contributing to its corruption and ruin with our own immorality.

Polluting the world with immoral deeds is analogous to polluting the world with trash: even when the effects of either act may not directly affect us, the result of setting the bad example, and contributing to the overall effect, is still exactly what makes for a dirty and unpleasant environment, in which we still must live. There is also a common advantage to be gained from working together with justice and compassion, as much on a project like maintaining a good aqueduct as on maintaining a good society. This remains equally true for Christians and Humanists, regardless of any God's existence.

1.2 What's Wrong with Secular Humanism?

As we have seen, every reason Moreland gives for us to be moral on Christian Theism is equally true, or corresponds to an equally true reason, on Secular Humanism. In fact, in most cases the Secular Humanist has more secure or more empirically verifiable reasons. So it appears to me that Moreland has nothing better to offer. Thus, to make Christian Theism look better, Moreland is forced to make Secular Humanism look worse. I will now address this argument.

1.2.1 Do We Live in a Sick Society?

Moreland opens his article in *Promise* with the old saw that "our society is in a state of moral chaos." This is meant to convince us there is some new problem that he can then blame on secularism. Other Christians who argue that our present woes are something new, and secularism to blame for them, include Charles Colson, *How Now Shall We Live?* (1999), and David Myers, *The American Paradox: Spiritual Hunger in an Age of Plenty* (2000). But these writers fail to mention that this has been said in every century of human civilization for which we have any appreciable amount of social commentary, going on four thousand years now. The 'depravity of society' lament is nothing new at all—for the oldest examples of this genre, see Samuel Noah Kramer, "The First 'Sick' Society," *History Begins at Sumer*, 3rd rev. ed. (1981, pp. 259-69). But most relevant for our present situation, see Stephanie Coontz, *The Way We Never Were: American Families and the Nostalgia Trap* (1992) and *The Way We Really Are: Coming to Terms With America's Changing Families* (1998).

In fact, when we look objectively at history, Americans are more moral as a society today than any society at any time ever in human history, apart from our free democratic cousins around the world, who tend to be far less religious than we, yet somehow enjoy far lower rates of crime, and sometimes even greater economic equity and social justice, contrary to the very thesis Moreland is defending. But focusing solely on America, what do we *really* see? We see an amazingly progressive culture that has crawled out of an age of violent expansion and bigotry, and is starting to show incredible promise as an enlightened society.

Never before the 20th century have tens of millions of people voluntarily supported international human aid projects on a vast scale, without regard for political borders or religious affiliations, such as the Red Cross or Second Harvest. We even give aid to our *enemies*. We lead not only the world, but our own past, in hours devoted per capita to volunteer humanitarian work. Education is regarded for the first time in history as an inalienable right, and is universally provided, even for women and minorities, and illiteracy is almost a thing of the past. The poor have government-funded medical care and millions of people are calling for more, out of pure compassion for their fellow human beings.

For the first time in history, women have full political rights. Free speech and freedom of religion are not only regarded as fundamental to our national character, but are ardently and thoroughly defended by all social classes. Slavery is defeated, and all forms of racism and hatred are almost universally loathed rather than accepted. For the first time in history, most people actually have compassion for the plight of animals—even the ones they eat must be humanely killed, and national organizations are devoted to their protection. For the first time in history, when men went to die in Viet Nam, millions of people actually protested an unjust war and stood for peace, despite being beaten and killed for it. In contrast, before our time, there is no recorded case, ever, of any comparable mass Christian action against a war. Today, even poor Americans live as well as Medieval kings once did, with more political power (in the right to speak, protest, and vote) than most of their counterparts in history ever had, amidst a sea of free parks and libraries, subsidized public transportation and 911 response teams, with luxuries galore, from sanitized water to portable orchestras in a box, all because people *cared*, and were honest enough to make the government work and the economy prosper, very much unlike the hopelessly corrupt governments and societies of many third-world nations today.

Let's compare America today with past centuries. Despite complaints about high rates of crime now (which have mostly declined steadily since the 1990's), compared to any other century our crime "problem" is extremely minimal, and has never been greater than it was eighty years ago anyway, when the murder rate skyrocketed under Prohibition. No surprise that today it is a new Prohibition that fills the vast majority of our prisons and drives young men to crime. In past centuries bandits and pirates threatened most rural routes and seaways, and travel was often a risky venture. Kidnapping was so routine that many people prayed they wouldn't be snatched. Robbery was so routine many people hid their money in mattresses or socks. Even the most effective civilizations, at their peak, could stave off only some of these ills, compared to the safety of our current society, and only for small regions and for short spans of time...or for select social classes.

Children were legally beaten and very often exploited or abused. The rights of women and minorities, not to mention the incarcerated or mentally ill, were all but nonexistent, and slavery (and the oppression of native peoples) was defended as God's Will. Indeed, millions of God Fearing Christians actually fought to the death for their right to abuse and enslave millions of their fellow human beings. Whereas now people protest even if a single soldier might lose his life, in all past eras war was accepted as the natural course of things—indeed, until recently, it was hard to find any adult in history who did not suffer through one, or lose a member of his family to one. And there was never before our era much concept of a 'war crime' or any real concept of 'workplace safety'. Think of the horrors of days past. McCarthy. Jim Crowe. Wounded Knee. Salem. Dare anyone claim we were better people then?

I could go on, but the point is clear. The true picture of life in past times is always bleak. No one today, who really knew the facts, would trade this century for any other. No sensible woman would dare do without feminine hygiene products and civil equality. No sensible man would dare walk into the disease-infested cities of the past or risk maiming or death in a pointless war or a gruesome industrial accident, or worse, risk finding himself a slave in the Antebellum South, or an aborigine on the Trail of Tears.

And if you hate political corruption today, you haven't seen corruption until you've flipped through a history book—the scandals of our past are quite revolting. But even in this century we have witnessed unprecedented progress. Whereas hundreds were killed and entire towns burned to the ground in the race riots of the Red Summer of 1919, even the worst riots today are a walk in the park by comparison. Once, minorities lived among us abused and mistreated, most living in a level of poverty that would disgust any human being today, while affluent white families virtually ate off their backs. Millions of blacks lived every day of their lives in fear of being lynched, or worse.

Yet things have changed. Now, in nations that are free and secular, only pockets of resistance remain tainted by this past evil, such as inner city slums, or Appalachian hovels, and scattered individuals or neighborhoods. But men like Moreland don't seem to call upon these real problems in their condemnation of society. Rather, his best example is the staged-and-scripted buffoonery of the Jerry Springer Show, which is as harmless as any other circus, and hardly indicative of reality for the average American. Why, then, are men like Moreland so eager to see our time as somehow more immoral than ages past? Apart from the general fact that people tend to falsely idealize their past, I would suggest two other factors are at work. On the one hand, we have a common error of reasoning: more weight is placed on sensational cases than on overall statistical facts. For instance, despite outrageous school shootings like Columbine, school crime has declined nationally and continues to do so. On the other hand, people like Moreland simply do not have their priorities straight. Personally, I will not accept any form of oppression, such as of gays or women or religious or racial minorities, as signs of a 'moral' society. Nor, in contrast, do I accept open sexuality or vulgarity as immoral.

Signs of morality are simply this: compassion and integrity, on a wide social scale. But many among the religious right see gays being openly gay and being treated as equals, they see offensive art, people challenging authority, women deciding for themselves what their place will be, and so on, and they don't like it. I cannot say whether or to what extent Moreland has such motives for condemning society, but they are certainly not uncommon among Evangelical conservatives generally. But none of that is immoral. To the contrary, these are signs of a *moral* society, where freedom, tolerance, equality, and compassion are the order of the day.

This is not to say modern society is all peaches and cream. We have a lot of progress to make, a lot of maturing left to do. Humanity is a young race, only four thousand years civilized. That's a mere 200 generations. If you knew all your ancestors, back to the day the first human city was built, their names would barely fill a single sheet of paper. Now humanity is slowly reaching young adulthood, petulant and naive, faulty and prone to missteps, but not the child it has been through all its past. Human beings are generally more conscientious now, more educated now, more freethinking, more cosmopolitan and more compassionate now than ever before.

Though we are centuries yet from where we ought to be (and where Secular Humanists want to take us), it is quite dishonest to portray our current existence as somehow a moral *decline*. We look in vain for the 'Golden Age' we are supposed to have declined from. And though I agree with Moreland and others that progress is being hampered by a sort of spiritual aimlessness, I do not believe they have the solution. They are only partly correct. We do need people to be more philosophically skilled, introspective, and spiritual (in the sense described in III.10.4, "The Nature of Spirituality"). We do need them to be less shallow and materialistic. But I believe Secular Humanism does far better at providing this, given our current state of knowledge and enlightenment, than any other worldview. As we've seen, the Secular basis for morality parallels the Christian, yet in each case, as I see it, there is empirical evidence to support the Secularist, but very little, if any, for the Christian. And Christianity comes with far too much baggage.

We must observe that, historically, Christianity even at its height, in the Middle Ages or the American Colonial Era, has always failed to produce a moral or enlightened society. While today, of the surveys that have been done, all show that Christians are no more moral than non-Christians. For instance, see Richard Scheinin, "Not as I Preach," *San Jose Mercury News* (September 11, 1993) and Steve Chapman, "Praise the Lord, Pass the Ammo: If teen violence is the question, religion isn't the answer," *Slate* (June 30, 1999). A Roper survey showed Christians more prone to driving drunk than non-Christians, as reported in *Freethought Today* (September 1991, p. 12), any issue of which catalogues the routine horror of crimes committed by clergy (of all sects, not just Catholics). So even when adopted Christianity does not have the curative effect Moreland claims for it. Perhaps one might call for a study wherein devout Evangelicals are compared with others, but in such a study I would ask that devoted Secular Humanists also be singled out for comparison and not lumped with the apathetic majority. I will bet good money we will find no winner in this match either. Indeed, though anyone could name a dozen famous Christian criminals, I doubt they could name even a single self-described Secular Humanist to match.

In contrast, Secular Humanism as a value system isn't getting the press it needs in order to have the extensive influence on people that Moreland mistakenly thinks it already has. Hardly a soul can give a correct definition of it, and few have even heard the phrase. And what people *really* need is not being given to them, nor is Moreland asking for it. Philosophy is not taught every year in school, as it ought to be. Churches are on every corner, not Freethought Houses. Every Sunday, believers go to be preached to in silence, not to actively discuss and debate the important issues of philosophy or policy. No one is being given the tools to think analytically about life and morality, or to critically examine and make an informed choice about spiritual direction, and no one is being encouraged to practice these skills.

And yet, instead of encouraging and fighting for this, people like Moreland are selling what appears to many of us like a dubious quick-fix: believe on Christ and all will be well, a claim that sounds a lot like the naive rally cry to hang the Ten Commandments in schools and society will improve itself, to just pray and you'll cure every social ill. This has the ring of talismanic superstition. It has consistently failed in every previous century. Why should it work now? Instead, the Secular Humanist's call for a universal education in the skills of freethought and the value of self-examination is surely a much better prescription.

> On what Secular Humanism *really* means and *really* entails, see the Secular Web's library on the subject (www.infidels.org/library/modern/nontheism/secularhumanism). Books on it are numerous, including: Corliss Lamont, *The Philosophy of Humanism*, 8th ed. (1997); Antony Flew, *Atheistic Humanism* (1993); Paul Kurtz, *In Defense of Secular Humanism* (1983). There are also many anthologies of humanist literature, e.g. Norm Allen, Jr., ed., *African American Humanism: An Anthology* (1991) and Margaret Knight, Edward Blishen, and Jim Herrick, eds., *Humanist Anthology: From Confucius to Attenborough* (1995). There are

also at least three "Humanist Manifestos" in print: *Humanist Manifesto* I (1933), II (1973), and III (2000).

1.2.2 Does Believing in Evolution Make Us Immoral?

After launching with an attack on contemporary society in his *Promise* article, Moreland echoes a common polemic that we act like animals because "we are taught all week long in public schools that this is exactly what we are—animals," thus tying his imaginary 'problem' to its supposed secular 'cause'. But he presents no evidence or any real case for this supposed connection between teaching evolution and immorality.

In contrast, Moreland's thesis seems implausible from the start, in every possible way. Most criminals are high-school dropouts or otherwise poor scholastics, and thus could hardly have absorbed much in the way of what evolution really means. Scientists, who study evolution more than anyone, show no signs of being any more immoral than anyone else, including Creationists. Crime today is hardly worse than when the Scopes Monkey Trial shook the nation. Moreover, Humans far more often behaved like depraved animals before Darwin was even born (need I mention Torquemata? Custer? Southern slave masters?). And staunchly religious countries where evolution is *not* much taught are bastions of misery, violence, and evil (think of Sudan or Afghanistan, to say nothing of Serbia or Haiti).

Thus, rather than aligning his argument with the evidence, it appears Moreland would rather engage in the fallacy of "poisoning the well" by arguing that evolution must be 'factually' false because believing it makes people 'morally' bad. Such nonsense is reminiscent of Luddites predicting the automobile would bring on the Apocalypse. Moreland's slanderous nonsense flies in the face of the simple fact that society has actually morally *improved* since Darwin, in almost every measurable aspect of life. Justice and compassion are now far more pervasive than they were even when Darwin published in 1859. The brink of the American Civil War. Does Moreland really have the gall to claim things have since gotten *worse*?

So the evidence contradicts Moreland on this point, and certainly offers no support for his claim. But does evolution even *imply* the conclusion that we are 'just' animals and thus have no value and no reason to be moral? This sort of question is full of flawed assumptions. For instance, even animals have value, and exhibit moral behaviors (such as altruism, compassion, tolerance, and cooperation), so even if we were to conclude that humans were 'just' animals it would not follow that we have *no* value, or that we can throw morality to the wind.

But the biggest fallacy of all here, frequently used by Christian fundamentalists, as I have already noted before (in III.5.5, "Reductionism"), is a *modo hoc* fallacy, the argument that "*X* is just *Y*" when in fact it is not. Humans are obviously animals. We've been classified as such since Aristotle invented the science of taxonomy. But just as a cat is categorically different from a bird, so are humans categorically different than all other animals. Among less significant distinguishing attributes, we possess the ability of self-awareness and a capacity for abstract thought (see III.6, "The Nature of Mind"). In other words, unlike all other animals, we can understand the full consequences of what we do, and are aware of what other beings think and feel. Unlike all other animals, we can comprehend what it means to be moral and adduce reasons to be so.

Thus, far from being 'just' animals, we alone have the ability to adopt and live a moral life. Anyone who reasons that we should act like 'other' animals because we, too, are an animal, is forgetting that as every species acts in its own way, so should we: thus, even as animals, we ought to act like human beings. It is in this, our uniquely human qualities, that our unique value lies, as is more thoroughly argued by Christine Korsgaard in "The Sources of Normativity" (*MDAP*, pp. 389-406; see below). Indeed, humans have evolved to live in social communities where moral behavior is an essential component: everything from the benefits of cooperation to the conquest of loneliness requires it. To borrow Moreland's own words, "To live in disregard of morality and virtue is to live like a fish out of water, i.e., to live contrary to our proper functioning," or in scientific terms, contrary to the demands of our ecological niche. Evolution does not take anything away. It merely explains where it came from.

So the charge that belief in evolution (or naturalism, or Secular Humanism) leads to immoral systems of thought has no more merit than the charge that belief in God does so. Surely Moreland will not dispute that devout believers in God throughout history have developed immoral values and engaged in immoral acts and employed God as justification. Yet he would chastise me for the hasty generalization that belief in God must necessarily have this outcome, or that it even tends to do so. He should appreciate, then, being chastised for committing exactly the same bigoted fallacy with regard to naturalistic evolution and Secular Humanism.

1.2.3 Selfish Genes and Selfish Memes

There is another sense in which a critic might press this argument. Might it be that, if evolution by natural selection *is* our true creator, then humans will forever be selfish creatures, and therefore we will be immoral no matter what we believe?

As we will see below, this is simply not so. Since the 1970's sociobiologists have proved that natural selection can favor, and thus produce, both selfishness *and* altruism, cruelty *and* compassion. And humans have the genetic propensity for both. But it is a mistake to equate our genes with our selves anyway. Our genes might be 'selfish' in some metaphorical sense, but there are many strategies for them to use for their own benefit, and promoting in us a strong moral compass is one of them, one humans as social animals really do develop and need to survive. But even more importantly, we are minds, not genomes (see III.8.4, "Memetic Evolution"). Our mind is built on top of the basic structure of our genes, incorporating rather what we ourselves have learned and decided and acquired from our family and friends and community and culture. Thus, no matter what natural selection is doing to our genes, we are now more the product of our society and our own thought instead. We (as *persons* rather than mere bodies) are constructed of memes even more than genes—especially with regard to the most profound and complex features of our minds, where reasoned moral thought occurs.

Our genes simply cannot control how our environment and the generic deployment of reason will affect us. Genes can limit or encourage a range of effects, but rarely decide the precise outcome. This is especially clear with the introduction of reason, which must work only one way or it will provide no advantage, thus constraining what 'selfish genes' can do to manipulate us in the face of rational thought, which they cannot control (see III.9, "The Nature of Reason" but also III.6, "The Nature of Mind" and III.8, "How Did We Get Here?").

And though memes can also be 'selfish' in some sense, this is not always a bad thing. A selfish 'truth' meme, for example, would promote itself to the detriment of all 'untruth' memes, and if there were such a thing it would only benefit us, for if the idea of truth were selfish, once it infected us it would lead us to love truth and abhor untruth. And that would be a good thing. Of course, reality is more complex than that, but the point remains. There are even 'altruistic' memes, which promote themselves by aiding other memes. And yet still none of this defines who we are. That will ultimately be decided by our will and our reason, which do most of the producing, mutating and selecting of the memes that form us.

We are limited only by two realities: by genes that have given us a template to build on as well as a largely (though not completely) blank slate, genes that define our nature yet grant us boundless adaptability; and by a memetically driven but equally malleable culture that we can analyze, modify, and accept or reject, in whole or in part. And even within these constraints, a rational mind in possession of the truth will be a moral mind. Neither theism, nor atheism, nor selfish memes or genes will ever change that fact.

For more, see section V.2.2, esp. 2.2.1, "Evolution of Moral Values." On why genes are not always 'selfish': Mark Ridley, *The Cooperative Gene: How Mendel's Demon Explains the Evolution of Complex Beings* (2001). On the scientific basis for how evolution can select for kindness and other moral qualities, see the landmark paper by John Collier and Michael Stingl, "Evolutionary Naturalism and the Objectivity of Morality," *Biology and Philosophy* 8 (1993), pp. 43-50.

On how genes and memes, nature and nurture, biology and environment, interact in complex ways to produce good and compassionate people, see: Dale Jamieson, *Morality's Progress: Essays on Humans, Other Animals, and the Rest of Nature* (2004); Matt Ridley, *Nature via Nurture: Genes, Experience, and What Makes Us Human* (2003); Shelley Taylor, *The Tending Instinct: Women, Men, and the Biology of Nurturing* (2002); Leonard Katz, ed., *Evolutionary Origins of Morality: Cross-Disciplinary Perspectives* (2001); Andrew Brown in *The Darwin Wars: The Scientific Battle for the Soul of Man* (1999); Jane Maienschein and Michael Ruse, eds., *Biology and the Foundation of Ethics* (1999); Larry Arnhart, *Darwinian Natural Right: The Biological Ethics of Human Nature* (1998); Elliott Sober and David Sloan Wilson, *Unto Others: The Evolution and Psychology of Unselfish Behavior* (1998); Matt Ridley, *The Origins of Virtue* (1997); Frans de Waal, *Good Natured: The Origins of Right and Wrong in Humans and Other Animals* (1996).

2. Morality in Metaphysical Naturalism

I have critiqued Moreland's arguments and shown that he has nothing better to offer than Secular Humanism, and that Secular Humanism may even be better than alternatives such as Christian Theism. Now I shall develop what Moreland claims to be impossible: a complete theory of natural ethical value, a form of virtue-based objective realism, built and defended from the ground up, which is empirically testable and rationally justified. Throughout, the central outstanding issues in moral theory will be addressed, explicitly or implicitly. For this I shall take as the central reference to current disputes in metaethics the landmark work of editors Stephen Darwall, Allan Gibbard, and Peter Railton: *Moral Discourse and Practice: Some Philosophical Approaches* (1997), hereafter referred to as *MDAP*.

2.1 Outlining a Moral Theory

Moreland's objection to naturalism (more precisely, *physicalism*) is that it does not accommodate value statements. For values have to be, he says, non-natural properties of things, i.e. "by a non-natural property I mean an attribute that is not a scientific, physical characteristic of physics or chemistry," so values "are just not scientifically testable, physical properties" and since naturalism cannot easily explain how they came to exist, naturalism is false. But none of this is true.

To begin with, a 'value', Moreland says, is intrinsically normative, meaning it is "valuable in and of itself" and "something we ought to desire." This is misleading. First, of course, there are non-normative values: we each value some things because they are special to us, while others couldn't care less. I doubt my one and only oil painting has any intrinsic value, but it has personal value to me. But even when we limit discussion to normative values, Moreland doesn't have it quite right, for all 'normative' really means is 'true for everyone' (in his words "something we [i.e. everyone] ought to desire"), as it derives from 'norm', referring to what is normal or standard. Even if something had value for some reason apart from itself, so long as it 'ought to be' valuable to everyone, it would have normative value. For instance, funding a court system for administering justice is normatively valuable, but not in and of itself: it has value because *justice* has value. So the value of a justice *system* is both normative *and* derivative.

Now, once we define values correctly, it is an easy matter to see how countless more values can be deduced and derived from them, even on naturalism. So what really concerns us are fundamental or *core* values, values that are not derived from any others but are basic and primary. Naturalists understand core values in different ways. I will present only my own personal view here, even though there are naturalists who disagree with me, just as there are Christians who disagree with Moreland.

> For views very similar but not always identical to mine, which nevertheless complement what I say here with even more rigorous analyses, see: Richard Boyd, "How to Be a Moral Realist," *MDAP*, pp. 105-35, plus a further list of relevant works on moral realism, ibid. p. 106; also Peter Railton, "Moral Realism," *MDAP*, pp. 137-6.
>
> Excellent defenses of naturalistic moral realism are legion, and I list here the best: David Owen Brink, *Moral Realism and the Foundations of Ethics* (1989); Joseph Daleiden, *The Science of Morality: The Individual, Community, and Future Generations* (1998); and Paul Bloomfield, *Moral Reality* (2001).
>
> Also relevant to my perspective: Jason Casebeer, *Natural Ethical Facts: Evolution, Connectionism, and Moral Cognition* (2003); Richard Taylor, *Virtue Ethics: An Introduction* (2002); Michael Martin, *Atheism, Morality, and Meaning* (2002); Owen Flanagan & Amelie Oksenberg, eds., *Identity, Character, and Morality: Essays in Moral Psychology* (1993); David Carr, *Educating the Virtues: An Essay on the Philosophical Psychology of Moral Development and Education* (1991); Barry Arnold, *The Pursuit of Virtue: The Union of Moral Psychology and Ethics* (1989); N. Dent, *The Moral Psychology of the Virtues* (1984).

2.1.1 The Goal Theory of Moral Value

As stated in *The Blackwell Encyclopedia of Social Psychology* (1995, s.v. "values"), "values are trans-situational goals that serve as guiding principles in the life of a person or group" which "serve as standards of the desirable when judging behavior, events, and people (including the self)" (for more on this, see the first bibliography in V.2.2, "How Naturalism Accounts for Value").

In the simplest parlance, a value is a latent, ever-present desire, to be distinguished from fleeting, momentary, or incidental desires (see III.10, "The Nature of Emotion"). When anyone harbors in their character an enduring desire for something, that is a value, as the term is understood in the social sciences. The object of this desire is then said to 'have value'. So when *everyone* ought to hold such a desire for something, that desire produces a normative value, a value that everyone *ought* to have. In this I am following several prominent experts, including Gerald Gaus (see the first bibliography in V.2.2) and Peter Railton, who both argue for the reduction of values to desires. Railton also defends what I argue below, that the meaning of "ought" language is a reference to implied goals, and defends the very conception of the normative that I adopt below (*MDAP*, pp. 142-8).

On close analysis, I believe there is only one core value: in agreement with Aristotle and Richard Taylor, I find this to be a desire for happiness. I believe that all other values are derived from this, in conjunction with other facts of the universe, and that all normative values are what they are because they must be held and acted upon in order for any human being to have the best chance of achieving a genuine, enduring happiness. When we say "you ought to value *x*" we mean that, if you do, you will improve your chances of enduring happiness, and if you do not, you will decrease those chances. This is not a novel theory. Even Taoists have argued this, for whom a moral life is needed to live in harmony with nature; and Confucianists, for whom a moral life is needed to live in harmony with society; and Buddhists, for whom moral life is needed to live in harmony with oneself; and many ancient Western philosophies agree, such as Stoicism and Epicureanism, even the philosophy of Aristotle, which became a bulwark for Christian Theism for many centuries.

I believe this core value entails two particular values, which have the highest order of rational importance among the derived values: compassion and integrity, which are essential to a genuinely happy life. The Secular Humanist's moral credo could rightly be stated: cherish integrity in yourself and compassion for all. How people come to have these values ingrained in their character is a different matter from why they ought to ingrain them. The first story involves human psychology, socialization and parenting, and mental development in general, and is a story about becoming a mature, healthy person. The second story involves the logical and factual connection between having those values and achieving happiness. The two stories are interrelated, but I will only focus on the latter here. What follows is a scientific hypothesis, but I believe it is factually true—from observing numerous human lives, today and throughout history.

2.1.2 Happiness and the Moral Life

To understand the connection, we must understand happiness. By happiness I do not mean mere momentary pleasure or joy, but an abiding contentment, a persistent, underlying sense of reverie that makes life itself worth living, in the absence of which life becomes shallow, unsatisfying, and ultimately meaningless. As David Myers puts it, real happiness means "fulfillment, well-being, and enduring personal joy" (see bibliography to chapter III.7). This happiness is rarely possible, and certainly impeded, amidst loneliness, fear, purposelessness, destruction, misery, insanity, or chronic anxiety or stress, among other things. In contrast, happiness is found, secured, and improved amidst love, good friendships, security, purposefulness, creation, joy, sanity, and peace.

These are by no means complete lists, but they will suffice to make the point. It should be obvious to anyone of any experience that these connections are all true: the more you have from the second list, the more often you will know true happiness, whereas the more you have from the first list the less often this will happen. Even if one gets lucky and finds a little happiness amidst misery, there is nothing secure in this, nor is it a lottery anyone really wants to play. And all the while it is tainted and troubled, and thus incomparable to the genuine good life, the perfection of which is unachievable, but is surely approachable: it is a goal one always benefits from aiming at, and would never much benefit from avoiding.

It is this fact that, in conjunction with a defining convention, allows us to identify what behaviors are moral or immoral. By 'defining convention' I mean whatever convention a human society has established for itself to categorize actions as belonging to a moral sphere. For instance, eating ice cream usually does not belong to the moral sphere—it is neither moral nor immoral. Most human conventions keep the categories of 'moral' and 'immoral' behavior even narrower than the category of all normative statements—for instance, it is generally not regarded as immoral to smoke, even if it happens to be normatively wrong to do so. This is merely an issue of language, and does not affect my position: on what I shall call the Goal Theory of Moral Value, smoking is just as wrong whether we call it immoral or not. What we *do* choose to categorize as immoral, however, is a full sub-category of normatively wrong acts. In this fashion I avoid any controversy over whether moral language refers to human conventions or natural facts: for it clearly does both. However, only natural facts shall concern us here.

This means the Goal Theory encompasses all conventions regarding what is moral or immoral, insofar as those conventions only include truly normative propositions. For instance, Moreland's ethics would categorize homosexual sex as immoral. But on my theory this is not normatively wrong and thus cannot be immoral as he claims. To the contrary, since it hinders no one's happiness, while suppression of homosexual emotions has been empirically proven to be destructive of human life and happiness, it is actually *immoral* to denounce or repress them. Likewise, though Moreland thinks a popular human lust for retributive punishment is proof that such behavior is morally good, on my theory we can easily see that it is cruel, pointless, and barbaric—hence Hell is immoral, a doctrine Moreland obviously cannot accept. For instance, see *Scaling*, p. 124, where Moreland elaborates his reasons for thinking punishment that does not rehabilitate, repair, protect, or deter, is nevertheless morally good. I think it is clear to any compassionate person that this is really nothing more than torture, the very "cruel and unusual" behavior the U.S. Constitution rightly forbids.

In contrast, most people do not regard smoking as immoral (just stupid), but Mormons (and those of some Muslim and Buddhist sects) actually do categorize it as immoral. But whatever one calls it, it is still wrong on my view. The use of the term 'immoral' here may have a different rhetorical effect, but is otherwise only an arbitrary distinction. This does not make it an invalid or pointless distinction, however. There is an obvious utility for modern democratic societies to classify as immoral only those behaviors that are potentially harmful to *others*, and these behaviors are usually far worse, far more wrong. But this should not permit the assumption that other behaviors that are normatively wrong are acceptable. And, naturally, all this assumes the obvious fact of scaling: some things are more or less moral than others. And in the grand scheme of things smoking is a relatively trivial sin.

Having established what is moral or immoral on the Goal Theory, one conclusion becomes clear: immoral behavior is risky. Like playing Russian Roulette, having unsafe sex, smoking cigarettes, or driving drunk, you might get away with it, but it is a gamble, and you can never to your dying day be sure of escape. Now, I will emphasize from here on only the category of the malevolent, deeds that threaten to harm others, and thus bring the greatest threat of harm to ourselves. Some of what follows still applies to self-destructive immorality, things we do to ourselves that undermine our happiness, but it is the other category that we usually discuss, and to that I now turn.

When you undertake an immoral life in the malevolent sense, everyone else will be gunning for you—not just society in general or your victims in particular, but everyone who, by observing and assessing your character, loses respect and trust in you, or emulates you, making *you* the victim. Not only that, but the repercussions of your actions will make your life difficult. They can wreck everything from your friendships to your health. And yet they produce nothing but fleeting and momentary benefits—few significant feelings of personal achievement, or of self-worth, hardly anything that can be called contentment or satisfaction with life. Even if you don't outright, and foolishly, risk or ruin your life or health, with a significantly immoral life you will come more and more to live in fear, you will more and more find yourself alone and unloved. And even when loved, it will not likely be for who you really are, and thus it will not be truly satisfying.

In short, immorality is bad for you. Hence, I predict just what I myself have observed: that we will have a very hard time finding any bad person who does not live with chronic anger, disappointment, difficulty, depression, or paranoia. They will try to replace the hole in their lives with luxuries and distractions and power trips, but it is never enough, and they ultimately can know only fleeting and hollow pleasures, never the genuine happiness of which I speak.

In contrast, moral behavior is beneficial. Though people often complain of how morality places restrictions on them, in actual fact all that truly normative morality prevents people from gaining are momentary pleasures, not happiness. Of course, bogus morals, such as repressive sexual mores, do not bring happiness, but we are concerned with genuine morality, which is rooted in compassion and integrity. This morality is actually conducive to happiness, like any other discipline that trains people to avoid excess or needlessly dangerous risks. Like brushing your teeth or exercise, the moral mindset teaches you to think of the long term, rather than creating even more problems for yourself later by acting on impulse now. Moral people will naturally make more and stronger friendships, win more genuine love from others, have less to fear, and find more in themselves to love and appreciate. They will set up fewer traps for themselves, live less self-destructive and more creative lives, and enjoy greater security. Their life will have meaning and value, from their own perspective, as well as that of others.

Indeed, the moral life is often easier. Every lie, every crime, requires a never-ending labor of cover-up and evasion, telling lies to cover the lies, committing crimes to hide the crimes, to the point where one can easily get lost in the complexity of one's own devices, and sink into a quagmire of ever-magnifying ruin. But if one acts morally, everything is done. No cover up is needed. It is easy to remember what you've told people when you know you've always told the truth, and you never have to fear exposure when you know you've done no wrong.

In the end, what sacrifices you must make to be good are usually in the long run trivial, such as the loss of material goods or eating crow, things which have no relation to real human happiness anyway. They can in fact have over-all positive benefits, as one can see from, on the one hand, the plethora of self-help books advocating a reduction of material things, things which only increase our anxiety and take away from our time and resources, and, on the other hand, the commonplace wisdom that those with the courage to admit their mistakes learn far more from them than those who do not.

Finally, since cultivating a genuine sense of compassion and integrity will lead to habitual moral behavior, these character traits, these values, these *virtues*, are good for you. For you need habitual moral behavior to know real, consistent happiness. Thus, even if you don't realize it, even if you don't quite understand it now, you want these values. In truth, everyone would want these qualities to be a part of their person, if only they came to see how their pursuit of momentary satisfaction is futile, and discovered instead how wonderful real happiness is in comparison.

These habitual behaviors are especially desirable, since having these values makes moral behavior enjoyable, thus removing a lot of what people dislike about it. If you really have compassion for others, then you share in their happiness and sadness, and thus anything you do to alleviate their suffering or bring them joy makes you feel good, through simple human empathy. And if you really have a passion for your personal integrity, then you will be happy when you stick to your guns or preserve your self-image, even under torture or severe loss. It will feel good that you made your personal values meaningful by making them manifest and unconquerable, defying those who would work evil upon or through you. You will be in your own eyes a success, not a failure. But if you have no empathy, you eliminate a huge source of joy from your life, leaving you more vulnerable to depression. And if you falter, you will suffer to one degree or another from self-disgust or disappointment. In other words, merely *being moral* in and of itself contributes to a lasting happiness.

2.1.3 Self Worth and the Need for a Moral Life

It is important to stress the role of self-worth in this picture, for that is more important and more potent for maintaining happiness than any other factor. Amidst all other forms of misery, fear, and pain, a strong sense of self-worth can preserve happiness like a sturdy ship in a storm. Though happiness would be greater with the impediments removed, the man of genuine self-respect will always be happier than he would be in the same circumstances without it. This has been confirmed scientifically (see section bibliography below). In fact, every recommendation Christian psychologist David Myers makes for achieving the happy life (see bibliography to chapter III.7) is entirely secular—even the role of spirituality, differently conceived (per III.10.4, "The Nature of Spirituality").

Philosophers have long known this—like Immanuel Kant, who foresaw the same conclusion in chapter III of his *Groundwork of the Metaphysic of Morals*, where he argues that the only reason to adopt his moral point of view is that it will bring us a greater sense of self-worth. In particular, he says that we should "hold ourselves bound by certain laws in order to find solely in our own person a worth" that compensates us for every loss, for, as Kant believed, "there is no one, not even the most hardened scoundrel who does not wish that he too might be a man of like spirit," for only through the moral life can he gain "a greater inner worth of his own person." As Kant understood, a strong sense of self-worth is not possible for the immoral person, but a matter of course for the moral one. This is largely because immoral behavior inevitably leads to overt self-loathing—unless this is psychologically suppressed, and yet even that is damaging, the cause of numerous mental disorders from depression to all manner of delusions and neuroses. In contrast, only moral behavior leads to a sound self-respect, an essential component to mental health and human happiness.

The dynamics of this are not hard to perceive. Even Moreland seems close to seeing this, when he argues that "if the depth and presence of guilt feelings is to be rational, there must be a Person toward whom one feels moral shame" (*Scaling*, p. 123), and he is right: that person is our self, the one whom we can never really escape. We naturally hate those who lie, cheat, murder, and steal; we hate those who are intolerant or insolent or malevolent in some way, who cause misery rather than happiness. This is simply a natural emotion arising from the human condition, for we could never really be happy ourselves if we did not loathe the enemies of happiness. If we tried to admire them instead, we would inevitably get burned by them repeatedly and grow to hate them anyway. Such people represent a threat to our own survival and well being, as well as to that of our family, friends, and all those whom we care about—and this includes not just people, but ideals and institutions. No other attitude toward them would remove this threat, therefore only a hostile attitude is rational.

This hatred is often felt, consciously or not, and not merely for those who threaten what we hold dear. It is felt for anyone who embodies malevolence, the very *type* of person. For we react this way even to fictional characters who can never harm anyone in the real world. Hence, the very idea of villainy is repugnant to us. Even the real villains among us try to paint themselves as heroes, more often I believe to deceive themselves than to deceive others, just as most of us can only ever admire villains when we delude ourselves into seeing them as heroes. In the very same manner, we love those—in fiction as in reality—who are benevolent, honest, or otherwise virtuous, who create happiness rather than misery, we love the very *type* of person who embodies the good.

Because of this natural, acquired human sentiment, whenever we act like those we hate, we will be faced with a psychological dilemma: we will be forced, on some level of our being, to hate ourselves. With feelings of self-loathing, someone who hates himself, in any sense, will always be handicapped, even sabotaged, in his own quest for happiness. Even if we try to take steps to hide from this fact, as I believe most villains in the world do, as well as their admirers, we cannot avoid deeper psychological ramifications. Self-hatred, self-defeating hypocrisy, perpetual dissatisfaction and disappointment with the world and ourselves, even outright madness can creep upon us, as history, psychology, and personal experience demonstrates. The only place to flee from this hatred and self-loathing is into every delusion or distraction we can conjure, which inevitably compromises our rationality or mental health, our very ability to pursue happiness wisely and effectively, wasting time and resources on the useless, even the harmful.

And if we have seen the truth about ourselves, even the option to hide from it no longer exists. Our self-loathing will then become direct and profound. It is those who have achieved this state of mind who truly know what it means to ask others how they can sleep at night, how they can live with themselves, after doing something personally loathsome. In contrast, when we become a good person, we can more easily come to love ourselves—in the way we ought to, with respect and humble pride.

It follows that the only way to achieve self-respect, to truly love ourselves, to be content and happy with our lives, is to love and respect the worthy, and then be like those we love, and embody the virtues we respect. The reason virtue is said to be its own reward is that merely embodying the ideal, merely knowing that you are the vessel of creation, of justice, of compassion, gives your life greater meaning, and hence value. To fall short of this is to waste your life. To abandon virtue is to make ourselves useless, cancerous, loathsome. Only the monster can find pride and joy in being the vessel of destruction, injustice, or cruelty. And it is a far cry from humanity to want to be a monster rather than a man. Even living as a hunted and hated monster can never be an efficient or reliable path to happiness. No matter where you turn, no matter how you try to hide or escape, in order to avoid the misery of self-loathing and to enjoy the elation and fulfillment of profound self-respect, the moral life is a necessity.

It follows that the Golden Rule, the most universally recognized moral standard, is merely an expression of a basic fact of human psychology: if we embody what we already hate, we will hate ourselves, and be hated by others, but if we embody what we love and respect, we will love and respect ourselves, and be loved and respected by others in turn. One might thus restate the Golden Rule most simply: be a hero, not a villain. For this is the way to happiness. Our very meaning and satisfaction in life depend on it.

Quotations from Kant are from *Groundwork of the Metaphysic of Morals* or *Grundlegung zur Metaphysik der Sitten* (1785) § 3.4 (by Kant's arrangement), or § 4.454 (in the standard edition of the Royal Prussian Academy in Berlin), or pp. 112-3 in Kant's 2nd German ed. (1786), or p. 122 of H. J. Paton's English translation, Harper Torchbooks ed. (1964); see also, Robert Wolff, *The Autonomy of Reason: A Commentary on Kant's Groundwork of the Metaphysic of Morals* (1986), § 3.5, p. 211.

Regarding the link between (secular) morality and happiness, see Ruut Veenhoven, *Conditions of Happiness* (1984), a masterwork of social psychology, wherein happiness is treated in fairly great detail, e.g. defined, scientifically measured, etc. However, his work did not explore distinctions between happiness-defeating and happiness-improving moral systems, and such research is greatly needed.

See also: Martin Seligman in *Authentic Happiness: Using the New Positive Psychology to Realize Your Potential for Lasting Fulfillment* (2002); Stephen Braun, *The Science of Happiness: Unlocking the Mysteries of Mood* (2001); David Lykken, *Happiness: The Nature and Nurture of Joy and Contentment* (2000); and Michael Argyle, *The Psychology of Happiness*, 2nd ed. (2001), which comes with an extensive bibliography on the issue. Books recommended in III.10.2, "Reason as the Servant of Desire" are also relevant here.

Also important is philosophical literature that draws on the same kind of data and concepts, such as: A. C. Grayling, *Life, Sex, and Ideas: The Good Life Without God* (2003); David Cortesi, *Secular Wholeness: A Skeptic's Paths to a Richer Life* (2002); Richard Warner, *Freedom, Enjoyment, and Happiness: An Essay on Moral Psychology* (1987); or Russell Gough, *Character Is Destiny: The Value of Personal Ethics in Everyday Life* (1997). Even some Christian writers, like David Myers, agree with this view (see bibliography to chapter III.7).

2.1.4 The Futility of Secret Violations

Since morality is rooted in character-based virtues, not in rules *per se*, the benefits of a moral life come mainly from *being* moral, far more than from merely acting moral. Though the latter has benefits, these are not sufficient to secure a genuinely happy life. For the enjoyment and understanding of the behavior is missing, making it unfulfilling. It is also risky, for without the habit of it, you are setting yourself up for a tragic fall. These factors eliminate the rationality, in some respects even the possibility, of 'secret violations', and how this is so will further illuminate the content of my ethical theory.

First, this is a theory based on risk and probability. One can never be sure of secrecy anyway. And the risk is always unwise, because it is so unnecessary. There is no such thing as a perfect crime, and the possibility remains to your dying day that you will be found out—and the odds of this improve with every immoral act. Likewise, though one can sometimes get hurt being moral, or not gain the rewards expected, the odds are always in favor of the opposite, and they increase continually the more natural and habitual your moral character becomes. The whole point is one of shifting your chances: to aim at happiness rather than risk misery. Morality is a lifestyle choice, and it is the lifestyle as a whole that rewards, not individual actions *per se*. And vice versa: opting out of that lifestyle is harmful to happiness, even when this is not apparent with each and every immoral deed. Inevitably, 'secret violations' are as pointless as sneaking a smoke: you still destroy your health—or, like the one-time drunk driver, risk ruining a life, especially your own.

Second, though a moral life interspersed with secret violations can reap some benefits, the gain is not enough to make it a good road to happiness. Your conscience will always convict you, as it is pretty hard to hide a crime from yourself. So the consequences to self-respect and mental health discussed above cannot be ignored. And the effect of becoming the sort of person who doesn't care, or who actually admires villainy, is negative: such a character is innately self-destructive and incapable of finding any true satisfaction in life apart from fleeting glimpses. In contrast, it is only through cultivation of a heroic character that life can be given a satisfying meaning and become fulfilling. In short, for us there is only one reliable road to happiness.

Third, a person needs moral character to live a good life with ease, yet a fully-informed moral character has side-effects that actually eliminate many fleeting urges for immoral action. For example, a moral person rarely has much interest in theft, since material possessions are of relatively little value to them, and the feeling of having earned what one has is too enjoyable and satisfying to forsake. A person of informed moral character knows it is detrimental to happiness to have too great a care for material things—or even worse, to base one's happiness on their possession—and knows that it is beneficial to happiness to care less about objects than more substantive sources of pleasure, such as peace and friendship and satisfying work. Thus, this theory, when pursued, actually has the causal effect of reducing the desire for secret violations. They eventually cease to have any point.

2.2 How Naturalism Accounts for Value

Moreland argues in *Promise* that "intrinsically normative, non-natural properties are not known by the methods of science." But by now we should see he is begging every question here: values need not be intrinsic to be normative, nor need they be non-natural. And when we start talking about natural human values, such as a value for happiness, instead of the pie-in-the-sky somethings-or-other Moreland has conjured up, we see at once he is entirely wrong: such things are not only known to science, but are extensively measured and studied by psychologists and sociologists (see section bibliography below). Even moral statements ("you ought to do *x*") could be scientifically investigated and verified or falsified, as we shall see.

But Moreland is also building a straw man by assuming that something has to be an object of scientific study to be accepted by naturalists. To the contrary, a Metaphysical Naturalist, like any other philosopher or reasonable theologian, can use her worldview to fill in the blanks with speculation, or even direct experience, until such time as a methodologically rigorous investigation can determine the truth. But she fills those blanks with ideas that are derived from and rationally consistent with the sound findings of the sciences, and she never pretends such proto-scientific ideas are firmly established facts (see II.3, "Method"). More than this, much of the naturalist's knowledge comes from essentially nonscience sources, such as daily experience, or narrative and analytical history. We merely place the highest authority, not the sole authority, in the findings of science. For in all matters of fact science is far more thorough and careful than any other mode of investigation.

Still, Moreland complains that he "cannot find out that mercy is a virtue by some laboratory experiment" and he declares that values could not "come to be present by strict physical laws." But they can. Physical laws entail that specific objectives can only be efficiently obtained by certain behaviors. So one could demonstrate scientifically, in the lab or in the field, that mercy is an inevitable by-product of compassion, and that compassion is essential for human happiness, and therefore ought to be sought by all human beings, for all human beings desire happiness. Thus, science can indeed demonstrate that a value is 'normative' (and thus conventionally designated a 'virtue'), contrary to Moreland's doubts. In other words, just as science can demonstrate proper 'normative' behaviors for growing corn, so it can for enjoying human society.

We can also explain how values physically exist and come about: just like memories or character traits, they are chemically and mechanically represented in the structure of the human brain, physically arranged to generate certain emotional and emotive responses and behavioral tendencies (see III.6 and III.10, "The Nature of Mind" and "The Nature of Emotion"). They are both coded genetically and modified by physical interaction with the environment and internal brain activity. And the capacity for all this was developed through billions of years of neural evolution (see III.8, "How Did We Get Here?").

> On the nature of 'values' essential reading is Nicholas Rescher, *Introduction to Value Theory* (1969) and William Smart, *Introduction to the Theory of Value* (1966), but more recently and most in accord with my conclusions is Gerald Gaus, *Value and Justification: The Foundations of Liberal Theory* (1990).
>
> On the study of 'value' by science see Gaus, as well as: Daniel Kahneman & Amos Tversky, *Choices, Values and Frames* (2000); Herbert Hyman, *The Value Systems of Different Classes: A Social Psychological Contribution to the Analysis of Stratification* (1993); Andrew Reid Fuller, *Insight into Value: An Exploration of the Premises of a Phenomenological Psychology* (1990); etc.

2.2.1 Evolution of Moral Values

Moreland doesn't even consider that humans evolved to be moral animals. He thinks the fact that human happiness is dependent upon moral behavior suggests a god arranged it that way. But if a god were arranging things like that, as I've suggested before, wouldn't the *universe itself* exhibit moral affinities? In other words, why doesn't evolution favor survival of the kindest? Why don't evil people become sick, or their bullets miss more often? Why don't innocents become immune to permanent or painful injury or heroes gain supernatural powers with which to fight crime or natural disasters? Why doesn't our love physically heal others, or crops grow in accordance with a society's neighborliness? Why aren't bibles that contain the correct moral guidance, the life-saving truth, indestructible—or written in a universal language that we are all born to read? (see IV.2.3, "The Universe is a Moron")

In fact, the universe exhibits *zero* value affinity: it operates exactly the same for everyone, the good and bad alike. It rewards and craps on both with total disregard. It behaves just like a cold and indifferent machine, not the creation of a loving engineer. The only place any sort of value effect is ever seen is in human thought and action, and only when humans are psychologically developed a certain way. It thus stands to reason that values do not come from the design of the universe, but the adaptation of *Homo sapiens* to that universe—and, in particular, to a social ecology. After all, the only place values are ever found are in human thought, so it seems an obvious conclusion that values are a product of human thought.

I would even argue that the link between morality and happiness is all but inevitable, for two reasons. First, no highly-sentient animal has a good chance of survival if it finds itself in a hostile social environment of its own creation, and so it is natural that it would evolve the means to avoid creating such an environment, and would evolve instead the ability to create a friendly and beneficial environment, which means one of peace, reciprocity, cooperation and trust. Perfecting this environment is in the interest of all individuals who benefit directly from living in it, providing an obvious evolutionary basis for moral progress, very much in line with Railton's thesis of social rationality (see section bibliography below), and similar to the inevitable utility of evolving limbs capable of holding and manipulating tools, without which, like social harmony, no civilization is ever likely to develop.

It is not by accident that humans have mastered the Earth and are the only species to go beyond it. We are highly adapted for social cohesion and mutual aid, and that makes us nearly unconquerable. Of course, as our 'design' is the result of an essentially blind, uncaring, and imperfect process, our innate moral abilities are as flawed as our bodies, but like our bodies they are perfectible and improvable, especially with technology. And morality is basically a technology of happiness and social well-being. It doesn't take much to see that we aren't helping ourselves by polluting and dragging society down with our immoral deeds, and we have much to gain by training our moral sentiments, as I've noted already.

Second, the moment you become self-aware, you adopt a whole new array of problems, from loneliness and difficulties with self-image, to a realization of your own mortality. Some means must be developed to cope with these problems if the advantages of self-awareness are to be enjoyed. The most obvious, and clearly the best trait for this is a moral outlook: through such, one can enjoy the benefits of sympathy among happy people, establish meaning and purpose by adopting a stable and satisfying self-image, and even accept the prospect of death without anxiety—by simply loving life, manifesting and establishing those values the universe would not.

And that's just the short list. There are many adaptive advantages of morality, and thus it makes sense that this should arise in accord with the needs for human happiness. Indeed, I cannot think of anything superior to it. The adaptive benefits of moral action, and the human-cultural disadvantages of immoral action, have long been understood and documented by cultural anthropologists, as any standard college textbook on the field shows. Though criminality and deceit can find an adaptive niche within a society, just as diseases can hitch a ride on otherwise healthy bodies, deceit and criminality are a precarious niche, the least effective and most dangerous over-all, and ultimately the feeblest path to genuine happiness.

> For Railton's theory of social rationality, see *MDAP*, pp. 150-4; Allan Gibbard, "Wise Choices, Apt Feelings," *ibid.*, pp. 179-198; David Gauthier, "From *Morals by Agreement*," *MDAP*, pp. 341-61.
> On the evolutionary basis of moral sentiments and values, see the bibliography to section V.1.2.3, "Selfish Genes and Selfish Memes." For a broader survey of all the relevant science: Michael Shermer, *The Science of Good and Evil: Why People Cheat, Share, Gossip, and Follow the Golden Rule* (2004). On cultural evolution in general: e.g. Carol & Melvin Ember, *Cultural Anthropology*, 9th ed. (1998); for more specific treatment of morality: e.g. John Cook, *Morality and Cultural Differences* (1998).

2.2.2 Human Nature

Moreland might respond by repeating his attack in *Promise* on the very concept of 'human nature':

> Darwin's theory of evolution has made belief in human nature, though logically possible, nevertheless, quite implausible....no reference to biology can be made to support one's claims about 'human nature'.

That isn't true. Whatever he thinks he is talking about, when I use phrases like 'human nature' (and I haven't much), I mean, in the most fundamental connotation, the qualities *sine qua non* a human being, and in the more colloquial connotation, the qualities *sine qua non* a normal, healthy human being.

Regardless of variations within and among species and the nature of animals as evolving, chemical systems, it is obvious that in order to be called a 'human' one must possess certain qualities, all the things we call 'human beings' clearly possess those qualities, therefore a 'human nature' exists. Q.E.D. Even the idea of a normal, healthy member of a species is neither incoherent nor implausible—it is a regular component of medical science and psychology. And though human nature is more defined by culture than biology, this very dependence is in turn biologically determined: humans are helpless without a culture. It is thus *human nature* to be a cultural animal, and it is in the nature of cultures to promote a certain way of life and thought. All cultures have language and morals, for example. Thus, contrary to Moreland's assertion, we can indeed base claims about human nature on biology—as well as culture.

> This has been aptly defended by Boyd, who articulates a more thorough defense of the use of 'natural kinds' within naturalism (*MDAP*, p. 111, 115-18), even though no doctrine of essentialism is necessary to the concept of human nature. See also Paul Ehrlich, *Human Natures: Genes, Cultures, and the Human Prospect* (2002) and Steven Pinker, *The Blank Slate: The Modern Denial of Human Nature* (2002), both of whom address many other important issues we have touched on before (e.g. III.4, "The Fixed Universe and Freedom of the Will," and V.1.2.3, "Selfish Genes and Selfish Memes").

2.2.3 Personhood

Moreland also challenges the naturalist to define 'person' or 'personhood' in a manner acceptable to an ethical worldview. On Metaphysical Naturalism, Moreland says, there "can be human non-persons (e.g., defective newborns, people in comas) and personal non-humans (e.g., orangutans) and the latter have more value than the former." Against this he argues that "being a person is to being a human as being a color is to being red." In other words, "There can be non-human persons (angels) but there can be no human non-persons just as there can be colored non-red things (blue things) but no red non-colored things."

But Moreland has made a crucial mistake here. For most defective newborns and people in comas are still persons on Metaphysical Naturalism. As long as the memories, character traits, reasoning faculties, and other mental properties that define someone's personality exist, the *person* exists. That none of these faculties are active is irrelevant to the question of whether they exist. Though dormant, they are there. A naturalist would only hold that a person does not exist when a defective newborn or someone in a coma was brain dead, or otherwise physically lacked the architecture of personality. And this should be obvious: even Moreland would not regard a corpse as a person, yet surely he would admit it was human. Thus, his whole argument falls flat. Being a person is *not* to being a human as being a color is to being red. Personhood is defined by properties that are not innate to a human, but that a human is capable of producing. So being a person is to being human as being red is to being a tomato: though a tomato can produce redness, it only does so when ripe.

Moreland also argues that on Metaphysical Naturalism, "the features that constitute personhood can be possessed to a greater or lesser degree," which means "some individuals can be more of a person and, thus, have more rights and value than other individuals." Of course, I am certain Moreland actually agrees with this to some extent. Surely, he would not argue that children should be allowed to guzzle beer in bars, visit strip clubs, vote, or sign up for the army. It is obvious that rights are attenuated to abilities. And though a person's value will vary according to their potential, and what they actually make of themselves, this does not mean some persons have no value, or that anyone has a trivial value. To the contrary, even non-persons have some value: a great many animals are deserving of compassion and respect. And a newborn human baby, deserving even greater compassion and respect, has more value than any animal on Earth, with the possible exception of adult apes or dolphins (or, perhaps, elephants).

Moreland might still object to placing an ape before a human baby, but I would again argue that he does not have his priorities straight. Koko is a gorilla who has proven human-like awareness and understanding (see www.koko.org), even showing compassion and abstract insight. She can have actual conversations with people in sign language. If in a dire circumstance I had to choose between saving Koko and saving a newborn human baby, it would be hard to justify saving the baby—only the baby's value to someone else, and its potential to develop into a fully-effective human being, would weigh against Koko. Yet if I had to choose between Koko and an adult human psychopath, that would be a no-brainer for Koko. So the fact that different persons have different value is a fact that we all assume to be true already. This does not discredit our ethics in any way.

2.2.4 Speciesism

Moreland in turn responds to the idea that humans are special on Naturalism by calling this "speciesism," the belief that humans are more valuable than animals based solely on "a racist, unjustified bias towards one's own biological classification." But this is a bogus charge. The special value of humans is not based on our commonality of species. As my ape example above proves, and as anyone who is familiar with the *Star Trek* universe knows, we would extend the same value to all species, even machines, who shared our special qualities.

Our value is based on objectively distinct properties possessed by humans that are the most valuable things to us, or to any fully-sentient being—far more valuable than anything possessed by other animals. Each human mind is an entire universe unto itself, a marvelous architecture of personality, of thoughts, dreams, memories, values, of knowledge and ideas and creative drives and skills. No animal comes close to possessing this. The loss of a human mind is a truly profound loss to the entire universe, and the development of a human mind is the greatest, most marvelous thing the universe may ever realize. But more importantly, each human shares our awareness of being, our understanding, our capacity for perceiving happiness, and agreeing to help each other achieve it. And it is by virtue of our ability to truly comprehend happiness in this way that our happiness is so valuable. It cannot be said that we disvalue the comprehension of happiness in animals out of bias, for they do not possess such a thing. They can experience a kind of happiness, and thus they have value, but they cannot comprehend it, truly perceive and savor that happiness, nor can they join society and act among us accordingly. Yet if being happy is good, *knowing* happiness, and how to produce it, is necessarily better. That has little to do with what species you are.

2.2.5 The Meaning of Normative Propositions

Moreland also criticizes the naturalistic metaethic in another way. He argues that "it confuses an 'is' with an 'ought' by reducing the latter to the former" and that "every attempted reduction of a moral property to a natural one has failed" (*Scaling*, p. 112).

But moral statements *only* have meaning if they are reducible to some factual state of affairs, and this is so even on Moreland's theory. Since his whole argument is that moral statements can only be true if Christian Theism is true, it follows that moral statements must be reducible to some fact that is true on Christian Theism. And he tells us what that fact is: the existence of intrinsic values, which are immaterial somethings created by God. On Moreland's theory, had God not created an intrinsic value for humans, and linked it to his system of salvation and damnation, then the statement "you ought to value humans" would be false. Thus, Moreland's 'ought' is also reduced to an 'is', and so he has no right to complain that naturalists do the same thing. There is no relevant difference here between a reduction to immaterial objects and a reduction to patterns of matter-energy.

Moreland happens to be referring to the so-called "naturalistic fallacy." But naturalism does not commit this fallacy if it answers both Hume's call for stating the relation entailed by an 'ought' statement (cf. "Of Morals," 3.1.1, in *Treatise on Human Nature*, 1739) and G.E. Moore's call for resolving the "open question" (*MDAP*, pp. 51-63, with p. 3 and p. 35, n. 2, for sources on the nature of the fallacy Moore claimed to have discovered). My theory answers both.

The Goal Theory of moral value reduces every 'ought' to an 'is' in the same way Kant did for what he called "hypothetical imperatives." For example, when someone says "You ought to regularly change the oil in your car" they may believe they are issuing a normative statement—that it is true everyone ought to do this—but in fact it is only *hypothetically* or "conditionally" true. If I want my car to run well and cost less to maintain, then it would be true. But if I want my car to break down, then it would be false. For instance, if I am a scientist testing the effect of oil deprivation on my car's engine, then to tell me that I "ought to change the oil" in my car would be a false statement. The fact that this statement has exactly the same grammatical and logical form as a moral statement implies that moral statements operate the same way, with one single difference:

they are normative, i.e. true for everyone (and, if we are not discussing the category of *all* normative statements, there is a second difference: a defining convention, per V.2.1.2).

This is similar to a position proposed by Philippa Foot ("Morality as a System of Hypothetical Imperatives," *MDAP*, pp. 313-322). Her original argument was debated by John McDowell and I.G. McFetridge ("Are Moral Requirements Hypothetical Imperatives?" *Proceedings of the Aristotelian Society Supplementary Volumes* 52, 1978, pp. 13-42), where confusion between the roles of facts and values, and what I regard to be the necessary role of desire in motivation, muddles their conclusions. Enlightenment is gained when Peter Railton weighs in on this issue ("Facts and Values," *Philosophical Topics* 14:2, Fall 1986, pp. 5-31), drawing a correct distinction between intrinsic and absolute value that Moreland would do well to heed.

Following this analysis, an "ought" statement is actually the *apodosis* (the "then" clause) of a conditional proposition, where the *protasis* (the "if" clause) of that conditional is implied or suppressed. So a full statement would be "You ought to regularly change the oil in your car, *if you want your car to run well and cost less to maintain*," which is equivalent to "Regularly changing the oil in your car will advance one of your goals, if the effect of regularly changing the oil in your car is one of your goals." Thus, if the 'protasis' of a correct imperative proposition is true ("since x is the case"), then the 'apodosis' is true ("then you should do y"), and hence the imperative statement can be empirically verified as constituting a true proposition.

However, it should be apparent that if the protasis is false, it is not necessarily the case that the apodosis is false. For there may be some other reason to desire the same outcome: there may be another valid protasis that is true, another goal that is satisfied by the same action. This is probably why the protasis is generally suppressed when people state imperatives, normative or otherwise: if they included it, then the statement could be false simply because the wrong motive was stated, whereas if they omit it, then the statement remains true so long as any equivalent motive exists, even if we are not sure what that is in any given case, and even if there are numerous motives, too many to state.

My view of what an "ought" statement means accords in part with that of J. L. Mackie (*MDAP*, p. 98-9). Or as Stephen Darwall puts it, once we are aware of all the facts, regarding how moral actions fulfill our goals in ways nonmoral actions don't, we will act accordingly. In other words, so long as we are rational and sane, "what we *would* do *if* we knew all the facts" is literally equivalent to "what we *will* do *when* we know all the facts." Darwall assumes that this cannot be true by definition (*MDAP*, pp. 306-7). Yet I cannot imagine how it could *not* be true by definition. The word "ought" must be given an established meaning that grounds the truth-value of any statements that incorporate the word, before any other analysis, ethical or otherwise, can even begin. And study of human linguistic behavior for hypothetical imperatives makes a strong case for my analysis being the conventional meaning of imperative vocabulary. The word "ought" is simply meaningless unless it factually connects personal goals with actual (or at least predicted) outcomes.

Analyzing the distinction between my view, called 'internalism', and the opposite view, called 'externalism', Bernard Williams comes to the same conclusion I do, finding that externalism must either be incoherent or a disguised redux of internalism, or else there is no effectively universal motive to be moral ("Internal and External Reasons," *MDAP*, pp. 363-71). So it seems internalism must true by definition, before we even start looking at any facts. Thus, my understanding of the meaning of "ought" statements must be correct.

It follows that the most common, ultimate appeal in all ethical systems is: "You must adopt our values to be happy," and since you want to be happy, once you see the entire truth of our claim you will have every motive to comply. This is true even for Christians: for example, "you must adopt our values to go to heaven" necessarily implies that going to heaven will make you happy, and not going will thwart your happiness. It also assumes there is such a place, and that you will actually get there by adopting the stated values. And so on. Whatever reason you offer for anyone to be moral, when we keep asking "Why?" eventually you will land on "because it is necessary for your happiness." If it wasn't, no one would really care.

Consequently, when challenged ("Why should I do what you say?") people then fish for protases (which describe goal-oriented motives) that verify the truth of the statement that one should do what they say. In effect, metaethics is this very occupation: the quest for a thorough accounting of the motives (as conditional protases) that make moral imperatives universally and thus normatively true. This is why all normative statements are reducible to facts: in particular, factual propositions about human motives and the causal relationship between ends and means. And Moreland cannot really disagree. For if someone had no motive whatever to care about those "intrinsic values" that God is supposed to have scattered here and there around the universe, then they would have no reason to listen to God's moral statements, and thus those statements would have no truth value for that person. It would *not* be true that they "ought to do" what God says. For instance, what if I *want* to go to Hell?

The Goal Theory is based from the start on identifying a universal motive for moral action. So it can easily establish certain moral imperatives as true or false. Moreover, by using the same meaning already established for auxiliary verbs like 'ought' or 'should' in nonnormative cases (and this is the same for other idioms: e.g., 'right' and 'wrong' can be used equivalently), the Goal Theory makes more natural sense—it does not render morals bizarre. So, in general form: "You ought to be moral, *if you want to find true, lasting happiness and to avoid various forms of misery*." The form for any particular moral statement is complex but always derived from this, e.g. "You ought to tell the truth, because doing so has effects $A, B, C, \ldots$ etc., and not doing so has effects $a, b, c, \ldots$ etc., and you desire $A, B, C, \ldots$ etc. and do not desire $a, b, c, \ldots$ etc." Or in a more complete sense: "You will regularly tell the truth when you have integrity and compassion" hence "You ought to cultivate integrity and compassion, because possessing them has effects $A, B, C, \ldots$ etc., and not possessing them has effects $a, b, c, \ldots$ etc., and you desire $A, B, C, \ldots$ etc. and do not desire $a, b, c, \ldots$ etc."

It is this that gives moral statements a standard hypothetical truth value, allowing them to be verified or falsified. All that is needed to make them normative is the fact that (1) everyone "desires $A, B, C, \ldots$ etc. and does not desire $a, b, c, \ldots$ etc." and (2) doing what is stated tends to have "effects $A, B, C, \ldots$ etc.," and not doing so tends to have "effects $a, b, c, \ldots$ etc." Hence, an empirical investigation can ascertain whether lying, or the failure to cultivate compassion and integrity, does indeed have effects $a, b, c, \ldots$ etc., and whether truth-telling and so on has effects $A, B, C, \ldots$ etc., as well as precisely when. For instance, one can account empirically for white lies and other forms of moral deceit, or for the actual level of risk or reliability in producing said effects (an element of the *degree* to which something is moral or immoral), and so on. This sort of research would be a task already well-suited to sociologists and anthropologists. Likewise, an empirical investigation can ascertain whether everyone actually (*when fully informed*) desires the one outcome and not the other (which is the physical fact that establishes normativity), a task already well-suited to psychologists. Thus, ethics falls within the umbrella of scientific investigation, and I suggest that "ethicology" should become a *bona fide* branch of scientific research, the last great scientific field yet to be endeavored.

So, like Thomas Nagel, "I conceive ethics as a branch of psychology" (*MDAP*, p. 323). We are not alone. There are many examples of the rising use of psychology in ethical theory (*MDAP,* p. 47, n. 147). There must be a call for more. Ethicists have been in the armchair too long, and must start collaborating with scientists and gathering actual facts. Ethicology would answer that call, leading to a "real revolution in ethics stemming from the infusion of a more empirically informed understanding of psychology, anthropology, or history" (*MDAP*, p. 35).

As to why such a science has not already begun, I note two facts. First, it is the most controversial and ephemeral subject of all those undertaken by the sciences, and thus the most difficult. It is politically and emotionally charged, so bickering and debate and irrational commitment to beloved dogmas have prevented us from seeing the necessary research programme. Cultures also conceal the truth of morality behind customs and traditions, making moral facts difficult to get to. By comparison, it has taken a long time for science to enter other difficult and controversial domains, as is shown by the histories of evolution theory, scientific cosmology, and consciousness research. Our science would also have to be a specialization of psychology, which is already one of the youngest of the sciences, so ethicology must necessarily be younger. It depends on psychology getting its act together first, which it only recently started doing. In a way, ethicology is to psychology as electronics is to physics. A certain perfection and completeness of the one is necessary for the other to begin.

Some object that the field of ethics, which is prescriptive and involves finding "a valid reason *for acting*," is wholly unlike other sciences, which are descriptive and involve finding "reasons for belief" (*MDAP*, p. 9). But that is not true. First, ethics is a logical deduction from described facts, from what we believe about the facts of the universe and our values, and thus it *is* about reasons for belief—belief about what we ought to do. Second, engineering, agriculture, medicine, and all the technological sciences, even language (no less a technology than anything else), involve prescriptive conclusions about what we ought to do (to achieve x in circumstance y). The Goal Theory places morality as a technology—the technology of social harmony, from a more restricted cultural perspective, but from a broader and more fundamental perspective the technology of individual human happiness. In the one case, we would perhaps be discussing morality within a modern democratic context (wherein "the moral" is conventionally limited to behavior affecting others), and in the latter case we would be discussing the whole set of normative propositions.

2.2.6 Moral Relativism and Moral Controversy

Moreland might then resort to accusing Metaphysical Naturalism of being doomed to a feeble or self-destructive moral relativism, which leaves no foundation on which the moral critic can stand. But this is not the case here. For the Goal Theory does not root moral value in popular sentiment or majority vote, or even in current beliefs or attitudes, but in the actual facts, particularly in the sentiments that *every* person *would* have if they were both (1) fully informed of all the true facts about themselves and the universe and (2) cognitively accurate in their analysis of these facts.

In other words, when we know what is true, and logically deduce from this what is actually valuable to us, we will all find ourselves in agreement about a certain set of objects of value. And this entails a certain set of behavioral principles, and by virtue of this universal agreement, those objects would have objective value, and those behaviors would be objectively right and wrong. This is because their value or rightness or wrongness would be true for everyone, and this truth would not be based on what anyone actually values at any given time, but on what *every* person *would* value given the proper conditions for making an accurate judgment about the truth of any statement.

Of course, since values, by definition, refer to attitudes held by individuals, there seems no sense in which values can be objective. This is because values refer to an individual's experiences, and all experiences are ultimately subjective—they are only shared to the extent that the same world, and the same physiology, sensory organs, and nervous system are shared. But the fact that all values are 'subjective' in the sense of belonging to an individual does not entail that none of these values are 'objective' in the sense of being universally shared (besides being manifest physically in the brains of valuers).

Thus we can explain all moral controversy and progress. For when we try to argue that someone is wrong about a moral proposition we always find ourselves arguing in either of two ways: either we point out how their understanding of the facts is wrong, or we point out how they have deduced the incorrect values from the actual facts. Both can be valid arguments. For instance, we tell the Nazi that his beliefs, like that Jews are not human beings and that they are plotting to take over the world, are *factually* false, and therefore his morals regarding Jews are in error. We also tell the Nazi that even if his belief that Jews are not human beings were true, it does not logically follow that their lives have no value, since nonhumans (even nonliving things) can have value, and the special value assigned to human beings is not based on their species but on qualities they can in principle share with other species and that, as a matter of fact, Nazis clearly share with Jews, even if Jews really were a distinct species. In the first case, errors of logical reasoning from the facts, or false beliefs about the facts, produce a wrong moral conclusion even when our values are correct. In the other case, errors of logical reasoning from the facts, or false beliefs about the facts, lead us to adopt an incorrect array of values, which produces wrong moral conclusions even when those values are correctly applied to true facts in any particular case.

There are thus at least four types of moral error: errors of reasoning in particular cases, and mistakes of fact in such cases, then errors of reasoning in the construction of our value system, and mistakes of fact in such constructions. And these four kinds of error are inevitably frequent and intertwined: for we are often ignorant of or incorrect about relevant facts or deductions in any given case of moral decision (as many of us are in the construction of our worldview, from which we derive our value system). It is the role of the ethical philosopher and moral critic to spot and correct these errors, pointing out how individual human happiness is not served by certain defective values, and how beliefs that are correctly aligned with true facts entail different conclusions, not only about the proper values to adopt, but about all moral subjects, no matter what values one has adopted.

The causes of such errors are multiple, ranging from ignorance and immaturity on the one hand, to confusion and prior commitment to false dogmas on the other. Moral progress is the outcome of overcoming these errors and their causes, and evolving a common value system ever closer to what is normatively best for everyone. Boyd (*MDAP*, pp. 128-30) also gives a fairly extensive list of the natural causes of moral disagreement and controversy, which is just as pertinent within the Goal Theory, adding to or elaborating on what I have just said.

2.2.7 Defining Good and Evil

The Metaphysical Naturalist is also accused of using the vocabulary of "good" and "evil" without justification, since this language only makes sense if God exists, or so the argument goes (Moreland, *Scaling*, pp. 240-248). But when we examine how people actually use these words we find the case to be quite different. What things people actually call 'good' all have in common can be described most succinctly as 'benevolent', 'beneficent', or simply 'benign' (in the distinctly positive sense), whereas for 'evil' the equivalent vocabulary is 'malevolent', 'maleficent', or 'malignant'. It is an undeniable fact that 'evil' is a word humans coined to refer to "anything causing injury or harm," i.e. that which is "harmful, injurious; characterized or accompanied by misfortune or suffering"; likewise, 'good' is a word humans coined to refer to that which is "healthful, beneficial; free of distress or pain; agreeable," per the *Random House Webster's College Dictionary*, McGraw Hill Edition (1991).

In simple terms, the good is that which helps people, heals or nourishes or improves or protects them, while evil is that which hurts people. This is simply how human convention has categorized things, and logically so. It is vain to argue that people can't talk like that if there is no God. Obviously they can use language any way they want to. The distinctions being described are real, and human reaction to those distinctions is naturally explicable.

However, a distinction must be made here between moral good and evil and 'mere' good and evil. For instance, cancer is an evil, but not a moral evil: no immorality produces it, nor is cancer in any sense 'immoral' (at least not on naturalism). In contrast, moral good and evil is that which can be related in some significant way to moral propositions. As we've discussed above, this is usually a subset of all normative propositions, according to some local convention, but even if a society should elect not to call some evil a moral evil, it remains evil nonetheless. For no matter how a society defines 'moral' good and evil, this would remain a subset of all goods and evils, which are defined simply, by human convention everywhere, as benign or malignant. That which has benign effects overall on human beings is good, that which has malignant effects overall is evil, even as many things combine both in different proportions.

It furthers the merit of the Goal Theory that it explains why there should be any connection between moral good and general good, or moral evil and general evil. Not only do the obvious social consequences of an individual's behavior relate to his own personal good or ill, but compassion and integrity relate to human good directly: for the compassionate want to be benevolent (benevolence brings everyone, even the agent, pleasure; malevolence, pain), while discarding one's integrity has malignant effects on society as well as the self (producing self-loathing or mental illness) in contrast to the benefits of self-respect and social order that integrity provides.

Likewise, when we identify which things are truly *normatively* good and evil we can distinguish them from those things that are merely likable or unlikable, or praiseworthy or contemptible, by reference to less universal values, such as personal principles and aesthetic tastes. And so we can intelligibly limit our use of such weighted words as "evil" to what is indeed universally harmful, and also proportion that attribute accordingly.

This also leaves us with an answer to Moore's "open question argument," which is that, where *x* is "ultimate human happiness" and *P* is "good," one might ask: "Yes, I see that *x* is *P*, but isn't there still room for me to wonder whether *x* is genuinely good, whether I ought to regulate my life accordingly?" (*MDAP*, p. 29). In other words, does it still make sense to ask "whether my own ultimate happiness is genuinely good, whether I ought to regulate my life accordingly?" Clearly it makes no sense to ask this. If you should *not* regulate your life according to your own happiness, what worth could life possibly have? What motive could you have to follow such a self-defeating path?

Even Moreland must agree that if faith in Christ did not lead to one's ultimate happiness (e.g. eternal bliss), then there would be little point in being Christian. Indeed, if everyone knew for a fact that Christians burned in hell forever, it would be impossible to argue that being Christian was good, that one should regulate one's life accordingly. So it is self-evident that happiness is the ultimate good, without which life can have no plausible meaning, and therefore things can be described as 'good' only as they relate to happiness. And human convention proves the case, for 'good' is always used in reference to what is believed to be benevolent, because it happens to contribute significantly to happiness in the universe.

For more: see Aristotle's *Nicomachean* and *Eudemian Ethics*; Russ Shafer-Landau, *Whatever Happened to Good and Evil* (2003); and Richard Taylor, *Good and Evil: A New Direction* (1999).

2.2.8 Moral Reason and Moral Intuition

Finally, how does naturalism account for the facts of human moral experience? For not only are we capable of reasoning through ethical issues, but of intuiting the correct solutions. Where does this innate moral sense come from, and can we trust it? The naturalist answer (as described in detail in III.9.1, "Reason vs. Intuition") is that any form of intuition is a skill, like any other, honed through long enculturation, habit and experience. So its trustworthiness will depend entirely on whether it has been built up correctly, and how developed and practiced the skill is. But though a skilled carpenter might be able to intuit the right way to build a house, a carefully reasoned-out plan of construction can correct or improve upon his judgments. Thus, there is a role for moral reason as a check upon, and as a practice for, moral intuition.

Still, moral intuition is necessary. The question of what is good, what is moral, in any particular case is complicated by the fact that we are ignorant of most of the things we need to know to answer the question. Our capacity to predict the future is greatly limited, yet entirely essential to any decisive answer as to what is right and wrong. Our ability to know the secret thoughts of others is also limited, and just as essential, and so on. Thus, the ability to do the right thing, even to know what the right thing is, will depend upon one's wisdom and knowledge, which will never be complete. The degree to which you really know the consequences of what you do, and the significance of what you embody when you do it, will determine the degree to which you can ascertain what is right or wrong in any given case, and that is hard to put down on paper. This is all the more true when we realize that moral decisions usually have to be made quickly, often leaving little or no time for thorough reasoning, much less fact-checking.

The complexity of moral thought, like the complexity of other crafts and enterprises, is thus often replaced with convenient, easy-to-follow rules that various experts believe to be the most useful or universal. But just as no man can be good at anything simply by learning the rules, true morality cannot be found in them. Rather, it is found in wisdom and intuition. Even a chess master must know much more than the rules of chess if he is to be a skilled player. But in morality, the rules cannot even be fixed. Any set rule can fall upon an exception. Thou shalt not murder—but what if you must kill a villain to save an innocent? And any set rule suffers from the flaw of ambiguity. What if you kill by mistake? Rules are useful because they allow us to act quickly when we lack the time to think something through. And when we practice at the rules long enough, they become instinctual, and thus even more effective—assuming the rules were good ones in the first place. For there are such things as bad ideas that seemed at first to be good ones, and these can become bad habits that are hard to break even when we discover their faults.

Since "intuition" is just another word for skill, and though skill can "think" faster than reason, it does not always think as well—especially if it was trained incorrectly to begin with. So having reason as a check on ourselves is essential, and equally necessary for guiding us in the development of our moral intuition in the first place. But when we need to engage in a cognitive examination of our moral beliefs, we should not try so hard to seek moral truth in rules, which are merely artificial expedients, "rules of thumb" devised for those cases when we must act without thinking. Moral truth should be sought in broader principles, such as the so-called "Golden Rule," which has been found in nearly all religious and ethical philosophies in all cultures throughout history. Jesus was repeating an old Jewish proverb when he said "Do unto others as you would have them do unto you," and Confucius was recording an old Chinese saying when he wrote "Do not do to others what you would not want done to you." Both were repeating a very rational concept derived from the two key virtues essential for human happiness: compassion (putting yourself in someone else's shoes) and a sense of integrity (being in ourselves what we expect from others). From principles like these we can then derive moral rules and then rationally arrive at appropriate exceptions to those rules, relying on reason and a well-trained intuition working hand in hand.

2.3 Eliminating Some Metaethical Defeaters

Though Moreland doesn't emphasize them, others have: the inevitable 'what if' questions, posing bizarre circumstances as challenges to a moral system of thought. I will deal with two examples of this genre, to show how the Goal Theory of moral value can generally meet such challenges.

2.3.1 What About Moral Suicide?

For instance, the inevitable question is always pulled out and dusted off, "But what's the evolutionary or even personal advantage of jumping on a grenade for your buddies?" Moreland touches only briefly on this (in *Scaling*, pp. 128-9). That is, since many people believe suicide is sometimes not only morally proper, but even morally obligatory, and suicide seems to provide no advantage whatever, evolution cannot account for this view.

But suicide *does* provide advantages, and we know many animals that commit suicide or engage in self-sacrifice. It is not hard to fathom why: preservation of the species is aided by it. Humans have more than a species to preserve: we have built a monumental cultural structure, of knowledge and ideals, of political and economic systems and artistic creations, that is in some ways even more valuable than the bodies that carry it around. It is not hard to understand why some people would be willing to die for it. This is, after all, what makes the human race so precious and valuable, perhaps unique in the cosmos.

But personal psychology must be looked at to really understand the value of moral suicide. A life of cowardice and failure is hardly worth living, and when we realize that a situation confronts us in which the only choices are death or perpetual misery, death is preferable. There are indeed things worse than death, such as endless torture, destroying those you love, or abandoning the very ideals that make your life meaningful. And there are things worth dying for, to preserve the life of someone or something you love more than life itself.

Ultimately, the moral suicide makes sense within naturalism, since death is the end of all fear and pain, of all sensation whatever, and there is nothing to fear from the end of fear itself. One must certainly dislike and oppose the prospect of death, because a happy life is so valuable and the prospect of attaining it almost always remains for the future. But our love of life and our fight against death need not entail chronic anxiety over dying, whether we choose death or not. When we have exhausted all options, and still conclude there is no longer any prospect of happiness, death becomes an acceptable alternative. And indeed, in cases like the grenade, where you are going to die anyway but have the choice either to save lives or not to save lives in the process, the moral suicide results in no additional loss at all, but a net gain.

Finally, one should not assume everyone's moral beliefs are correct, least of all about suicide. Very often an act of suicide is held up as moral when in fact it is not—very often it is, at best, an act above and beyond the call of moral duty (a 'supermoral' act); or worse, pointless; or worst of all, immoral. The circumstances obligating total self-sacrifice are extremely rare. Even the circumstances that merely justify it, without obligating it, are not at all common. Metaphysical Naturalism can accommodate the concept of the moral suicide well enough, and it does so to a perfectly rational degree, without demanding or expecting too much from people.

This is a point that can apply to many irrelevant objections to natural morality: if something does not make sense as a moral belief on this view, then it probably *isn't* a moral view after all. Evangelical Christians like Moreland would have us believe that early-term abortion or homosexual sex are immoral. But we hold they are not. Jesus, as depicted in the Gospels, would have us believe that letting people rob and beat us is moral (Matt. 5:38-42), or forgiving a criminal five hundred and thirty nine times is moral (e.g. Matt. 18:21-2), or that surrendering *all* your wealth and time to succor the poor is moral (Matt. 19:16-24). But we hold that these are at best supermoral, and that it is immoral to expect such behavior from anyone.

2.3.2 What About Weird Aliens and Psychopathic Robots?

Consider another question in this genre. I have argued that values become normative (and thus 'moral') by being essential or conducive to individual happiness, which all humans desire. But what about people who don't want to be happy? What about psychopaths? What about weird aliens that evolved differently than us? What about heartless sentient robots? And so on.

Of course, there may be some race of sentient alien monsters who could never be persuaded by anything we had to say. One thinks of monstrous races in movies like *Aliens* or *Starship Troopers*. But this has nothing whatever to do with how we as *human beings* ought to live. Morality is necessary for our happiness, even if it isn't for other beings. Still, my theory of the evolution of a moral species (section V.2.2.1, "Evolution of Moral Values") would also apply to alien races. So we can expect most alien races to have evolved similar basic moral values as our own simply due to convergent evolutionary forces—all the more so since a space-faring civilization could not be built from scratch, much less maintained, by a heartless, treacherous society. Likewise, robots could be modeled on humans, avoiding any question of an alienating difference. But apart from such things, rare exceptions, which are clearly abnormal, do not invalidate a normative rule. For instance, there are certain normative procedures for growing corn, but given some truly rare and abnormal climatic conditions these rules might not apply. So the fact that there might be some people who don't want to be happy would not affect the normativity of moral values for the rest of us, only for them.

Are psychopaths just this sort of person? Psychopaths (also called 'sociopaths') certainly seem to be an exception to the rule: conscious beings lacking any values in common with other human beings, by which the truth of moral statements would be established. Hence, I allow for the "monster" as an exception to my theory of moral meaning, granting the hypothetical case of a conscious being so alien to normal human existence that it is beyond good and evil, in the same way as a spider or a shark. But this being would then be worthy of no more compassion or respect than such dangerous animals—hence the way they are treated in *Aliens* and *Starship Troopers* is more than appropriate. They would be incapable of earning the full status of moral dignity, and they would be the closest thing to genuine demons as there can be. Boyd has even more to say on the psychopath and his irrelevance to the normativity of natural values (*MDAP*, pp. 130-2). Many systems of Christian theology already accept the existence of such beings: demons are condemned to be evil, and are, unlike us, incapable of escaping the company of Satan by an act of free will.

But does such a monster really exist? What, after all, is wrong with psychopaths? Do they not care if they are miserable? Or do they for some reason actually achieve happiness through immoral behavior? Several leading experts who have worked with and attempted to treat psychopaths report that their patients put on a front, that in fact they hate themselves and existence and live generally miserable lives, constantly wracked with paranoia and loneliness. Some could even be brought to openly admit this—and not only that, but admit that they hated the fact that they couldn't escape their sociopathic behavior patterns and couldn't share the emotions that others enjoy.

Though several psychopaths have said they would gladly accept any treatment that would fix them, no such treatment is known. To this day there is no cure for the disease, which is partly genetic (it is shared by more twins than chance predicts, and is far more common among men than women) and partly produced from certain inept parenting environments (for those who are curious, the key cause appears to be a sustained environment of inconsistent rewards and punishments). But this means my theory holds even for psychopaths: moral behavior is essential even to their happiness, but due to a mental disorder they are incapable of developing the mechanisms that produce moral thought, and so are trapped in misery.

> See Dennis Doren, *Understanding and Treating the Psychopath* (1996); and William Reid, John Walker, & Darwin Dorr, eds. *Unmasking the Psychopath: Antisocial Personality and Related Syndromes* (1986).

3. Moral Conclusions: Tying it All Together

Throughout his attacks on secular metaethics, J. P. Moreland fails to disentangle the words *meaning*, *purpose*, and *value*, yet they have slightly different connotations. In my vocabulary, life has meaning to the degree that it has any significance or importance to anyone, life has purpose to the degree that it is employed to any creative end, life has subjective value to the degree that anyone values it, and life has objective value to the degree that every sane and informed person would value it if informed of all the facts, and their cognitive faculties were functioning without error. It is special pleading to insist that meaning, purpose, and value must be cosmic or ultimate or transcendent or whatever jargon one wishes to throw at it. So long as life has any meaning, purpose, and value, it has meaning, purpose and value. Q.E.D.

Contrary to Moreland's claim, I have shown that Metaphysical Naturalism can explain and justify a moral view of life more than adequately. I conclude that naturalism provides the most sensible, most confirmable, most adaptable, and most universal value system, one that is reasonable and effective, and ideal for progress into a new age of enlightenment. Secular Humanists have always said that a practical, philosophical education and lifestyle, and a healthy family life, are essential to healing moral deficiencies in any society. This is entirely true. Believing on Christ has nothing to do with the matter, as history demonstrates. Some even argue that Christian Theism creates more problems for an ethical theory than it solves (see section bibliography below).

Finally, philosophers might ask how we should classify this moral system, and I will conclude with a list of names and jargon that the layman can be excused for skipping. Within the Goal Theory, morality is most like a technology, and one could call it a technology of happiness. Like techniques for farming corn, there is more than one way to do things, but there are better ways and worse ways, possibly even one *best* way, and this varies by circumstance, but in all cases it is the nature of the universe itself that determines this, and not human opinion. That a good moral technology is better at securing human happiness than a poor one is a fact independent of my or anyone's beliefs, feelings or wishes. Thus, the Goal Theory is an objective and not a subjective ethical system, even though it derives entirely from the nature of conscious existence, and thus ethics does not exist apart from the existence of conscious beings. Though it is based on essentially subjective human desire, it only rests on a *universal* one, meaning it is objective in a relational rather than an absolute sense. It thus has the merit of accommodating both subjectivist and objectivist theories of ethics, obtaining the benefits of both and the faults of neither.

Likewise, this is a realist theory, since moral propositions describe real facts about humans and the universe, which can be discovered by objective empirical inquiry. I do not accept Simon Blackburn's straw man of "realism," where it is said to entail that values are external to humans ("How to be an Ethical Antirealist," *MDAP*, pp. 167-78). Objective realism does not require this. It only entails that values exist independently of human opinions and beliefs. Values are patterns of relationship between our happiness and actual and potential things, actions, and consequences in the universe, patterns that we must train ourselves to detect, just as we must train ourselves to detect complex notes in a piece of music. But the fact that we think they adhere in objects (rather than in our brains, in our perception of complex relationships involving ourselves) is indeed a perceptual illusion that Blackburn's "projectivism" adequately captures and explains.

In accord with John McDowell's analysis, my theory presents value as a secondary rather than a primary quality (to borrow a rather antiquated distinction). He draws an apt analogy between legitimate and illegitimate attribution of the terms "valuable" or "right" or "good" to valuable phenomena, and legitimate and illegitimate attribution of the term "fearsome" to fearsome phenomena (John McDowell, "Values and Secondary Qualities," *MDAP*, pp. 201-13). There is nothing innate in something that makes it fearsome: its fearsomeness is entirely relative to the person observing it. Nevertheless, the attribute "fearsome" is a property really possessed by an object when placed in relation to someone who is in fact threatened by it. So it is with value terms, which are goal-oriented character-innate desires inherent within *us*, which happen to focus on certain external facts that people like Moreland then mistakenly take to be inhabited by our projected values.

The Goal Theory also accommodates moral intuition, which is like any other skill, and so this theory unites the best features of cognitivism and noncognitivism. A farmer can intuit correct actions in corn farming, in direct proportion to his skill and experience and the correctness of his over-all knowledge, and so can a human with regard to moral reasoning, without engaging deliberate reason or being aware of the truth-value of particular moral propositions. And yet all moral propositions have a cognizable truth value. The Goal Theory also unites both deontological and teleological ethics under the umbrella of a virtue-based theory, since integrity relates to the categorical concerns of theorists like Kant, while compassion relates to the utilitarian concerns of theorists like Mill, yet in neither case do we fall into their extreme views.

Finally, the Goal Theory can easily incorporate Richard Boyd's "homeostatic consequentialism" (Richard Boyd, "How to Be a Moral Realist," *MDAP*, pp. 105-35). In fact, the only aspect of his theory that I reject with certainty is his denial of egoism. Not only is the Goal Theory fundamentally egoistic, I have argued that no successful metaethics could ever be anything but. However, the Goal Theory still eliminates selfishness as a valid moral motive and is thus not ego*tistic*. Selfish behavior (as in caring *only*, or mostly, for oneself) is a different sub-set of self-interested behavior than that appealed to by the Goal Theory, and there is certainly no equivalence. If I engage in self-sacrificing and selfless acts, if I maintain concern about others in my decisions, I am not selfish, even if I act this way out of self-interest. I am still a good person even if I am kind because I enjoy it or if I am honest because I value my reputation. Compassion and integrity definitely conflict with egotism and selfishness in numerous ways, yet everyone always has self-interested (or ego*istic*) reasons to be compassionate and consistent (as shown in section V.1.1, "Do Secular Humanists Have No Reason to be Moral?"). So though one can call my theory egoist, it still upholds compassion and integrity as the most overriding values any human can have, essential for that ultimate of all values: the pursuit of happiness. In effect, I wholly reject Thomas Nagel's conclusions with regard to metaethics, at least concerning egoism, as wrong in almost every conceivable respect (Thomas Nagel, "From *The Possibility of Altruism*," *MDAP*, pp. 323-40; on egoism: pp. 336-8).

This ability to unite the bulk of all competing ethical theories into one coherent view is a strong argument in favor of the correctness of the Goal Theory. For if there is anything objectively true in ethics, it follows that significant existing metaethical disagreements must be based on genuine perceptions (or misperceptions), and therefore the correct theory, whatever it is, must be able to explain all those perceptions and show how they do not lead to disagreement after all. The Goal Theory does precisely this, and better than any other existing theory. It thus has the strongest claim to being correct that I have yet seen.

> A wide-ranging critique of traditional theistic ethics and metaethics appears in Joseph Daleiden, *The Final Superstition: A Critical Evaluation of the Judeo-Christian Legacy* (1994). Regarding how theism actually creates new problems for the ethical philosopher, see: Michael Martin, "Copan's Critique of Atheistic Objective Morality," *Philosophia Christi* (Series 2) 2:1 (2000), pp. 75-89; Richard Gale, "Freedom and the Free Will Defense," *Social Theory and Practice* 16:3 (Fall, 1990); Theodore Schick, Jr., "Morality Requires God... or Does It?" *Free Inquiry* 17:3 (Summer, 1997); Graham Oppy, "Is God Good by Definition?" *Religious Studies* 28 (1992), pp. 467-474; Adolf Grünbaum, "The Poverty of Theistic Morality," *Science, Mind and Art: Essays on Science and the Humanistic Understanding in Art, Epistemology, Religion and Ethics, in Honor of Robert S. Cohen* (1995), pp. 203-242; Richard Gale, "R. M. Adams's Theodicy of Grace," *Philo* 1 (1998), pp. 36-44.

VI. Natural Beauty

1. Beauty as Emotional Appraisal

What is art? The word "art" has many uses and connotations, and all debates about what it means only expose this versatility. Every definition is correct in its own context. But "Art is the skilful production of the beautiful," in the words of the *Oxford English Dictionary*, is the primary meaning, which raises the fundamental question: What is beautiful? And why? That is something any complete worldview must have an answer for. Here is mine.

"Beauty" is something we feel, an emotional response that humans (and no doubt other animals) have when confronted with a particular experience, whether auditory, visual, tactile, intellectual, or whatever—even complex experiences involving several senses at once. This feeling tells us what we like, and it reminds us, in a way, of what is good for us, or good about us. When you consider what this means in my worldview (III.10, "The Nature of Emotion"), you will naturally wonder what exactly it is that the beauty reflex is appraising, and whether it can be right or wrong, and if so, when. To answer that question we must explore what sorts of things typically have beauty, or at least could have it.

Art, for instance, is usually a work, something made or done, and the skill involved in doing this is also called an "art." The kinds of things we regularly call art are either of a performance or material. Performance art includes theater, dance, and music, where the work our mind appraises is something in the moment and then, when done, is gone. Material art includes sculpture, painting, and poetry or prose, where the work our mind appraises is something enduring. It can be stored, moved around, revisited. It is static and unchanging. But there is also a kind of art that combines both: film, for instance, or recorded music. These arts produce enduring works that can be stored, moved around, and revisited, yet they involve active performances that are anything but static. What do all these things have in common that might move our mind to feel their beauty?

Moreover, works of art are usually human creations, distinguished from nature's products. Yet we also experience beauty when beholding works of nature. In fact, our most frequent and powerful experiences of beauty probably involve just such things: the beauty of an animal's coat or gait, of a flower, a landscape, a cloud, a person's face or body. The list is endless. What do these things have in common with works of art? We can see at once that they share the same categories: there is beauty in something happening, and beauty in something static and enduring. In every case it would seem to be certain patterns of perception that trigger the beauty response.

On naturalism, whatever this mechanism is that causes us to see and feel a beauty in things, it must either be a product of evolution, genetically encoded in us for some purpose conducive to our species' survival (or the accidental side-effect of such a thing), or a product of culture, solely and entirely taught to us, the mere passing on of an arbitrary tradition, which may or may not provide a benefit to that culture or society. Or, as is often the case, it could be some combination of both. Either way, the beauty response must be, in whole or in part, either good for us, or frivolous.

As it happens, scientists have only just recently begun to explore this issue, plumbing the biology behind the beauty experience. Their results so far are basically this: the feeling of beauty is an appraisal mechanism that is useful for survival and is thus inborn, but it keys on certain patterns in experience that can be reproduced in many different ways, and how humans exploit this response in their appreciation and production of art is largely cultural.

2. Eight Rules of Beauty

In a 1999 issue of the *Journal of Consciousness Studies*, Drs. Ramachandran and Hirstein elaborated eight natural criteria that trigger our brains to an emotional feeling of beauty (see section bibliography below). Their work focuses on visual media, so the other senses might key on some different properties besides these, but many of the same principles can be adapted to other sensory domains.

In general, Ramachandran and Hirstein argue that "what the artist tries to do (either consciously or unconsciously) is to not only capture the essence of something but also to amplify it in order to more powerfully activate the same neural mechanisms that would be activated by the original object." In fact, "to abstract the 'essential features' of an image and discard redundant information is essentially identical to what the visual areas themselves have evolved to do" (p. 17). They identify eight features along these lines. Logically, we should expect that the degree of beauty response—from mere distraction to amusement to awe—is related to how many of these features are present and how strongly. These evolved triggers are as follows.

2.1 The Peak Shift Effect

The peak shift effect is ... a well-known principle in animal discrimination learning. If a rat is taught to discriminate a square from a rectangle (of say, 3:2 aspect ratio) and rewarded for the rectangle, it will soon learn to respond more frequently to the rectangle. Paradoxically, however, the rat's response to a rectangle that is even longer and skinnier (say, of aspect ratio 4:1) is even greater than it was to the original prototype on which it was trained. This curious result implies that what the rat is learning is

> not a prototype but a rule, i.e. *rectangularity*. [p. 18; this relates to what I discussed in III.5.4, "Abstract Objects"]

Thus, the first rule of art is that observers abstract from particular cases more fundamental rules about certain patterns of experience, and the beauty response is a reaction to peak manifestations of likable patterns in experience.

I discussed human beauty a little in III.10.3, "The Nature of Love," where I noted the scientific fact that a ratio of 0.6 to 0.7 between a woman's waist and hips is most widely regarded as the most beautiful, for that is the "peak" point of convergence between our brain's keying on a particular hourglass form, which loses its pattern if it is too straight or too curved. But I also noted two other facts there. First, nature thrives on variation, and this is only the statistical average result from surveys of admirers of the female form. Many are keyed to like slightly more or less curvy women. Second, this is only half the story—for much of what we find likable in the first place is the psychological result of our upbringing and past experience, which is heavily influenced by our culture. It is also very individualized, as I noted in hypothesizing why I liked brunettes.

But even after including all this (some triggers are innate, but many triggers are acquired, and all triggers vary from person to person around a common average), the peak shift effect is still relevant: what we will regard as most beautiful will be the peak example of what we like—in other words, as much of it as we can experience without it becoming something else (as if we stretched the rat's rectangle so far we ended up with a mere line). Notably, this means Plato's theory of the Ideal Form stems from a natural, inborn human beauty response: we *like* idealizing patterns in our experience. And this keeps us striving toward the best.

2.2 The Correlation Effect

> One of the main functions of 'early vision'...is to discover and delineate *objects* in the visual field...and for doing this the visual areas rely, once again, on extracting correlations. (p. 21)

This is the origin of the correlation effect, the pleasing nature of certain correlations in perceived patterns. For example, "In 'colour space' the equivalent of this would be wearing a blue scarf with red flowers if you are wearing a red skirt; the perceptual grouping of the red flowers and your red skirt is aesthetically pleasing—as any fashion designer will tell you."

They hypothesize that "there may be direct links in the brain between the processes that discover such correlations and the limbic areas which give rise to the pleasurable 'rewarding' sensations associated with 'feature binding'." I would put it this way: we are hardwired to like harmony, wherever it manifests, whether visual, auditory, even in living and working. After all, harmony is usually indicative of health, as well as order, which is usually safer and more useful than mayhem.

One way science can inform our study of beauty is in just this respect. Our brains, for instance, are known to have twenty or thirty areas dedicated to recognizing different visual attributes, like motion, color, depth, form, etc. Ramachandran and Hirstein describe the way these brain areas work by employing physical distance. For instance, we know that objects close together in our visual field are registered in cells close together in the brain, and objects distant from each other are registered in cells farther apart in the brain, and this physical mapping is employed by the brain to remember these distinctions. But separate areas devoted to other attributes like color probably do this, too: "points that are close together in *wavelength* are mapped close together in the colour area of the brain *even though they may be distant from each other physically*" (p. 21). When such closeness is detected (by keying on the time it takes to send a signal between the related cells), in certain cases it may trigger a pleasing feeling, since such matches are very often important.

Thus, understanding how the brain breaks down and analyzes the attributes of what we see will inform us about what it is the brain uses to identify visual beauty, and it may be so for all other forms of beauty. Even our love of simplicity might be explained this way: simple ideas involve fewer components, possibly stored closer together in the brain's cells.

2.3 The Stand-Out Effect

All of our senses are especially keen at isolating and calling attention to phenomena that stand out amidst the background, things out of place or otherwise stark. This is not only useful, but often a necessary outcome of having a finite brain, which can only do so much at once, working much better when it needs only to focus on one or two interesting features or properties of an observed work or scene.

As Ramachandran and Hirstein put it, "extra redundant information can actually distract your limited attentional resources away from the defining attributes of that object. Hence the aphorism 'more is less' in Art" (p. 24). They mean, also, not merely a single object *per se*, but even one or two qualities: that is, given that there are some twenty and more separate areas of the brain involved, each dealing with one type of quality (such as color in one place, form in another), the brain does better when it doesn't over-use too many of these. Thus, a stand-out object that is interesting in twenty or more ways will be less attractive, less potent, than one that is interesting in one or two ways: a stark monocolored object, vs. a stark mishmash of complex patterns.

There are many ways this can generate a pleasing sensation. It is possible that, for instance, an attraction to smooth and unblemished skin marked by the stark contrast of eyes, lips, and eyebrows, is a manifestation of this effect. This would be heightened by our already inborn attraction to the organization of a human face (babies, for instance, find objects with features in a face-like pattern to be more pleasing than others, and some studies have shown that beautiful faces conform to predictable geometric patterns).

Likewise, it is common for stark features to make a scene more beautiful: observing a grassy hill may be pleasing, but plant a single oak tree in the picture and it becomes awe-inspiring. The ability to bring a subject to the fore, and contrast it with its background or periphery, is one of the most common skills artists in all fields focus on. They often seek new, creative ways to accomplish it. And when they can do it with as few tricks as possible the better—the more impressive, the more alluring.

2.4 The Contrast Effect

Related to the Stand-Out Effect is the brain's obviously useful deployment of resources on distinctions and changes in the sensory field, rather than on the more mundane and complex background noise. Our visual cells, for instance, respond far more strongly to "edges" and other visual transitions than to "homogeneous surface colours" and "what the cells find interesting is also what the organism as a whole finds interesting and perhaps in some circumstances 'interesting' translates into 'pleasing'" (p. 25). Ramachandran and Hirstein give a perfect example, which draws upon even more complex perceptual associations and contrasts than mere color transitions:

> For instance, a nude wearing baroque (antique) gold jewellery (and nothing else) is aesthetically much more pleasing than a completely nude woman or one wearing both jewellery and clothes, presumably because the homogeneity and smoothness of the naked skin contrasts sharply with the ornateness and rich texture of the jewellery.

In fact, it may be even more profound than that, as the brain enjoys *intellectual* dissonance as well: the contrast of antique artifice with the raw beauty of nature and youth, for instance.

The beauty response's triggers may seem contradictory, but pleasure still arises when they function without conflict. Again, Ramachandran and Hirstein:

> Grouping on the basis of similarity [i.e. the Correlation Effect] is rewarding, but if so how can contrast (the very opposite of grouping) also be rewarding? One clue comes from the fact that the two mechanisms have different spatial constraints; grouping can occur between similar features (e.g. colour or motion) even if they are far apart in space (e.g., the spots on the nose and tail of a leopard). Contrast, on the other hand, usually occurs between dissimilar features that are physically close together. Thus even though the two processes seem to be inconsistent, they actually complement one another in that they are both concerned with the discovery of objects—which is the main goal of vision. (Contrast extraction is concerned with the object's boundaries whereas grouping allows recovery of the object's surfaces and, indirectly, of its boundaries as well). It is easy to see then why the two should be mutually reinforcing and rewarding to the organism. (p. 27)

2.5 The Symmetry Effect

As physicists have noted over the past forty years, symmetry and the breaking of symmetry is an underlying pattern manifest throughout the universe in even its most fundamental structure. Most physical laws can be defined in terms of the symmetry or asymmetry in certain patterns. Symmetry has also become fundamental to the overall construction of living organisms, with many obvious advantages. It happens to be simpler than coding for asymmetrical forms, and simpler things are always easier for natural selection to hit upon and preserve. As a result, symmetry quickly becomes a good indicator of physical and genetic health.

It is thus not surprising that a natural liking for symmetry should end up in our brains. Symmetry is a good indicator of a living organism as opposed to a random natural landscape, or of a healthy rather than an unhealthy organism. This is distinguished from the Correlation Effect by the fact that symmetry is not defined by a mere nearness of attributes, but by an actual inverse reflection of a particular pattern of matter-energy in space-time. In other words, a pattern paired with its mirror pattern is naturally beautiful.

2.6 The Counter-Symmetry Effect

Ramachandran and Hirstein note how "your visual system abhors interpretations which rely on a unique vantage point and favours a generic one or, more generally, it abhors suspicious coincidences" (p. 30). This is an inborn perceptual feature that actually keeps the Symmetry Effect in check, just as the Correlation Effect can counter the Contrast Effect. There appears to be a natural inclination in some contexts to prefer experiences that are at least a little asymmetrical, a little random. Too much order, and for some reason the response is to feel jarred rather than pleased.

This is why irregularities, imbalances in the composition of artworks and natural scenes, can sometimes be pleasing to us, while unnatural or "surprising" symmetries are not. Ramachandran and Hirstein analyze this in terms of the natural logic of preferring what appears to be a general point of view as against a unique one, for the former typically contains more information, and more objective information, while the latter is more limited to subjective circumstances, which are less relevant to the object being observed. A preference for the former is therefore naturally useful to survival. The example they give is of a tree accidentally matching the terminus of two mountains. We prefer to see the tree off-center, and thus are often stimulated to move our head to make sure the scene is not artificial.

But this might also be a perfectly practical inborn love of nature itself: nature exhibits certain fractal and chaotic patterns, nevertheless merged in orderly ways, and we find pleasing what reminds us of natural surroundings. The human love of greenery, for instance, manifests this tendency. And consider trees as a particular case—the organization, the pattern of their branching and of their roots and trunk and bark is not thoroughly symmetrical, nor yet truly random. As it is useful to be comforted by a tree—a sign of water, shade, wood, and of life in general—it is useful to be comforted by its particular sort of natural pattern. And so we have evolved to be.

2.7 The Analogy Effect

Humans love analogies. They are a useful way of thinking about things, a primitive and often practical way of applying knowledge of one thing to another, a crucial first step toward reason. And though analogizing thought is more flawed than any precise logic, it is more probable that a pre-rational brain would evolve a less-accurate but easier-to-deploy logic of analogy. And this would remain useful even as a rapid fallback when we lack the time, ability, or information to take apart a problem piece by piece with some tool of formal or informal analysis.

The human love of puns and symbolic myths is an early reflection of this natural passion, but as our minds become more sophisticated in their reasoning ability they grow bored with simpler pleasures and seek more profound challenges to their higher cognitive talents. Hence, we are charmed, sometimes deeply moved, by profound and well-chosen symbols or metaphors in art. Not only does analog logic improve, even set the foundation, for the ability to comprehend more about things and to convey such information efficiently, it is also challenging. And our brains love to be challenged, at least when they are healthy and have been stimulated properly in childhood. This instinctual love of challenge is what drives us to excel, an obvious survival advantage.

Ramachandran and Hirstein describe the mechanism this way:

> A metaphor is a mental tunnel between two concepts or percepts that appear grossly dissimilar on the surface. When Shakespeare says 'Juliet is the sun,' he is appealing to the fact that they are both warm and nurturing (not the fact that they both reside in our solar system!). But, again, why should grasping an analogy of this kind be so rewarding to us? Perhaps the use of a simple concrete example (or one that is easily visualized, such as the sun) allows us to ignore irrelevant, potentially distracting aspects of an idea or percept (e.g. Juliet has nails, teeth and legs) and enables us to 'highlight' the crucial aspects (radiance and warmth) that she shares with the sun but not with other women. (p. 31)

More even than that, this focus on the essential detail to be communicated by employing a simple, well-understood symbol, is a remarkably efficient way to communicate. For it contains within it a whole galaxy of emotions and ideas, which we can immediately grasp by drawing into mind the sun and a woman and extracting everything they can have in common, leaving a great deal to impress the mind, all triggered by a mere four words. This all taps into the basic skill of being able to take a series of specific perceptions of a pattern and then extract the features of that pattern that are common to them, assigning *that* metapattern a name and a place in the mind. In other words, herein lies the fundamental machine of abstraction (cf. III.5.4, "Abstract Objects").

2.8 The Anticipation Effect

Ramachandran and Hirstein identify a phenomenon in human beauty response that they also attribute to our brain's love of challenge: art that presents a puzzle or difficulty the brain presumably must solve, is pleasing (and useful, since it keeps us at trying to figure out something that has caught our attention). However, the evidence they present for this is of a rather different nature altogether—for they do not offer solvable puzzles. For instance, they cite the example of "a model whose hips and breasts are *about* to be revealed" which is widely regarded as "more provocative than one who is completely naked" (p. 33). This is true. But the brain can't actually undress her, except imaginatively, which is not really a solution, nor very challenging. Rather, this is an example of the pleasure of anticipation. One can think of other examples, such as Myron's *Discus Thrower*, where the peculiar beauty of the sculpture lies in the way the athlete is wound and ready and just about to release the discus. In both examples, it is *anticipation* that initiates the pleasure response: our brain can almost touch, almost see the next thing happening, and its attention is seized by this. Consequently, many forms of art employ features of suspense and mystery.

3. Beauty in Human Life

I have not used the same order of principles as Ramachandran and Hirstein, nor have I done justice to all the details they explain in each case with regard to the underlying neurophysiology, the natural advantages of generating a pleasure response to each effect, or the particular ways the principles manifest themselves in practice and in theory. Ramachandran and Hirstein themselves point out how their eight principles are not exhaustive: there are many others one might explore (rhythm, balance, depth, weight, or more subtle patterns, such as evocativeness and originality). But the overall picture should be clear: the beauty-sense is a natural, inborn instinct in us all, whose principles are shared by us all, which have practical and useful reasons for being there.

3.1 Is Beauty Bunk?

Does this mean, then, that all beauty is just a set of rules that stimulate cells? Have we robbed the world of its beauty by explaining what beauty is? Hardly. This is the same *modo hoc* fallacy I have confronted before (see index). Knowing *why* beauty feels the way it does and *why* we find certain patterns beautiful does nothing to change the fact that beauty exists and is pleasing and makes our life more enjoyable. It is all the better that the things we find beautiful will more likely than not be good for us as human beings. As Ramachandran and Hirstein put it:

> If a physiologist were to publish a paper explaining the neural basis of sex, i.e. in terms of the activity of neurons in the hypothalamus, septum and other limbic circuits...would [people] suddenly stop having orgasms or stop engaging in sex? Or would reading about the detailed physiological

mechanisms and evolutionary origins of digestion suddenly stop you from enjoying or digesting food? (p. 73)

The obvious answer is "No." Kittens don't cease to be adorable just because we know we are programmed to find the features of infant mammls disarming. Beauty remains even when we know why. This holds true even when we factor in the cultural and idiosyncratic molding of our beauty analyzers. As I described in III.10.1, "Emotion as Appraisal," we can mold our values. Likewise, we can mold our beauty evaluators, and in fact they are very easily influenced unconsciously by our own life experience. Growing up in a particular culture, a particular geography, a particular household, with a particular sequence of happy and unhappy experiences, changes what we notice and what we like, what brings comfort and pleasing associations, and what not. All this is meaningful, and therefore worthy of appreciation. That is not only why these things are beautiful, but why they should be.

3.2 The Subjective Nature of Beauty

It follows from the above that our idea of beauty will be different from everyone else's. It will be different when compared across cultures, and different even in comparison with neighbors in our own culture. We can also "acquire tastes" through repeated exposure to new environments or objects, especially under happy circumstances. This is easier still when aided by others who help point out the salient and beautiful features of some particular work of man or nature.

Herein lies the difference between imperative values, and opinions (II.2.2.3, "Opinions" vs. II.2.2.4, "Moral Imperatives"). The crucial difference is that, though in each case there might be universal (and thus objective) truths about the causes and nature of beauty, just as for moral good, only the latter is always, by definition, tied to what is good for people—to the practical achievement of happiness (per V.2, "Morality in Metaphysical Naturalism"). And so it makes sense to say you *ought* to have certain moral values. But it makes much less sense to say you ought to have certain *aesthetic* values. Though the beautiful has a relationship to what is good for us, it is not a necessary relationship (especially in the long run), nor does it relate in any logical way to proper *action*. Beautiful things can be bad for us, or indifferent, while ugly or plain things can be good. Moral reason thus supercedes artistic taste when it comes time to decide what to do. And whereas talk of moral value is about values that motivate actions that are good for all people, regardless of everyone's differences, beauty is always highly personalized, by one's own culture and unique life experience.

Therefore, beauty *is* in the eye of the beholder. We know there are, for humans at least, universal underlying principles that generate commonalities even amid differences. But despite this, people value different things artistically, and it is rational for them to do so. No one can argue with my reasons for being charmed by kittens, for the associations they bring to me from my personal background, and the features of their personality that I admire, and other qualities that are pleasing (like their cleanliness and furriness), are perfectly rational reasons to feel charmed, even as they might not be for someone else who has different experiences and interests. We can influence each other, by drawing attention to features of the pattern "kitten" that someone did not notice before, or by giving each other certain experiences that change the associations we make or the interests we have. But there is no sense in saying "Kittens are charming as an absolute inviolate principle, and so they ought to be charming to all people." That would simply be a false statement. It would be false even if everyone on earth just so happened to find kittens charming, since this congruence would be purely accidental.

But this does not mean art criticism is all just hot air. Though aesthetic values pertaining to the content of art and its presentation are largely (though not wholly) subjective and non-universal, the art critic still aims to do three reasonable things. First, as we've already noted, she calls attention to the merits or flaws, the features deemed beautiful or ugly, which an audience might otherwise not detect or appreciate. Second, a critic may write for a particular audience that shares certain aesthetic values in common, or for several such audiences at once, thus aiding like-minded people in seeking out what they enjoy, letting them decide whether they would like to view the piece, or support other artists of similar style, and so on. And third, by explaining why the critic likes or dislikes it, or what is likable or unlikable about it and why, the reader will gain insight into the aesthetic values of the *critic* and whether he would agree or disagree with her, hence expanding our understanding of humanity and human variation, and of particular aesthetic, cultural or subcultural groups in particular.

3.3 The Higher Virtues of Art

Though I am persuaded by the analysis of Ramachandran and Hirstein—even if they are wrong in some details they are very likely right about their general programme and its significance—I see three "higher" principles at work in what we might call "high art," truly moving masterpieces, classics, the pinnacle of the beautiful in human creation. These criteria I draw from my own experience, and I have confirmed them again and again not only in analyzing my own appreciation for certain works of art, but in discussing and learning about how others view them, too. The relevance of these lie in the same place as for the eight principles, and in the same qualified way: to find these qualities beautiful is often a good thing, good not only for human survival, but for achieving the even more elusive and valuable object of human *happiness*.

3.3.1 Art as Communication

One of the three highest virtues of art is that when the human brain is trained to comprehend art's subtlety, this becomes an effective means of communicating emotions or ideals. This is mainly achieved by "The Analogy Effect" (VI.2.7), but any of the other effects can enhance a symbol's power or meaning. Through art we can capture the essence, and the rich depth of emotional and symbolic associations, of the heroic, the good, the villainous, the bad. In short, art can carry deep cognitive meaning. And truly good art, art that is genuinely good for the human race, art of which we can be most proud, will do this. This is especially true in the art of cinema, which can deploy the most media to a common end, merging the visual with the auditory and written, the moving and the static.

But even on a simpler level art has such a virtue. Art can move people, it can set a tone, capture and convey a mood, and that makes it very useful. We lack the power to directly stimulate emotions, and thus need tools and techniques to do so. Art is one of the best—the safest of all drugs. Even furniture can convey a social or personal meaning, when made to elicit an attitude or to otherwise soothe, even as it performs a utilitarian function. Art can also provide an emotional escape, or a challenge. And it can give our imaginations experience, through suspension of disbelief, and manipulation of emotion and reaction. This is art's first virtue as a tool of communication.

3.3.2 Art as Education

Related to its first virtue is art's second: its ability to aid the understanding of the real, even through the medium of the unreal. The contrast alone educates. We can learn a lot about ourselves and each other through an appreciation and immersion in the arts. Our universe becomes more colorful and vibrant, more interesting and valuable, when we can see in it patterns of beauty that were invisible to us before.

When an artist employs his medium to demonstrate his careful observation and comprehension of the natural world, or people or society, when he captures and conveys the essential patterns of things and juxtaposes them in meaningful ways, he captures nature, he captures humanity, he captures society. And he presents it in a way no dry history book or photograph typically can. By commanding our interest from the integration of the elements of beauty (like the eight surveyed above), we are stimulated to learn from a work of art, rather than ignore it in boredom. And it does us credit when its message is true. For then we really learn, and learn something we really should learn, about ourselves and our world.

This virtue does not even require the artist's intention. The pure accidents of composition, or the subconscious effect of an artist's vision on his work, or the role of the audience's background and perspective, all can produce entirely new meaning worthy of our interest.

3.3.3 Art as Skill

Related to its second virtue is art's third: art is the exhibition of talent and skill. I do not merely mean creativity or originality. Plenty of bad art is creative and original. I mean the actual ability, the skill, required to carry out the technical demands of producing a good work of art. Humans naturally admire, often deeply, the achievements of their peers. Our own accomplishments awe us. But even more, they make us proud, proud that humanity can be so clever, so disciplined, and produce something so profound. It is one of the things that makes us so manifestly valuable to all other humans, and so worthy of each other's universal love. It is a great disappointment when humans abandon this merit in themselves and waste their lives in useless dissipation. Nothing is so ugly as apathy.

But place in contrast to this a man's master craft, and we see the final virtue of art. I confess I find little beauty in abstract art—or at best a trivial beauty, which might amuse but will hardly impress. Consider the onerous and exacting skill, the tireless accumulation of ability through trial and error over years of practice and dedication, that a painter exhibits when he creates a scene full of vibrant realism—as in the awe-inspiring glories of Classical and Renaissance art. Compared to them, the works of an abstractionist are an embarrassment. The one paid profound attention to the natural world, to the human form, to the qualities of perspective and light, to anatomy and geometry, and mastered them all, and brought them out in a creation unparalleled in nature. The other did little more than interior design on a canvass. Should an alien race encounter us, I would weep if anyone sought to impress them with Picaso. For surely nothing represents better the greatest of human achievement in learning and ability, in attention to detail and understanding of things, than the works of such true masters as Da Vinci or Donatello, Phidias or Myron.

As I see it, the abstractionist has abandoned and thus betrayed one of art's highest virtues: the manifestation of man's power to master and know his world, and create beauty of a complexity and depth that rivals nature herself. To revel in abstract art is like elevating Kabuki at the expense of Shakespeare, or to praise Marcel Marcaux over Sidney Poitier. But that's just my opinion.

> On the biological and sociological basis for artistic feelings and inclinations see: Caleb Crain, "The Artistic Animal," *Lingua Franca* (October, 2001), pp. 28-37, who summarizes the landmark work of Ellen Dissanayake in *What Is Art For?* (1988), *Homo Aestheticus: Where Art Comes from and Why* (1992) and *Art and Intimacy: How the Arts Began* (2000); see also: Margaret Livingstone, *Vision and Art: The Biology of Seeing* (2002); Semir Zeki, *Inner Vision: An Exploration of Art and the Brain* (2000); Joseph Goguen, ed., *Journal of Consciousness Studies: Art and the Brain: Controversies in Science and the Humanities* (1999); Joseph A. Goguen and Erik Myin, eds., *Journal of Consciousness Studies: Art and the Brain Part II: Investigations Into the Science of Art* (2000); and William Benzon, *Beethoven's Anvil: Music in Mind and Culture* (2001). See also: Richard Dawkins, *Unweaving the Rainbow: Science, Delusion and the Appetite for Wonder* (2000).

VII. Natural Politics

1. Morality vs. Politics

In Part V, "Natural Morality," I surveyed and defended the ethical philosophy of "Secular Humanism," which I defined there as any philosophy that holds to two basic doctrines: that humankind is the most important thing in the universe and the welfare and betterment of all human beings is a fundamental good, and that only secular solutions to our problems are credible, not religious or mystical ones.

As a Metaphysical Naturalist, Secular Humanism informs both my morals and my politics. This final part of my defense of naturalism will be the most personal and speculative of all, for it is in politics that we find the greatest ignorance and uncertainty, and thus should permit the greatest scope for honest disagreement. Nevertheless, what follows accords generally well with material in any current textbook on political science, and of course I shall explain why I believe my politics are rationally grounded, and why I believe it is what every reasonable, informed person should agree with. It should also become clear what it would take to change my mind (not much, really: just the truth).

As Aristotle said, man is a political animal. We congregate in complex societies that require governance and negotiation, and we thrive on this. In contrast, when alone we are virtually helpless, and hardly able to satisfy even our most basic needs, much less achieve anything in the way of genuine, enduring happiness. The success of human communities especially depends on two different aspects of their culture. The first is the sphere of morality; the other, that of politics. The difference between them is that moral science (and its underlying philosophy) is the study of how we should behave in order to achieve real happiness in harmony with all other beings. It informs our conduct at all times, and it relates to how we as *individuals* should decide and act. It is a practical art of everyday behavior.

In contrast, political science (and its philosophy) is the study of the use and distribution of *power*, and how this can best be achieved in any society so that all individuals may achieve their own happiness.

I have defended the ultimate aim of morality as individual happiness in Part V, "Natural Morality." My extension of the same goal to politics is a necessary product of that. For politics is ultimately the study of human behavior as it relates to power, for none of its recommendations can be achieved without human actions and decisions, and none of what it describes involves anything but, ultimately, how humans should act. Consequently, if the ultimate moral aim is happiness, then human action knowingly contrary to that end is immoral, while all non-trivial human action, to avoid that condemnation, must contribute in one fashion or another, directly or indirectly, obviously or not, to the achievement of happiness. Therefore, this must necessarily be the aim of politics, too, for politics is simply one set of human actions, those pertaining to the use and distribution of power.

However, that morality and politics have the same end does not make them identical. Not all moral thought relates to power, its use or distribution, but all political thought does so by definition. Politics is the extension of the technology of morality to the more specific, and very distinct, technology of government. To put it another way, ethics is the science of living, but politics is the science of governing. As a consequence, politics cannot be the vehicle for compelling moral behavior. It is easily shown to be immoral, if not impossible, to compel all people to agree on all things, least of all on moral philosophy. This is clear from the gross destruction of happiness all around, from strife and misery and the inevitable descent into crimes against humanity the effort has always entailed. As no use of force or the threat of force can *morally* be employed to get others to agree with your views, it follows that politics cannot be used this way.

Politics *must* concern itself, for example, with the practical prevention and correction of physical harm, as a matter of necessity. But neither law nor government can have anything to do with compelling *belief*, or dictating in just what way people are to be happy. Moreover, while morality is derived from one's philosophy, and not all people in a society share the same philosophy, *politics* concerns the governance of the whole. It must accommodate all people, even those very different from yourself. So the first task of the political philosopher is to work out how best to secure happiness for all without forcing universal agreement—indeed, to work out just what uses of force *are* legitimate given this aim. This is certainly a major difference between the objects of the sincere moralist and the honest politician.

2. The Rationality of the Moderate

In politics, as I see it, there should be neither liberal nor conservative. The fundamental nature of both ideologies is such that to divide them, and thus to exclude one and lavish attention on the other, is never a practical or rational method of achieving proper political ends. But it is perhaps inevitable that there should be such extremes of division, since the conservative and liberal are actually very much the same sort of person: both see the world in naive, black-and-white terms; both are easily offended, becoming outraged beyond reason; both are paranoid, and exaggerate every imagined threat; and both abandon empirical objectivity, aligning their beliefs instead with their ideology—skeptical of what conflicts with it, and credulous of what agrees with it, often without even *asking* what the evidence is, or even in spite of what it is. Perhaps as a consequence of their common defects of character, both struggle to impose their own moral vision on everyone else, rather than sharing a common civil society with mutual toleration and respect.

By contrast, the moderate is the most rational political animal in any society. It is most unfortunate that in the United States moderates have not seen the merit of uniting behind that principle, to form a power party to be reckoned with, one where disagreements are not only allowed, but encouraged, so long as members acknowledge the virtues of moderation in all things, the avoidance of extreme measures or allegiances, and the superior importance of facts and evidence in making political decisions, trumping any ideology that would argue the contrary. But above all, the moderate desires to cultivate a civil society, rather than impose his will upon the people.

Character defects aside, how can I say the conservative and the liberal are both right, and for that very reason both dead wrong in abandoning the other's ideals? To understand this we must examine their basic foundations *apart* from their personal flaws, their circumstantial allegiances, and their pet programs and platforms, which change with the fashions and the times, under the forces of money, influence and hubris, and either have little to do with rational adherence to any real political theory, or are but the means of the moment for achieving more fundamental aims. Hence, though people associate many ideals with liberals and conservatives, none seem historically essential to them but the following.

The conservative, in essence, is concerned with preserving and defending existing institutions and traditions (including the reigning economic regime—like, say, capitalism), and in some cases abandoning new failed projects for old successful ones. It is easy to see the value of this. Many institutions ought to be preserved and defended. Many aspects of the status quo can be, and often are, just right, and even many of those that aren't may require only minor tinkering or retooling. And it certainly can happen that a new project was simply a bad idea, while an old one, even if not yet the best solution, may nevertheless be better than anything else tried so far, and thus should never have been altered or abandoned. To deny that this can ever be the case is irrational. And to oppose the conservative goal simply because it is conservative, without reference to whether it might actually be *right*, is equally irrational. This is no less true of fiscal conservatism, as there are obvious advantages to spending and taxing less, and keeping free of debt.

The liberal, in contrast, is concerned with the social and institutional problems and failures that plague and worry society, seeing that new programs and solutions must be tried—which requires money, and thus some redistribution of wealth—while old approaches must be abandoned or greatly restructured. It is easy to see the value of this, too. It is obvious that we are not living in the best possible society, with the best possible government. Many traditions *are* obsolete, or were never enlightened to begin with. Likewise, it is obvious that what is wrong with our current social and political system might not be wrong anymore if we changed that system in relevant and effective ways. This is clear in economics, in science, in technology: progress is a good thing, and that means new, indeed often radically new things must be sought after and deployed. Everyone of sound sense acknowledges the benefits of this pursuit of the novel, even those who rightly criticize its excesses and seek to correct its less fortunate side effects. To deny that this is equally true in politics, in government and law, is irrational. And to oppose the liberal goal simply because it is liberal, without reference to whether it might actually be *right*, is equally irrational. And, again, this is no less true of fiscal liberalism, since there are many advantages to paying for a strong, effective, and supportive government that brings true security and equality of power to all its citizens.

Because of this divided view of proper political goals, liberal and conservative parties also tend to fall on different sides of another issue: the matter of personal freedom. This division of views is likewise irrational. So obviously irrational, in fact, that conservatives often conceal their aim to restrict freedom under the guise of increasing it, or employ justifications based on inflammatory rhetoric and exaggerated claims of danger, efforts they would not employ if their program would actually be approved on the honest and sober truth. Liberals are often equally guilty of the same trick when they attack *institutional* freedom. Still, in my view, it is by uniting the two ideas into a common, moderate program that we find the rational view.

The liberal usually desires to create, secure, and defend new freedoms, thinking 'the more, the better'. They have three valid lines of argument for this. First, the more a government's resources are tied up in needlessly restricting freedom, the less resources it will have for solving real problems. Second, the more power a government has beyond what it really needs, the more opportunity and scope there is for abuse of that power, and history proves it is only a matter of time before such abuses appear. And third, since happiness is the goal of all people and of politics itself, and freedom is essential to the effective and successful pursuit of happiness, and attempts to curb that freedom are often detrimental to happiness and conducive to greater misery and conflict, the government ought to procure more freedom for its people, not less. But there is a fourth reason that has some merit, too, which relates to the very fundamental basis of liberalism: to discover novel solutions requires the freedom to experiment. It is therefore not in any government's interest to be too controlling of a society, or too rigid in its own operation.

In contrast, the conservative usually desires to oppose the liberal quest for more freedoms, and sometimes attempts to take away freedoms that already exist. They have only one argument of any merit: freedom can be abused, to the grave detriment of others, and thus must be kept under good regulation. The liberals tend to agree with this when it comes to limiting the freedom of magistrates, professionals, and corporations, in order to protect individuals from their corruption, incompetence, and abuse. Yet it is irrational to regard ordinary citizens as less in need of such regulation—for they are no less human.

This is the basic foundation of almost any political theory: order is always better than anarchy. The question is how much. For certainly, though anarchy is the enemy of happiness, so is oppression. Thus, to decide on where freedom ends and regulation begins is the very matter that concerns all of political theory generally. The conservative tends to ask for too much control of individuals, while the liberal asks for too little. Their positions oddly *reverse* when addressing the rights and freedoms of corporations and governments, since tradition abhors interference with the privileges of wealth and power, while liberty demands restraints upon them. But in both cases the moderate seeks the best harmony between liberty and restraint, in both the private and public spheres.

3. Basic Political Theory

I have said that politics is about power: who has it and how it shall be used. Power ultimately consists of the use force. Force can be used not only to compel behavior, or to punish and reward, but to seize property and employ it for political ends. The wanton use of force is anarchy. But when power is conceded to an individual or group with a common purpose, which organizes power more and more into a hierarchy of authority, you have government. In its simplest form, government has no law, only the will of those in power. But humans are corruptible, fallible, ignorant, and limited, and governed by passions that all too often interfere with their reason. Thus, we need law.

Law is an agreement made (often only tacitly) by everyone in a society to use and distribute power only in certain ways. By accumulating, reforming, and perfecting the law over generations, it becomes less fallible and ignorant than any human can be, for through trial and error and the accumulation of knowledge and wisdom by a great many people over a long period of time, the law progresses toward a better overall model. Likewise, so long as it does not codify some particular commonplace passion (such as the abuse of unliked people), the law has no passions at all, and is thus more objective. And so long as the corruption of legislators does not corrupt the law, the law itself is free of corruption. You can bribe or trick a judge or a voter, but you cannot bribe or trick a law. Finally, the law can be reproduced and employed everywhere. It is not dependent on the limited time and resources of a single individual or committee, nor is it so easily changed as to be subject to passing whim.

Of course, it is not strictly true that the law is agreed to by everyone in society. People are rarely asked, and when they are, it is usually on a very limited aspect of the law, and even then (as in every democracy) they might not get their way. But there *is* an implicit agreement to the law made by all persons when they choose to remain within and a part of a society rather than outside and disconnected from it. And this agreement can encompass disagreement, which is the basis for sound political structures anywhere: we can all agree to abide by the law, even the laws we disagree with, provided we also agree to abide by legal procedures for changing that law.

This is one of the founding-stones of the success of democracy, and the primary reason it has so far proved to be the best form of government in terms of securing the greatest amount of happiness for the most people. Because we agree to allow the laws to be changed through persuasion and argument, and allow everyone to have some voice in that process, people are generally comfortable with an imperfect society, knowing they can work to improve it without resorting to violence. And this is essential, for it is impossible to create a perfect society from scratch. Any society, in order to make stable progress, while on the way avoiding the worst pitfalls of civil corruption, chaos, and war, must have citizens capable of coping with, and accepting peacefully, the inevitable imperfections in that society.

Naturally, this entails that the best form of government is what best achieves this end of consoling the most people in the face of existing imperfections in the law. And governments structured only to produce laws that enforce a civil society, rather than imposing a moral order, have been the most successful at this. It follows that the proper function of law is to create and maintain a civil society. Only in a totalitarian state is the purpose of law to enforce morality, whereas in a civil society this purpose is *rejected*. Instead of legislating anyone's moral vision, the law should ensure a common peace and welfare, so each individual can manifest her *own* moral vision, live her life according to her own moral beliefs—and exercise her freedom to change her mind. If you want to live by strict commandments in the Bible, if you want to shun abortion or birth control, or pray in school, you can. But you can't compel or intimidate anyone else into joining you. That is what it means to be free.

By thus removing from government the power to impress its will upon the people, including all the loopholes it can exploit to that end, all the seeds of a state's self-destruction are removed. Instead of a vastly expensive and cruel apparatus of control, fostering civil violence, discord, oppression, tyranny, and withering paranoia, which will merely end up the puppet of whatever ideology seizes power at any given time, our system bars such a monstrous institution from arising by placing in front of it the barriers of freedom and individual rights—which are, in fact, the same thing: without individuals being granted rights that *supercede* the will of the majority, freedom literally does not exist.

When governments and their laws are thus structured to promote only civil society rather than moral order, then no matter what ideology might gain power or influence, it will not have the apparatus to compel adherence among those who disagree with it. It is solely because of this that people with diverse religions and beliefs can live together in peace, achieving happiness and prosperity for all who labor for it. This is what it means to live in a civil society. That is why the tools of oppression, which lead only to violence and chaos, were replaced by liberty and freedom of speech. We war with words, not swords. And it is for that very reason, and no other, that we prosper.

It follows that political violence from oppressed or dissenting groups or individuals stems either from their rejection of the very idea of a civil society, or from their perception that they do *not* have the power to peacefully reform the government. In the first case we have nothing more than tyrants in waiting. But in the latter case we sometimes find a legitimate complaint—though far more often the complaint is unjustified, resulting in any of three errors that lead to rash action.

First, rather than see their oppression as a mandate to reform *how* laws are changed, thus securing them the peaceful access to power they lack, such radicals often choose violent means directly to their desired ends instead, and thus irrationally choose anarchy over order, which is self-defeating. Second, rather than acknowledge that they cannot get what they want because they have failed to persuade others to agree to it, and instead of seeking such agreement by improving their arguments and evidence and their diligence in communicating this to society, they often abandon the very principle that the law is what all agree to, and resort instead to compulsion, the very tool of the tyrant, the epitome of what they are supposed to be standing against. And third, sometimes, rather than acknowledge that they have failed to secure the agreement of their peers because they are *wrong*, or at least lack sufficient evidence and valid argument to know they are right, they abandon reason and take up arms instead, which is a fundamental rejection of even the *idea* of a social contract, a degeneration from "man the rational animal" to just "man the animal."

Those who use violence for political ends within any democracy, provided it is a genuine democracy (no matter how slow or flawed), are almost always behaving irrationally, for one of the three reasons above. And, sadly, political theory can only advise the rational man. Fortunately, this is where men and women usually place themselves, accepting a multitiered system geared toward limiting the use of force to just and proper ends. We obey traffic laws because otherwise we will pay traffic tickets, or lose our right to drive. And we submit to these punishments because to reject them entails even more severe consequences, such as garnished wages, seizure of property, or imprisonment. And if we refuse to abide even these, and evade them, the government's ultimate basis of authority is truly revealed in the use of violent force to compel compliance: to capture and detain non-compliant members of society, and to physically seize goods.

Force is therefore only appropriate against those who have rejected the social system. This assumes we are speaking of the moral use of force in the first place, when it is the *only* alternative to broader victimization. So long as the social system is genuinely aiming at and progressing toward universal happiness, it is a moral act of self defense to use that system to deploy force against those who reject such a goal and the common project to achieve it, and who use that rejection as a rationale to harm or abuse others. But this goes both ways. For it is equally appropriate to use force against those in power who ignore the law, or the enforcement of the law, in order to harm or abuse others. And when those in power are so uniformly corrupt that the law is all but a dead letter, armed revolution is a moral imperative.

Though some claim nonviolent revolution is possible, there is very little evidence this has ever been true in any uniformly corrupt society, much less an undemocratic one. Violence in your own defense is essential if you wish to survive and remain free in the face of real, unjust violence against you. In contrast, nonviolent civil disobedience is simply a means of persuasion, acceptable for the very fact that it abides by the social contract to seek peaceful means of reforming the law. This is made clear by the willingness of nonviolent dissenters to pay the penalties and serve out the allotted time for violating any laws in the carrying out of their protest. Thus, nonviolent protest *assumes* a civil society exists in the first place. It is futile otherwise. And as a matter of fact, nonviolent protest has only ever worked against genuinely civil, democratic societies (like the United States and the British Empire).

> For the thought that led to the creation of the most novel and important democratic republic in history see: Thomas Paine *The Rights of Man* (1791) and *Common Sense* (1776); and Alexander Hamilton, James Madison, and John Jay, *The Federalist Papers* (1788). And everyone should read the *Constitution of the United States*—it is far from perfect,

but it is the first of its kind: a social charter with general human happiness as its declared aim, and universal liberty its declared means.

For more on political science and philosophy in general, and as understood today, see: Austin Ranney, *Governing: An Introduction to Political Science*, 8th ed. (2000); Thomas Dye, Harmon Zeigler, and L. Harmon Zeigler, *The Irony of Democracy: An Uncommon Introduction to American Politics*, 11th ed. (2000); and Jonathan Wolff, *An Introduction to Political Philosophy* (1996).

4. The Politics of Metaphysical Naturalism

The most distinctive aspect of politics from a Metaphysical Naturalist perspective is dependence on evidence. As I described in detail in chapter II.3 ("Method") the only sources of truth are, in order of authority: logic, science, experience, history, expert testimony, and plausible inference. This is what determines how a metaphysical naturalist will approach political issues. Let's examine some consequences of this fact.

4.1 Political Method

First, it is foolhardy to make decisions about policies that will affect the lives of thousands or even millions of human beings based on merely plausible inferences, which is one reason why government has no business telling citizens what philosophy or worldview to adopt. Though plausible inferences will do in a pinch, when there is nothing better to go on and an emergency is pressing that demands immediate action, this is a problem that should usually only vex the executive branch of a government, and when such snap decisions go bad, the government should own up to the fact and right its own wrongs. Otherwise, there is rarely any excuse for the judicial or legislative branches of a government (nor much even for any other important institution, such as the media) to engage in snap decisions. These institutions are *designed* to allow time for review, inquiry, and deliberation. It is only the executive who is assigned the right to act on a moment's notice, and only when unavoidable. And this is not a bad thing: a good government must be able to handle all contingencies, to be slow and methodical in all things except those that demand more immediate action. But there must be competence and accountability, and above all caution and reserve, and respect for rights and law.

Likewise, to base decisions on the advice of experts is perhaps not always foolhardy but is still risky, since we are then brought back to the original problem of the corruptibility, fallibility, and ignorance of humanity that law is supposed to avoid in the first place. For it is difficult for anyone asked to give advice, affecting a national or even community policy, to always remain honest, unbiased, and careful. People do risk exploiting the power this gives them to further their agendas, people do risk following faulty inferences or educated guesses, concealing them in the guise of established fact. Expert testimony is nevertheless essential as a *component* of political processes: and we see this in the United States in congressional inquiries and debates, trials at court, the chief executive's cabinet, etc. But individual citizens should not rely too much on this source of information in deciding political policy, much less fundamental issues of the form and nature of their government. And even when they do, it should always be with a good dose of methodical skepticism (on the criteria for a proper argument to authority, see II.3.6 and IV.1.2.6, "The Method of Expert Testimony" and "The Criteria of the Good Historian").

We start to enter the realm of good evidence on which to base our political views when we get to critical history. Well-established historical facts are a useful and necessary source of data about what works and what doesn't, about causes and consequences. It is of paramount importance to make sure you are relying on a genuine, competent, critical history, and not the much more numerous tracts and texts that distort history to serve an agenda, or that simply treat history incompetently or negligently. But so long as your facts are secured by a sound and expert application of the critical-historical method, you are on good ground for building and defending a political view. However, it is not the best ground.

The best ground on which to base any decision regarding proper public policy is science. The facts of science are the most important resource for information about what is needed, and what will work and what won't, what the consequences of any policy will be—so long as it meets two criteria. First, which almost goes without saying, it must be science established by the whole scientific community, on sound facts and method, and not pseudoscience or half-baked science or the authority of a few expert mavericks being passed off as scientific fact rather than scientific *opinion*. Second, one should, in every case possible, rely on specifically relevant research. That is, once you have a range of policy options, objective scientific research (especially experimentation) should be deployed specifically to address the concerns related to those policies.

So long as the scientific method is honestly, correctly, and competently used to that end, you won't be able to find any better facts to base your decision on. And this applies to every branch of government.

What about experience and logic? First, personal experience, critically examined, is less reliable than science generally (which is just personal experience rigorously and methodically perfected and repeated), but in politics in particular it is a double-edged sword. On the one hand, this is very important. When you know first hand that what politicians are telling you is false, because you are in the midst of the very effects or problems created by a policy, or you can see first hand that the facts are other than they are letting on, then you know whom to side with. On the other hand, this is risky. Your singular perspective is always biased even beyond your control. No matter how objective you are, you can only have the experiences you've had. If you live in a slum or a posh neighborhood you will not know what the rest of the country is really like, at least nowhere as well as you do your own unusual slice of it. At the same time, personal enmities, passions, and preferences are naturally built up in you by being so close to certain causes and consequences, and these can easily blind you from seeing the forest for the trees. Hence the importance of science—and, incidentally, democracy.

Last but not least is logic. Though there is no beating logic, it is easily abused in the name of politics. In fact, there is very little that logic can tell you about the best policy to adopt. That requires knowing things about the world and about people, and logic tells you nothing about such things, at least nothing you didn't already acquire from science and history. All logic does is ensure that you are thinking clearly about the facts you *do* have. This does make it very important. Logic is certainly needed to see through the routine use of fallacious reasoning and deceptive rhetoric by politicians and pundits, just as a decent command of statistical mathematics is essential for seeing through the abuse of numbers in the political arena. But logic can never and *must never* be used to replace the need for facts and evidence to support any position. When it comes time for proof, arguments are hollow. Evidence is king.

4.2 The Best Polity

There is probably no perfect state, even in principle—only better ones. And in our endless quest to improve, there are three questions of primary interest in politics, to which the above methods and advice must be put by any Metaphysical Naturalist: (1) how best to design a system for making and revising the laws, (2) how best to design a system for enforcing and carrying out those laws, and (3) what the best laws are.

The first question is of the fundamental structure of government, and as history has proven, it must center around the best way to form and organize some form of democracy. Everyone must have a voice, so as to ensure the confidence of all citizens that there are peaceful means to reform the law when sufficient facts and reason warrant. In this respect, the Founding Fathers had good reasons to believe a republic (like that of early Rome) was better than a pure democracy (like that of classical Athens), and they were probably right then as now. But there are serious questions about whether the republican model now employed in the United States or any other country is a good one. It is certainly not the best—though it remains to be seen what will be. Still, this is an empirical question, which cannot be answered in the ivory tower of theory (though I make a suggestion below in VII.4.3, "Choosing Our Leaders").

The second question, of the system needed for enforcing and carrying out the law, is mainly a matter of fiscal and administrative engineering, which is just as empirical an enterprise. But for this we have a lot of data to work with—if only we could gather it honestly and analyze it objectively. We should endeavor to study and learn from our past successes and failures, at all levels of government, and organize in some central way the clearest advice regarding what is best and what is worst, what works well and what doesn't, and why. Only then could we steadily improve our government in efficacy and efficiency, instead of remaining hamstrung by ineffective bureaucracies that never learn from each other or themselves (see VII.5.3, "A Commitment to Executive Reform").

The third question is one we more often face. How do we tell good laws from bad? The first test is whether the laws allow more people to find happiness than would be the case without them, or with different ones. The second test is whether the laws foster rather than hinder a civil society (see VII.1 and 3, "Morality vs. Politics" and "Basic Political Theory"). And in both tests we must always consider not hope or theory, but the actual facts: we must know whether the law will even work, and what *all* its effects will actually be. In other words, reality must guide us, not ideology.

4.3 Choosing Our Leaders

Finally, related to all of the above, there is a fourth question to answer: whom to vote for. To be honest, I believe We the People should only ever vote for laws (at the very least, *tax* laws), and not for people. Legislators should instead be chosen by lottery from among a pool of qualified and willing candidates, ensuring a completely fair system, as well as genuinely proportional representation—both by ideology *and* demographic. For example, a lottery would give us legislative bodies wherein roughly half the membership would always be women, for the first time in history. Moderates would rule, beholden to no one, and special interests and ideologues would no longer pull the strings. The qualifications to stand for this lottery could be anything we are comfortable with, as long as any citizen can achieve those qualifications with suitable effort. Everyone else in the government (like, say, the President) should be nothing more than an employee hired or fired at will by the relevant legislative body—although judges must be immune to dismissal for any reason other than impeachment. That's what I suspect is the best possible government, in rough outline.

But Americans live now in an electoral system. We are stuck having to vote for our leaders. So whom do we elect? To decide that question requires having reliable and accurate information about the options, which is not easy to get. You need to read what both sides are saying about both candidates, and not let your own biases cloud your judgment. You need to think critically and see who actually cites sources and evidence for their claims, and who doesn't, and who tells you the whole story, rather than only half-truths. Don't trust the media. Don't trust the campaigns. Don't trust anyone. Look at it all, but assume most of it is false or deceptive—and, sifting through it all, try to identify on your own what is most likely true, and what most likely isn't. Once you have some kind of trustworthy information to go on, there are four considerations that should govern your choice, in the following order of practical importance.

The first consideration is that of rational success: we must go with a candidate who might actually win, at the very least to prevent one worse from achieving the office. And though the parties and pundits like to fool you into thinking all candidates are the same, they are not—of any two matching off, no matter how despicable both may be, one of them is *always* worse. Hence it is as foolish to vote for a noncontender as it is not to vote at all. In practical effect on your future, on the exercise and distribution of power, they are exactly the same. And rational action demands we do what will *actually* produce the best outcome, even if that 'best' still sucks.

The second consideration is that of simple honesty: a candidate must mean what she says and say what she means. She must not be a self-serving criminal or two-faced ideologue. This does not mean the candidate must be a saint—she can be a complete bastard in her personal life—rather, it means that when it comes to performing public duties she won't cheat, deceive, or double-cross *the people*. In other words, she will actually serve the proper end of politics (negotiation for universal happiness), instead of abusing her power for personal or ideological aims. She will make reasonable deals. She will be responsive to public concerns. She will admit her mistakes, and correct them. But above all, she will serve the people, not use them.

Naturally, apart from those two matters, for any specific position, competence should be the determining factor. We will never find perfect, flawless men and women, but we can find competent ones. Just as naturally, you should vote for rational moderates before all others, or those closest to that ideal, for the very reasons I laid out in chapter VII.2. Everything else is secondary. For instance, most people vote for two *other* reasons, which ought instead to be given a back seat: they vote for the charismatic candidate (whoever they 'like' or who 'feels right') or they vote for the candidate they believe supports all the same positions they do on major issues. Both are grave mistakes, especially if they guide one's vote to the *exclusion* of the criteria of competence and moderation—or worse, lead us to act irrationally by backing a liar or noncontender.

True, there is utility in charisma: a charismatic person will make a good leader, they will be heard, and get things done. Unfortunately, however, charisma does not always correspond to competence or moderation. Plenty of liars, tyrants and bunglers have been quite charismatic. We should be suspicious of charisma, not swayed by it. We need to look past it, to see if the candidate really has what counts: prospects and honesty, competence and moderation. And though you certainly want to place those in power who share your plans for improving law and society, there will almost never be any candidate who agrees with you on every important issue, much less one who actually has a chance of getting elected and is demonstrably honest and competent. It is thus irrational, first, to hold out for some hero who may never arrive or, second, to let this romantic quest leave you open to the manipulation of any liar or dissembler who is good at *pretending* to be that hero.

Once again, you will most advance your political goals by backing a practical rather than a 'perfect' candidate. A moderate will always have good ideas, or at least not terrible ones, on almost every issue anyway, whether they are your ideas or not. Even when it comes to cases of serious disagreement, as a moderate she will not support any policy that is too radical or driven by ideology over fact and sense anyway, and that is far more important than any complete agreement with your interests.

When we have absorbed all of this advice, and grasped the rationality of political choice, we can then keep our eye on two final issues. Above all considerations, every candidate in a truly free and sensible society should at least *pretend* to be someone who would fight for the liberty and welfare of every law-abiding citizen, even one who opposed his most sacred beliefs. But if someone doesn't even put on this pretence, who makes plain instead his disinterest in the liberty or welfare of some people he disagrees with or dislikes, who intends to replace the civil society with his own moral order, then he is an enemy of all that is good, and must be actively opposed. On the other side of things, all else being equal, I believe we should support candidates whose views are closest to the six humanist objectives I describe below. This will better our chances of making progress toward those goals, which I believe are essential to the good society.

5. My Politics

What follows are what I believe are the most rational positions to take on six general issues of pressing importance to American society today. I could bring up many others, but these will do as prime examples. There are few issues more central to creating the good society than these. In fact, most other issues are not yet so clear in their apparent solutions, at least not to me. For instance, I am undecided whether legalizing drugs would be better or worse for society overall. There are not yet sufficient facts to decide the issue, and there are worthy arguments both for and against. Still, it is clear the current system is not working, and the best solution will most likely involve neither the complete legalization nor the complete criminalization of drugs. But exactly what that middle ground must be is unclear to me at present. All I know is that few people are seeking that middle ground, but instead are wielding specious arguments and dubious facts in an attempt to win, and the genuine truths and good ideas are thus getting lost in the fog of war. There is nothing more contrary to how a Metaphysical Naturalist should engage in deciding political issues than such combative, obfuscatory, and factional politics.

5.1 A Commitment to Freedom

The first plank of what I believe to be the right political platform is a commitment to freedom. Freedom is typically defined under the rubric of rights. As I stated above, human rights are those rights all people must have in order to best pursue their personal happiness. It is therefore immoral to oppose or suppress them, and essential for any polity to defend and respect them.

In particular, limited solely when in conflict with the rights of others, all individuals should enjoy freedom of speech and thought, freedom of labor and trade, access to impartial justice, and equality before the law. That is only the short list. The only freedoms that should be restricted are the uses of freedom to deprive others of freedom, and even then, when conflicts arise, they should be resolved, in every case possible, by peaceful negotiation and mutually agreed compromise rather than blunt force or blind fiat.

Consequently, I am doubtful of the merit of 'rights' to goods or other demands on the labor of others (like a right to 'have food' or 'medical care'), since that is tantamount to slavery. Such rights entail the use of force to compel others to serve you. We can still have government food and medical care, but they should be regarded as privileges, which like any economic transaction are entirely negotiable and ought to promote *mutual benefit*. Even if it is agreed that a government ought to ensure everyone has some particular good or service (like food, medicine, or education), this should be by fair economic transaction, not compulsion—and that includes compelling others to pay for them. In light of this, we should note that laws always create legal 'rights' of a purely formal variety. But those are not *human* rights in normal parlance, and certainly not in the sense I employ here.

This is not to say that we should have no preventive or regulatory law. Some measure of this is always necessary to prevent the infringement of human rights, by removing clear risks and promoting alternative behaviors. There is no doubt that safety laws, traffic laws, licensing laws, and similar legal phenomena are an unqualified good, securing far more widespread happiness than would otherwise be the case, and averting far more misery than such laws themselves create. But even in this, the aim should be the protection of human rights, with as much accommodation as possible for those rights otherwise being compromised as a result. In short, the government should interfere in our lives as little as possible.

For the whole story of human rights from a variety of perspectives (not all of them mine), see: Robert Maddex, Mary Robinson, and Desmond Tutu, *International Encyclopedia of Human Rights: Freedom, Abuses, and Remedies* (2000); Charles Black, *A New Birth of Freedom: Human Rights, Named and Unnamed* (1999); Tara Smith, *Moral Rights and Political Freedom* (1995). On regulatory law, see next section.

5.2 A Commitment to Social Reform

The second plank of what I believe to be the right political platform is a commitment to social reform. Though I do not agree with socialists, who wave around slogans like "from each according to his means, to each according to his needs" and so on, who in their compassion are inadvertently giving fascism the veneer of a noble society, I also do not agree with hard-line capitalists who believe the free market will solve every problem—both are baseless superstitions, obviously wrong to any clear-headed observer of the facts.

As it is an expression of our ever-essential freedom, capitalism is a good thing. We should protect and encourage all aspects of a free market and individual ownership of self and property, and employ market forces and similar principles even in public organizations, as much as is reasonable and effective. Though corporations are not angels and should be policed, they should not be derided as some sort of Satanic evil they are not. They are the heart and blood of our welfare.

Besides that, any attempt to put shackles on capitalism is flirting with human rights violations. There is simply no way to put the control of people's labor and property into the hands of a collective. Individuals and committees must always manage that labor and property anyway, and they will be no more compassionate and reasonable than corporate profit-seekers. Worse, in order to interfere in the free exchange of labor and goods, one must violate virtually every human right we have a name for. Controlling labor is little better than slavery, and the means that will ultimately be necessary to enforce compliance to collectivist property laws will be little less than tyranny. This is why every communist state that has ever formed has ended up a fascist state under Marxist guise—and then stagnated in militant poverty.

However, "hands off" is bad policy, too. For instance, there is no doubt whatever (as has been proven again and again by abundant sound science) that poverty begets crime. There is also no doubt, as history proves, that a free market does not cure poverty. To the contrary, as with the brutality of the natural world it imitates, a third part of the world will always be miserably pressed upon the bottom and paid just enough not to die. Likewise, it is often argued that education should be privatized, but there could be no greater recipe for disaster than that. I will discuss this more below, but just for starters, most people in the U.S. could never afford the real cost of even the poor education their children now receive for free. Forcing them to pay will only create a more ignorant populace, and that means more poverty and superstition, and less rationality and wisdom to go around. Then there is the obvious need and unqualified benefit of regulatory law. Fire safety laws have saved more lives than fear of lawsuits ever has or will. And so on.

So it is clear that the government must do more than maintain a status quo. Government should endeavor to reform society itself, especially when it is so wealthy that it can afford the luxury of greater public good. There are three obvious reasons for this that even Machiavelli would appreciate. First, any measure that makes people happier and reduces crime not only makes for a better society, the very goal of politics, but reduces the costs of government and, by producing more good citizens in the place of criminals, leaves more dedicated and competent labor, thus boosting the economy. Second, any measure that makes for a more rational, competent, and informed citizen body, as well as a healthier and stronger one, will eliminate many of the tensions, misunderstandings, and conflicts that bog society and politics down, and this will accelerate progress toward a better and more harmonious world, again the very goal of politics. It will also produce a more competent and therefore more productive labor force, and reduce crime even further, and hence, again, government costs.

Third, while it is easy to escape a lawsuit (through declaring bankruptcy if nothing else, though many other maneuvers are always available), it is not so easy to escape the police. The government obviously has far more well-trained personnel and far more resources to devote to solving crimes, than citizens will ever have to deploy in their own self-defense against abuses of their rights that result from corruption or neglect. Therefore, regulatory law is simply indispensable. And by preventing disasters, and compelling people to be honest when there is serious risk of harm, the economic and civil benefits are unrivalled.

There are two principles at work here. First, by establishing standards and enforcing them (one of the very first things governments ever did, essential to the success of any economy) society is made better at very little expense to personal freedom. Only when such regulation becomes onerous has it outlived its welcome and no longer serves the only end that justified it. One easy rule to go by is simply to mandate what people ought to be doing already: that is, if they would (in a perfectly just society) pay for their neglect or behavior in civil or criminal court anyway, it is certainly little harm to force them to avoid this fate in the first place. The idea must always be to outlaw only actions that are *likely* to bring harm, though such crimes should always be treated as less offensive than actually *doing* harm, since malice in the former case is less certain.

Second, while moral appraisal and statutory justice are concerned with identifying people with good or bad character, by observing the behavior their character causes them to commit, and is not at all concerned with what made that person good or bad, politics must take up the latter concern. Through the organization of the social system, which entails the negotiation and execution of power, we ought to seek to produce fewer bad and more good people. How to do that, without compromising human rights, is no easy question, and one that requires careful study before implementing any policy. The one case where the evidence is clear is education (see VII.5.4 below).

However, this should not be read as a recipe for throwing money at every problem we think we can fix. Government can rescue those truly at the end of their rope. But in most cases it would be much better to teach people to take responsibility for their own lives and give them access to the knowledge they need for achieving their social and personal goals through their own discipline, effort, and self-reflection. People can accomplish far more on their own than any government machine will ever do for them, and the will and the character they must cultivate to that end will benefit them in every aspect of life. As the saying goes: give a man a fish and you feed him for a day, but teach a man to fish and you feed him for a lifetime. Obviously the latter is better in every conceivable way—as long as there are fish for them to catch.

In contrast to the obvious benefits of such an education in private action, there is little to be gained by increasing taxation, certainly much harm by increasing it upon those who have little to spare. In contrast, ruthlessly taxing the rich is not such a good idea either: the poor would not have jobs without the rich spending their money. And only the massive capital investment the rich can provide produces those increases in productivity that are key to any nation's economic and technological advance. Of course, they can still afford to pay more than those lower on the economic ladder. And the rich deserve to, for they have benefited far more from the public order. But, as I will argue next, there are much cleverer ways to secure a luxurious public revenue without any taxation at all.

> Some worthwhile reading on the debate over how much government should or shouldn't do for us: Erich Fromm, *The Sane Society* (1990); Phillip Brown and Hugh Lauder, *Capitalism and Social Progress: The Future of Society in a Global Economy* (2001); Richard A. Epstein, *Principles for a Free Society: Reconciling Individual Liberty with the Common Good* (1998); Stephen Elkinand and Karol Soltan, eds., *A New Constitutionalism: Designing Political Institutions for a Good Society* (1993).

5.3 A Commitment to Executive Reform

The third plank of what I believe to be the right political platform is a commitment to *executive* reform. That is, the reform of how government works, both its administration and bureaucracy. So much can be gained, and so many problems solved, by simply redesigning the way we do things. And, likewise, there are many practical reforms of the law that would also be of great benefit to society at very little cost. So vast and yet so untapped is this well of resources that we really ought to have an entire government office established and staffed *solely* for the project of finding bad design in the rest of government and law and recommending, after experimentation, observation, and research, how the problem could be fixed or the situation improved. Call it the 'bureau of reinvention'.

In general, the government should operate on simpler, more flexible rules with a constant eye toward improving defects in any system. Paperwork and personnel should everywhere be reduced where it can, though in some cases, such as law enforcement and teaching, personnel *increases* are actually necessary, along with increases in pay and benefits. But in addition to this, government ought to seek a far more practical and beneficial reform: the abolition of income tax. There is little merit in seizing the property of citizens every year, and much ill comes from doing so, especially since the poor lack the education and resources to exploit the vastly complex tax code to their advantage and thus are harmed by the system far more than is just. Indeed, so much of the time of legislators and the resources of political action groups is wasted on endless yearly debates over taxation, and so much corruption and confusion surrounds it, that it can easily be called a major public evil.

It would be far more practical and justifiable to generate revenue from a hidden national sales tax. Such a cut of the profits of business is more than a fit price for helping free and fair trade to exist in the first place. If not for government, there would be no peaceful or reliable means of enforcing contracts and policing crime, and of basically creating the stability, peace and justice that gives consumers and businesses confidence in the market. Indeed, by any sound reasoning, businesses *ought* to pay such a market fee.

However, the most reasonable solution for generating public revenue is twofold: fiscal responsibility and public enterprise. The first is a simple matter that any economist will tell you is quite right and would be an unqualified public good: eliminating government debt and reducing or limiting the size and cost of government enterprises. Corporations do this all the time. There is a science to it that is taught in schools throughout the country. The government itself ought to get with the picture and do it, too. The bureau of government reinvention I just 'invented' earlier could also be tasked with rooting out and exposing government waste and fraud, and developing solutions to both. But the simple act of paying off massive public debt is undeniably useful: it would not only cut the huge costs of paying interest on that debt, but it would also free up billions of dollars of capital that would then become reinvested in the *economy* instead of the government. Until then, that capital is frozen and useless, doing no one any real good.

The second principle is a little more controversial. Rather than charge taxes, the government could simply generate revenue directly—by selling stuff. This would not require fascist requisitioning, since on a free market the government can just buy what it needs over time, and thus accumulate an economic branch, which makes its money the same way everyone else does. There are several obvious sectors that could best be exploited to this end, particularly all the natural resources. If the government bought up and then managed all the commercial forests, for instance, and rented this land out to lumber companies with an eye to maximizing public revenue, not only would we get a huge windfall of cash without violating anyone, but the forests could be managed more ecologically. Since the government would *have* to maintain a regular source of revenue, it could not afford to lose its forests. This is very much unlike corporations, whose boards and shareholders often don't care about any future beyond their own golden parachutes and the next big stock sale—making the rape of the environment a profitable venture. The mining and oil industry are also prime targets of government acquisition.

This is just one idea. Yet it appears quite practical, and the benefits obvious. Just imagine a tax free nation. What country could compete with us then?

> For more on all of this, see: David Osborne and Ted Gaebler, *Reinventing Government: How the Entrepreneurial Spirit Is Transforming the Public Sector* (1993); Philip Howard, *The Death of Common Sense: How Law Is Suffocating America* (1996); David Osborne and Peter Plastrik, *Banishing Bureaucracy: The Five Strategies for Reinventing Government* (1998); Al Gore, *Creating a Government that Works Better and Costs Less* (1993) and *The Best Kept Secrets in Government: How the Clinton Administration Is Reinventing the Way Washington Works* (1996).

5.4 A Commitment to Education

The importance of education to a successful free society cannot be overstated. We need a national education system that allows some local flexibility and control, but requires schools to adhere to standards established by a national bureau, one that engages in a perpetual review of school successes and failures, of effective and ineffective textbooks and teaching styles and curricula—basically, all the evidence pertaining to education, including scientific studies of what works in what areas, and what doesn't. Only then can we learn how to improve our education system and *actually improve it*. We can even learn by studying the school systems of other nations. And teachers' unions, the main stumbling block for all attempts at school reform, would not oppose any of this if you sated them with a single reform, one they would give all else up for, and that teachers already deserve anyway: outstanding pay and benefits for educators.

A national education system should aim at three goals: producing citizens who (1) have a high economic value, who (2) have a strong command of issues and concepts necessary for participating knowledgeably and effectively in the processes of law and government, and who (3) are capable of making their own informed decisions in matters of morality and faith. All three aims are equally important.

The first is a pragmatic benefit: a more competent and reliable workforce boosts the economy, and everyone gains from this—the government earns more revenue, communities get richer and more livable, and more people succeed, strengthening the middle class. And any political scientist will tell you a large middle class is the most important component of any democratic society—for that is the one group that knows what it means to work hard for what you have, yet still has something to lose. They are the foremost self-interested defenders of an honest free market—whereas the rich would benefit from a *dishonest* market, while the poor would benefit from getting rid of free markets altogether. Thus, we need to boost the knowledge, intelligence, and practical skills of all citizens. That will always be an unqualified good.

The second is a necessary benefit: an ignorant and uncritical public is a powder keg that will destroy any democracy in time. And until it does it will cause a great deal of chaos and misery as the mob is easily manipulated by factions and demagogues, resulting in power being gained and used irrationally, or to improper ends. An ignorant and uncritical public will also readily demonize and oppress minority groups, eventually bringing about their own downfall when the tide turns and they get as good as they gave. In short, this is a recipe for social strife. A democracy is only as clever and wise as its people. If you want facts and truth to prevail, if you want people to cooperate and listen to reasonable arguments and act upon what is genuinely the wisest course, what is genuinely demonstrated by real facts, if you want people to see through lies and fallacies and obfuscations, if you want people to act and vote wisely, and accept the flaws in the political system while peacefully working to correct them, you must educate them. There is no other way.

Finally, the third goal is one that is both practical and necessary: people who are educated well enough that they can actually make their own informed, intelligent decisions about what to believe and how to live their lives, will become better people. For reason and knowledge will lead them to see for themselves that the path to happiness lies in heroism, not villainy, in kindness and integrity, not cruelty or indifference. This will also reduce crime and increase widespread happiness, and is far superior to indoctrination in a belief or creed. For in the latter case people will have no clear idea why it is true or right, and will simply take it as rote, and not actually live it, not understanding it, and at best will adhere out of fear or fickleness, but at worst will reject it altogether and be left with nothing but self-serving apathy.

So if you want people to become genuinely good, you must show them *why* a certain creed will serve their happiness and *why* abandoning it will serve their misery, and you must do this with evidence and sound argument. But above all, you must be honest with them, and teach them how to find all this out on their own: for there is nothing more persuasive than self discovery. This means you must give them the facts and the tools, teach them how to use the one on the other, and let them come to their own conclusions.

In line with these three goals, any complete education program would emphasize throughout every student's academic career the fields of philosophy, logic and rhetoric, the methods—not just the findings—of science, and the contemporary real-world application of other vital fields, such as mathematics (learning how to understand and evaluate statistical arguments, for example, is far more important than trigonometry), economics (understanding capitalist principles of debt and investment, for instance, is crucial), and law (understanding what the law is, how it is interpreted, and why—a basic toolkit every citizen should have). Though not of comparable priority, the arts should have a place, too, as they teach us about humankind, how to find more beauty in the world, and learn new ways of expressing ourselves and our own creativity, all profound sources of personal happiness—securing for everyone the benefits of the higher virtues of art (per VI.3.3, "The Higher Virtues of Art").

Every school program should be geared with all these goals in mind. One can see where there is obvious overlap: the skills of critical thought are vital to accomplishing all three goals, and such skills require a thorough education in scientific method and practical logic, if not in the proper deployment (and typical pitfalls) of all six 'methods' (cf. II.3, "Method").

The importance of education is emphasized in another way by Collier and Stingl (see the first bibliography in V.2.2) who describe what is called the "Political Worry" in these terms: "We are obliged to be moral only insofar as we believe ourselves to be obliged; if there are members of society who do not believe themselves to be so obliged, the social system threatens to break down" (pp. 47-8). Of course, any rational agent in possession of all the facts can see that such a breakdown would be personally disastrous. Statistically, it is very likely to produce general misery, and very unlikely to produce genuine happiness for anyone. Happiness will then only be securable at a much greater cost than would be the case in a fair and well-ordered society. Therefore, the breakdown of the social system is in no one's best interests.

But this means there is no Political Worry so long as a social system encourages rationality and educates its members well. This is a very strong argument for giving a quality universal education a very high priority in any society, an education grounded in the promulgation of critical thinking and training in practical and theoretical reason. To this effect, Collier and Stingl cite Rawls, *A Theory of Justice* (1971), who persuasively argues "the idea that if a society's basic structure is fair and its citizens raised to be cognizant of this fact, the strains of social commitment will be minimal" since every rational person's interests will be secured (by the fairness of the society), and "trust is a correlate of cooperative behaviour, and fairness is likely to inspire trust" (p. 48). Thus, the whole success of any society hinges on the effectiveness of its educational system in making *all* people knowledgeable and rational.

And none of this will be gained by privatizing the education system. We have ample facts showing this. First, an unregulated school system will be self-seeking in its agenda. Rather than promote a uniform base of knowledge that will bind us all together with a common pool of ideas and experiences that make communication more effective, independent schools will teach different things to different people, dividing society further. And rather than promote universal tools of reason and critical thought, of philosophy, and scientific and critical-historical methods, most schools will suppress these things in favor of an ideology or dogma, thus destroying the most important reason we need universal education in the first place. Thus, we need to reform the public school system from the ground up, not abandon it to greedy businessmen, con artists, and cults.

Of course, deregulation will also degrade overall quality, as schools dumb their programs down, securing cheaper teachers and textbooks, in order to compete, indeed even to be affordable to most Americans, who would cringe at the very thought of forking over the actual $4000 to $8000 a year *per child* they get for free now. Vouchers would only be politically proper if they were restricted to open schools that respect the religious differences of their students, and who were regulated, forced to adhere to national standards of safety, equity, curricula, and achievement. Yet it is hard to see what the benefits of such a system would be that could not be gotten even more directly and securely by instead remodeling the public school system on free market principles.

Finally, a private school system would ultimately toy with the very lives of children. If you buy a defective product, you can return it (maybe—not all businesses are so conscientious). But how do you return a year of a child's life? A defective school, a scam, a failure, a bad product in a one horse town, all would be a disaster. It would not be a correctable economic bump in the road. You cannot play around with people's lives like that. Education is too important to become a mere profit venture. To see this, one need only look at higher education, where the large majority of private colleges are woefully inferior to those run by the state, which are heavily subsidized: students pay less than half the actual cost of their education there, which averages around $30,000 a year, more than five times the cost of secondary and primary education. The exceptions prove the rule: good private colleges are not only comparably rare, but outrageously expensive and beyond nearly everyone's reach. Only a national system of primary and secondary education—one completely subsidized so *all* people can acquire an *equally* good education, a system compelled by political power to meet minimum standards and unite citizens behind a common ideal of tolerance and liberty of thought, a power that can step in and directly take over a noncomplying school—only that will succeed in achieving the three goals of universal education in a free state.

> There are countless important works on education, but these at least are required reading: Joe Harless, *The Eden Conspiracy: Educating for Accomplished Citizenship* (1998); David Levine, Robert Lowe, Robert Peterson, and Rita Tenorio, eds., *Rethinking Schools: An Agenda for Change* (1995); David Berliner, Bruce Biddle, and James Bell, *The Manufactured Crisis: Myths, Fraud, and the Attack on America's Public*

Schools (1996); Gerald W. Bracey, *Setting the Record Straight: Responses to Misconceptions about Public Education in the United States* (1997).

Against the idea of public funding of private schools: Gerald W. Bracey, *The War Against America's Public Schools: Privatizing Schools, Commercializing Education* (2001); Edd Doerr, Albert Menendez, and John Swomley, *The Case Against School Vouchers* (1996); Edd Doerr and Al Menendez, *Church Schools & Public Money: The Politics of Parochiaid* (1991); and Art Must, ed., *Why We Still Need Public Schools: Church/State Relations, and Visions of Democracy* (1992).

5.5 A Commitment to Defense

If I could snap my fingers and make all the guns and bombs in the world turn into flowers, I wouldn't hesitate for an instant. But we don't live in that world. We live in a dangerous one, of predators, tyrants, and madmen, of confusion and anger, of ignorance and irrationality. There will likely *always* be need of a good command of violent means to defend oneself from the peaceless and unreasonable. It is precisely because violence cannot be opposed but by more violence that so many people resort to it to get their way, for most victims are too timid or unprepared to meet that violence with their own. Thus, it is irrational to call for the disarmament of any free democratic nation. After all, your nation will only remain free and democratic if you have the guns and bombs to defend it. History, even very recent history, proves this beyond any reasonable doubt.

Therefore, any rational political policy must include support for a strong and sophisticated military. Though we must winnow out the fraud and corruption in congress and the corporations that have made our military needlessly more expensive, and less effective, than it could be, we nevertheless need what we have, and more. And we should not be afraid to employ force to protect American property and welfare, and human rights here *or* abroad. Though we should be cautious, and never pick a fight we can't win, or where victory would do more ill than good, a strong and well-trained military is necessary and should be used, especially in cooperation with international coalitions seeking world justice. Cops, after all, would be foolhardy not to carry guns, and would hardly be effective if they didn't. And there are no police among nations but armies.

Finally, consonant with military strength is the fact that politicians should not interfere in tactical or strategic decisions. Once a mission's parameters and objectives are clearly set, if the supreme commanders then approve of the mission, they should be allowed to carry it out in accordance with their own knowledge and expertise with little or no micromanagement from the central civilian power. There are too many examples in history, from Viet Nam to Mogadishu, proving that disaster inevitably arises from political interference in military operations. The other side of the coin is that war and violence must always be the last resort. Diplomacy is still vital—for example, we must never treat other free nations any differently than we would expect to be treated in turn. And restraint is essential—for we can't kill every evil in the world without unacceptable cost to ourselves.

In the United States this whole issue brings up a unique reality: the saturation of civilian society with firearms. This is the outcome of the Second Amendment's liberal application over three hundred years. And regardless of what the Second Amendment once meant or can now be interpreted to mean, American citizens today probably have an established common law right to possess firearms, protected by the Ninth Amendment. And this is a state of affairs that could never be remedied. Even were we to outlaw them outright and engage in violent seizures of guns nationwide, which would itself become a disastrous slide into the abuse of power, we would hardly impact the problem. Guns would simply go underground, and we will have effectively armed the criminal class while disarming their innocent victims.

Thus, there is only one solution to the gun problem in this country: responsible gun ownership. This does not mean the right to bear arms is or should ever be unlimited. Owning a gun is even more serious than owning a car, and all the requirements of the one should certainly apply to the other: gun safety training and testing must be required, and no one should be able to buy guns, ammunition, or any lethal weapon or related parts or equipment without a license certifying that they have been trained in that class of weapon, are not a felon, and have obeyed the laws with regard to the bearing of arms. Likewise, only weapons suitable for hunting and personal defense against criminal attack should continue to be legally sold to civilians. Civilians have no just need of weapons of war, nor even if they had them would they be any match for a renegade modern army. In a civil war, every militia will be mere missile fodder. Gun fanatics won't have a chance. The only defense against tyranny in a modern free society is *participation in government*.

> On the nature and importance of war, and how to conduct it, there are several classic works that ought to be required reading for anyone who wants to understand the subject: Sun Tzu, *The Art of War* (c. 270 B.C.); Niccolo Machiavelli, *The Art of War* (c. 1520); and Karl Von Clausewitz, *On War* (1832). Related to the issue of military leadership is Oren Harari, *The Leadership Secrets of Colin Powell* (2002).

On guns in America: Michael Bellesiles, *Arming America: The Origins of a National Gun Culture* (2001); Gary Kleck and Don Kates, *Armed: New Perspectives on Gun Control* (2001); Terry O'Neill, *Gun Control: Opposing Viewpoints* (2000). On the subject of personal self-defense, see: Sanford Strong, *Strong on Defense: Survival Rules to Protect You and Your Family from Crime* (1996).

5.6 A Commitment to Secularism

Finally, as I already explained above in section VII.1, "Morality vs. Politics," government must never attempt to tell people what to believe, especially in matters of conscience. This is the grounding principle of what we call the "Separation of Church and State." It means that the government should never be in a position to decide approved from unapproved views in matters of religion, and the people's money should never be employed to support any church or religious mission. Our government must be religiously neutral, so it can be fair to all.

As the very first amendment to the U.S. Constitution puts it, "Congress shall make no law respecting an establishment of religion, or prohibiting the free exercise thereof," which in 1971 the U.S. Supreme Court interpreted with the three-part "Lemon Test" in *Lemon v. Kurtzman*: all laws must have a secular purpose (no laws can have a special *religious* purpose), every law must have a primary effect which neither advances nor inhibits religion, and every law must avoid the excessive entanglement of church and state.

This means, for example, no official prayer or creationism in schools, no public vouchers subsidizing church schooling, no religious tests for public office or for civil rights, no spectral or other mystical evidence in courts or as the basis of any law, and equal treatment of all churches and religious views by the government and the law. The reasons are easy to grasp: to give any religious institution or dogma *political* power is an invitation to corruption, intolerance and oppression (as history has proved without exception). Moreover, to base laws on anything but established facts is to replace truth with opinion, and agreement with oppression. Finally, to entangle the government with religion in any way has an inevitable crippling effect on the freedom of religion itself. Religious liberty is not possible unless the law and the government are blind to religious distinctions, neutrally treating all the same.

One mistake that is often made is to think that taking "God" (i.e. religious dogma) out of schools or government is the same thing as government support of atheism (or Secular Humanism). This is quite untrue. It would be just as wrong for the law or the government to coerce anyone to become an atheist or profess or agree with the philosophy of Secular Humanism, or to give special financial or legal support to organizations that promote specifically atheistic or secular humanist philosophy, or to otherwise interfere with anyone's religious freedom. But it is not wrong for the law and the government to support and promote the *facts*. What those facts might imply or entail about the correct philosophical worldview is not the government's business to dictate. But it *is* the government's business to be fair and honest, and that means agreeing with what is proven, with abundant objective evidence. It is also the government's business to treat all citizens equally, and that means not intimidating members of minority faiths with official proclamations of support for contrary religions just to pander to the popular, or to usurp public power as a vehicle for the promotion of your own personal beliefs.

And this is why schools only teach students what they need to *know*, rather than telling them what they must *believe*. For that, all are free to choose on their own. This is a distinction often lost in the debate over whether creationism should be taught in biology classes, for example. To be an informed citizen, much less someone who might be planning a future career in the biological sciences, you do not have to believe in evolution, but you certainly have to understand it and know a lot about it, especially since it is a well-established fact ubiquitously employed throughout the sciences. But there is no need for you to know about creationism, least of all in a biology class. It would be a good idea, and a legal one, to teach a world religions course to all students, wherein the ethical and creation and other doctrines of *all* religions are taught equally, but of course no creationists want that—for then their religion would have to compete with other religions, and that would deprive them of the unjust power over a captive audience they are really seeking.

Likewise, there is no valid ground for criminalizing abortion, for there is no evidence sufficient to convince any objective court of law that people can exist without a brain (see section III.6, "The Nature of Mind"), so elective abortion before the formation of a cerebral cortex (usually some time between the 20th and 24th week of gestation) does not violate anyone's rights by any standard except a solely religious one. Only educated medical professionals are capable of determining precisely when a cerebral cortex has formed, or when an abortion is necessary to save the mother's life, and these facts will vary with every case. Thus, it is not something that can be honestly legislated, without imposing religious beliefs on people, hence depriving them of their religious freedom.

These are just some of the issues surrounding church-state separation. There is a lot one could learn about it, and a good place to start is the Secular Web's library on the "Separation of Church and State" (www.infidels.org/library/modern/church-state/).

Important books include: Robert Boston and Barry Lynn, *Why the Religious Right Is Wrong About Separation of Church & State* (1994); Isaac Kramnick and R. Laurence Moore, *The Godless Constitution: The Case Against Religious Correctness* (1997); Marvin Frankel, *Faith and Freedom: Religious Liberty in America* (1995).

On the specific issues I raise: Robert Alley, *School Prayer: The Court, the Congress, and the First Amendment* (1994); NAS Council, *Science and Creationism: A View from the National Academy of Sciences* (1999); Ruth Dixon-Mueller & Paul Dagg, *Abortion and Common Sense* (2002); Robert Baird and Stuart Rosenbaum, eds., *The Ethics of Abortion: Pro-Life vs. Pro-Choice* (2001); Cynthia Gorney, *Articles of Faith: A Frontline History of the Abortion Wars* (2000); Rickie Solinger, ed., *Abortion Wars: A Half Century of Struggle, 1950-2000* (1998). On education, see VII.5.4, "A Commitment to Education."

6. The Secular Humanist's Heaven

Here is a picture of what Secular Humanists like me are aiming for, the society we are working to create—so you can ask yourself whether you want to help us achieve it.

Imagine a world rather like that in *Star Trek: The Next Generation*. I do not mean cavorting across the galaxy with warp drives. That is not likely. There is so far no plausible prospect for faster-than-light travel. But what *is* likely, indeed certain, if we keep our peace and prosperity long enough, is a world where human rights and freedoms are universally respected, and everything is free except human labor. All our needs will be satisfied by safe machines that can create anything, drawing on self-sustaining power sources like clean fusion or orbital solar. There will be no poverty. The mentally and physically ill will be cared for. The ignorant will be educated.

Everyone will be taught and encouraged to dedicate time and effort to some social good of their choosing: as teachers, builders, explorers, researchers, artists. Their week will be short, and their work fulfilling. All menial, dangerous, and unpleasant labor will be handled by machines. Everyone will be capable of understanding and appreciating themselves and their society, and will *want* to serve the public good, because a rich education will be freely given to all, throughout childhood and into adulthood, that encourages exactly this, and provides the skills for it.

This world will not be free of evil. There will still be criminals and loafers and attempts to abuse power or cause harm, there will still be accidents and mistakes and disasters. But the elimination of poverty and the reduction of illness and a universal humanistic education in reason and sense will make this much rarer than in any present society, as will our constant advance in knowledge and technology. And the government will have been honed and perfected by applying to its reform the scientific principles and findings of many centuries, until we have the most effective system for protecting human rights and managing the automated economy—and above all, for maintaining checks and balances, and providing quick remedies and effective preventive measures against the natural human impulses toward incompetence, negligence, corruption, and crime, as well as hubris. The government will be of the people, by the people, and for the people—a people who will almost all be healthy, happy, and civil, and a government that will be more self-critical, self-repairing, self-policing. Most natural evils will have been abolished. Genetic diseases and disabilities will be a thing of the past, illnesses cured, most injuries easily repaired. Natural disasters will be all but incapable of thwarting our countermeasures against them.

We will certainly have terraformed and colonized other worlds, at least in our own solar system, with fabricated worlds to live in as well in the very reaches of space. But at the same time we will have much more control over population pressures here on earth. Fertility will be so well managed that all men and women can turn it on or off without difficulty or side effects. With smaller population densities and cleaner technologies, a more comfortable and beautiful coexistence with nature will be possible, as cities and parks become more thoroughly and intelligently integrated.

We might even make immortality possible. It may even happen that, in the fullness of time, we will be able to transfer our minds, by transferring the patterns of our brains, into computer-simulated worlds that are in even more perfect regulation than the physical world, a true paradise. And this simulated universe, and the computers that produce it, would itself be a self-sustaining, self-maintaining, self-repairing, self-expanding artificial organism. It is possible it will never die. As the Third Law of Thermodynamics entails, it will take infinite time for the Second Law of Thermodynamics to dissipate all the available energy in the universe into unusable form. So there will always be an energy differential in the cosmos that a resourceful machine can exploit as a power source. Perhaps this will never be achieved in practice, but in theory it can be hoped for.

This is all science fiction, surely. But I hope one day to make it science fact. If it sounds like your dream of heaven, this is no accident. This is the society I want to work toward so that it may exist if not for us, then for our children, or our children's children—ultimately, so it may simply exist: so we can defy the coldness of space and the brutality of nature and create paradise in spite of them.

On humanistic futurism: Wil McCarthy, *Hacking Matter: Levitating Chairs, Quantum Mirages, and the Infinite Weirdness of Programmable Atoms* (2003); Douglas Mulhall, *Our Molecular Future: How Nanotechnology, Robotics, Genetics, and Artificial Intelligence Will Transform Our World* (2002); Gerard O'Neill, *High Frontier: Human Colonies in Space*, 3rd ed. (2000); Michio Kaku, *Visions: How Science Will Revolutionize the 21st Century* (1998); and see the Secular Web section on "Extropianism" (www.infidels.org/library/modern/lifeafterdeath/extropianism/).

VIII. Conclusion

There is one thing I have tried to make clear throughout this book. Metaphysical Naturalism is the only worldview that is supported by all the evidence of all the sciences, the only one consistent with all human experience, the established truths of history, and reason itself. No other worldview, including theism generally or Evangelical Christianity in particular, is supported by any evidence of any of the sciences. The only remotely plausible exception, 'fine tuning', is not very convincing evidence for the divine, and supports no doctrine of salvation (see III.3, "The Nature and Origin of the Universe"). Science doesn't necessarily contradict alternative worldviews, for one can adjust most of them to be compatible with almost any evidence. But no other worldview is directly and substantially supported by any scientific evidence, whereas all scientific evidence so far *does* support Metaphysical Naturalism, often directly, sometimes substantially. Though naturalism has not yet been proved, it is the best bet going.

Even the facts explained by Big Bang Theory are solely and entirely physical and natural. None are facts about spirits or gods or supernatural entities or powers, and the theory does not include any reference to such things. Insofar as anything is left unexplained by it (such as matters of cosmic order or first cause), there is only humble ignorance. Theories are never scientifically established on what we *don't* know or *can't* yet explain, but always and only on what we do know and can explain. To argue that science has not explained something, therefore *our* explanation (whatever that is) must be correct, is not a scientific argument. Such an argument might be good and persuasive, but not because it is scientific—though it may be well supported by science. This is the distinction between science, as a database of facts established by a methodologically sound empirical inquiry, and metaphysics, a speculative enterprise of interpretation and plausible hypothesis formation. You can reject all such efforts to go beyond established science, rejecting all worldviews, or you can adopt the most probable hypothesis: Metaphysical Naturalism.

The title of this book is "Sense and Goodness without God," because Metaphysical Naturalism is full of sense, and encourages nothing but good. Reason and acute thinking are its very bedrock, and the love of wisdom its main driving force. To be wise and practical is our motto. And this worldview provides adequate, if not strong reasons to devote yourself and your life to high moral ideals, to compassion and integrity in the pursuit of happiness. It is thus a good philosophy—good for you, good for all humankind. And all this without recourse to a god. Though we have found no evidence for any god, and no reason to believe there is one, the sense and goodness of our worldview stands as it is even if there is. It stands on its own terms, on reason and fact.

I have included bibliographies of recommended and related readings in most of the sections of this book. I strongly encourage all readers to pursue those readings, to understand our philosophy better, and why it is true. This is especially the case for those of you who have objections to what I have argued that are not answered here: the answers are very likely already in print elsewhere, in the very resources I mention. Re-read my Introduction for when and how to reach me if you still have objections not answered by these other resources.

Two aspects of this worldview that I have left out of this book so far are its advantages to humanity's survival and welfare, and its cultural manifestations. I will devote only very brief space to these here, to round out this important picture.

The adoption of Metaphysical Naturalism will benefit the survival of any society, by eliminating fatal or exhausting religious conflict and instead managing disagreement with reasonable debate, by stopping the waste of time and other resources on falsehoods and taboos, by encouraging humanistic cooperation and preservation (especially against extremism, apocalypticism, fatalism, and religiously-inspired apathy, bigotry or panic), and by instilling the proper values necessary for an enduring, contented culture, one actively interested in exploring and colonizing the universe and ending misery and want.

Metaphysical Naturalism is tailor-made for advancing human survival and welfare. Its rigorous respect for truth and caution, and reliance primarily on what is evident to all people and thoroughly demonstrated, will ensure stability in ideology, and thus in society. Our worldview is constantly attenuated to the facts, which never really change. Our approximations to the truth do change—though as they do, such changes become less and less common, and more and more trivial or esoteric, as the truth is approached.

This stands in stark contrast with religious ideologies, which are grounded in no permanent facts and thus subject to constant and profound alteration and upheaval as societies and whims drift.

History proves the case. The evolution of Christian ideology alone has been profound and unending, never finding any agreement at any time, but always fracturing into more and more sects with no way to reconcile them. So while there is no sign of Christianity converging on any common ideology, Metaphysical Naturalism clearly is. Indeed, naturalists throughout history, who arrived at their views wholly independently of each other, even in widely differing cultures, have all converged toward the same general conclusions and world picture, ensuring that our worldview, even if always a minority view, will still find more and more uniformity rather than division of views. Yet Christianity a thousand years from now will not be the same Christianity lived today, just as what we have today is not the same as that lived a thousand years ago. In all periods we meet hundreds of sects at fundamental variance with each other. Every other major religion faces the same story.

Cultural manifestations of our worldview can be briefly surveyed. The cultural effects of adopting Metaphysical Naturalism and its moral element, Secular Humanism, lie in our continuing desire to experience scientific and historical discovery, in journals and magazines and books, which continue to reinforce and realize our worldview, enriching the way we see and understand the world. They lie, also, in our personal enjoyment of a secular spirituality, an awe of life and all things, in our attentiveness to our lives and the universe around us, and in continual, passionate learning. Though there is no unique regimen of rituals, we do enjoy all secular celebrations and festivals, and make them occasions for joy and happiness, as well as personal reflection and meditation on the nature and meaning of our core beliefs.

Maybe one day there will be enough of us that Secular Humanists can congregate on symbolic occasions to evoke an affective response to these core beliefs in every town, every neighborhood. But as we are born of individual reason and conviction, not cultural upbringing or socialization, we are spread out geographically and thus lack the strength of united enclaves. Yet already many of our organizations hold annual conferences, getaways, summer camps, or local community meetings. And authors like David Cortesi, in *Secular Wholeness: A Skeptic's Paths to a Richer Life* (2002), discuss how we can cultivate as individuals something of a new "secular culture."

Still, we do not congregate to reinforce a dogma through indoctrination and peer pressure, but rather through the open sharing and debating of ideas and the enjoyment of common society. And we also gain a sense of community and conviction through fighting together against our common enemies—the foes of reason, truth, and liberty. There are numerous major organizations of atheists or other freethinkers, which aim at many goals like these (the Secular Web lists and links to them at www.infidels.org/org). Finally, the most obvious cultural manifestation of our philosophy is our story of everything: from the Big Bang to the Enlightenment, and thence to today, and on to the ideal humanistic society towards which we strive.

If you are persuaded by my defense of Metaphysical Naturalism and Secular Humanism, and like our story and goals, I would like to extend a plea: Join us. Accept a life of reason, a life of love for science, truth and humanity. Find spirituality in simply being and knowing. Seek to live your life by the true humanistic ideals of compassion and integrity, unpolluted and uncorrupted by the "exceptions" and distortions heaped upon these virtues by various religious or political dogmas. Base your beliefs on the evidence, and humbly admit what you don't know or hold less certain. And help us work toward a better future for all of humanity.

Failing that, if you'd rather pass, then I would like to extend another plea: for tolerance, acceptance, and understanding.

Index

A

abiogenesis. See 'biogenesis'
abortion, 342, 403. See also 'personhood'
Abraham and Isaac, 16, 295–96
abstraction and abstract objects, 30–31, 37, 119, 124–30, 133, 141, 174, 177–78, 198, 309, 353, 354, 359
ad hocness, 71, 72, 73, 75, 76, 77, 79, 80, 150, 239–40, 241, 250, 253, 256, 282–83, 284
aesthetics. See 'beauty'
afterlife. See 'immortality'
agnosticism, 11, 15, 50–51, 255
AI, 139–42, 145–47, 177–78
Alexander of Abonuteichos, 234–35
animals, 104, 135–37, 138, 142, 145–47, 149–50, 153, 171, 172, 173, 178, 179, 193–94, 198, 199, 219, 273, 276, 277, 284, 285, 304, 308–9, 328–31, 341, 343, 351, 353–54, 377
Apollonius of Tyana, 234
argument from evidence, 242–45
argument to authority, 58–59, 243–44, 246–47
argument to the best explanation, 238–41
Artificial Intelligence. See 'AI'
artistic criticism, 39, 363–66. See also 'beauty'

Asclepius, 233, 234, 236
atheism, 4–5, 9, 17, 18, 19, 200, 203, 253–89, 402–3, 412, 414. See also 'agnosticism', 'Secular Humanism', and 'God'
Azathoth, 281

B

balance of proof, 222–24
beauty, 37–39, 171–72, 198–202, 351–66
bible, 9–11, 15–17, 19, 205, 242–52, 259–61, 262, 263, 264–65, 270, 295–96, 298–99, 300–301
Big Bang, 74–81, 84, 90, 92, 105, 166, 411, 414
biogenesis, 166–67
black holes. See 'Smolin Selection Theory' and 'Big Bang'
blindsight, 147–48
brain, 33, 127, 135–59, 173–76, 177–88, 192, 193–97, 325, 330–31, 353–60
brute fact, 73, 82, 93, 122
Buddhism, 12, 13, 204, 205, 206, 263–64, 266, 267, 315, 317
burden of proof, 221–22

C

capitalism, 391–93
Cartesian Demon, 32, 45,

49–50, 51–53, 184, 191–92, 287–88
Cassius Dio, 228, 230, 243, 245
causation, 85, 92, 94, 98–99, 103, 106, 220
  agent, 100–117
  natural, 68, 72, 80, 83, 87, 129–31, 132, 273
  ontological, 85
  supernatural, 204, 211, 221–22, 227–28
Chaotic Inflation Theory, 75–77, 79, 80, 81, 82, 83, 84, 87, 90–91, 93. See also 'multiverse theory'
Chinese Room, 139–44
Christianity, 4, 9–10, 11–12, 13, 15–19, 44, 65, 156, 158, 204, 205, 206, 221, 225, 228–30, 234–35, 242, 245, 257–58, 259–68, 270, 274, 289, 293–302, 304, 306–7, 315, 331, 333, 339, 343, 345, 411, 413
Christmas Carol, 282
colors, 30–31, 94, 123, 125, 127, 130, 138, 146, 156, 187–88, 329, 355, 356
communism, 206, 262, 268, 391–93
compatibilism. See 'free will' and 'determinism'
computers, 32, 44, 106, 132–33, 136–43, 145–47, 155, 169, 177–78, 179, 186, 194, 254, 273, 406
consciousness, 119, 135–48, 149, 152, 156–57, 166, 173–74, 178–79, 180, 198, 201, 224, 327. See also 'personhood', 'AI', 'brain', 'knowledge', and 'reason'
contradiction, nature of, 42–43, 188–91
convergence, criterion of, 51–53, 59, 91, 128, 171–72, 215, 219, 230, 256–57, 413
creativity, 126, 139, 140–41, 174–76, 177–78, 214, 220–21, 330, 356, 365, 397–98
cultural decline, 3–4, 115–16, 199–200, 235, 269–70, 302–11, 365–66

D

death, 158, 161–63, 327. See also 'immortality' and 'suicide'
Deism, 11, 253–54, 260–61, 288
democracy, 11, 17, 31, 198, 263, 301, 317, 335, 375–78, 383–87, 396–97, 400
depression, 163–64, 318–22
design, 85–88, 104, 168–74, 184, 186–88, 227, 228, 254, 273–75, 276, 278, 277–78, 280, 284, 285, 286, 287, 326, 381, 383–84, 394, 411. See also 'evolution' and 'Cartesian Demon'
desire, 40, 60, 104, 109–14, 196–97, 315–16, 331–34. See also 'morality' and 'emotion'

determinism, 95, 97–99, 103, 104, 109, 113–17, 200–202. See also 'free will'
dreams, 137–38
drugs, legalization of, 304, 389

E

education, 149, 232, 270, 307, 345, 391–92, 393, 396–400, 402–3
Einstein, Albert, 88, 95, 214, 219
emotion, 193–208, 351–52. See also 'happiness'
energy. See 'matter-energy'
entropy, 129–30
epistemology. See 'knowledge'
ethicology, 334–35
ethics. See 'morality'
evil
definition of, 337–39
examples of, 15–16, 257, 261, 274, 276–82, 284–86, 287–88, 296, 309–11, 320–22, 326, 378, 405
evolution
biological, 47, 83, 132, 166, 167–76, 181–88, 192, 193, 196, 216, 219, 221, 223–24, 273–74, 287, 308–11, 325–27, 328, 335, 341, 343, 352, 362, 403. See also 'design'
cosmological. See 'Smolin Selection Theory'
cultural. See 'memes and memetic evolution'

F

facts, 40–42, 219–20, 331–32. See also 'brute fact, the'
faith, 60, 61, 217–18, 257–58, 269–70, 286, 287
fatalism, 97, 115–16, 200, 412
fine-tuning, 80, 86–88, 411. See also 'design'
first cause, 81, 84–85, 93, 411. See also 'time, theory of' and 'brute fact, the'
first principles, 24, 27–29
fossils, 170–71, 219
free will, 97–116, 196, 277, 284, 285, 284–86, 296, 343. See also 'determinism'
freedom, political, 303, 306, 370–74, 376–77, 389–90, 397, 402–4, 405. See also 'capitalism' and 'democracy'
freethought, 26, 270, 307. See also 'tolerance'

G

Gibbon, Edward, 231
Goal Theory (of moral value), 315, 317–20, 331, 334–39, 341, 345–48. See also 'morality'
God, 5, 14–16, 44, 49–50, 58, 88–89, 93, 104, 161, 171, 172, 185, 191–92, 203, 211, 213, 222–24, 227, 247, 294–97, 298–301, 309, 326, 331, 333–34, 337–38, 402. See also 'atheism', 'bible', and 'Jesus'

Golden Rule, 322, 340–41
good. See 'happiness' and 'evil'
gun control, 400–402

H

hallucination, 13–14, 32, 49, 139, 152, 155–56, 205, 233, 280
happiness, 11, 12–13, 15, 18, 19, 25–26, 53, 163–64, 197, 201, 202, 204, 206–7, 275, 277, 297–99, 301–2, 315–27, 330–31, 333, 335, 337–39, 341–47, 362–63, 364, 369–70, 373–74, 376–78, 384–86, 389, 390, 397–98, 412. See also 'love' and 'meaning of life'
Harnouphis, 229, 238, 241
Hayek, Salma, 200
heaven, 9, 14, 16, 73, 155, 162, 262, 268, 276–77, 278, 284–85, 297–99, 300, 333, 339, 405–6. See also 'hell' and 'immortality'
hell, 9, 13, 14, 18, 40, 260–62, 286–87, 297–99, 300, 317, 334, 339. See also 'heaven' and 'immortality'
Hermes, 229, 241
Hinduism, 263
historical method, 57, 223, 227–52, 382
homosexuality, 199, 305–6, 317, 342
human nature, 104, 205, 223, 240, 245, 328–30. See also 'personhood'
human rights, 329, 389–90, 391, 393, 400, 405, 406
Hume, David, 7, 14, 231–32, 331
hypocrisy, 11, 15, 255, 256, 283, 286, 299, 321
hypothesis. See 'prediction', 'fact', and 'truth'

I

immortality, 157–58, 254, 297–99, 406. See also 'death', 'heaven', and 'hell'
Inference to Metaphysical Naturalism, 68, 72, 166, 221–22, 227
inference, metaphysical or plausible, 50, 59, 68, 213, 216, 221–22, 227–28, 324–25, 381–82, 411. See also 'plausibility'
infinite regress, 73
intellectual or interpretive charity, 5–6
internalism, ethical, 332–33
intuition, 33, 34, 178–80, 183, 192, 339–41, 347
Islam, 19, 205, 206, 262–67, 270, 278, 289, 317. See also 'Koran'
is-ought dichotomy. See 'naturalistic fallacy'

J

Jehova's Witnesses, 18
Jesus, 9, 10, 16, 18, 19, 259–60,

261, 262, 268, 298–99, 300–301, 340–41, 342. See also 'resurrection'
Job, 15–16
justice, 114, 158, 263, 266, 279, 285, 302, 303–5, 314, 322, 390, 392–93, 394, 398, 400. See also 'responsibility', 'retribution', 'punishment', and 'law, human'

K

Kant, 320, 322, 331, 347
knowledge
cognitive, 33, 179, 198
noncognitive, 33, 179. See also 'intuition'
theory of, 21–61, 91–95, 115, 124–34, 144–50, 173–92, 203–6. See also 'truth'
Koran, 260, 262, 266, 270–71. See also 'Islam'

L

law, human, 106, 264, 265, 375–79, 383–404. See also 'justice'
laws of nature or laws of physics. See 'physical law'
liberty. See 'freedom'
life after death. See 'immortality'
life force, 35
Linde, Andrei, 77, 81
logic, 11, 12, 17, 18, 24–26, 28–29, 31, 45, 50, 54, 53–54, 59, 61, 67, 107, 125–26, 173, 177–92, 213, 214, 223–24, 271, 335, 337, 358–59, 381, 383, 397–98. See also 'reason', 'method', 'warrant', and 'knowledge'
love, 19, 23, 31, 102, 125, 151, 161–63, 180, 194–202, 205–6, 253–54, 262–63, 277, 281, 285, 294–96, 316, 318, 321–22, 326, 327, 365, 414

M

Marcus Aurelius, 228–31, 234, 239
mathematics, 31, 50, 53–54, 61, 67, 98, 125–26, 132, 133–34, 175, 181, 225, 383, 397. See also 'numbers' and 'logic'
matter-energy
discussion of, 68, 120–22, 124–25, 126–30, 134. See also 'reductionism', 'Big Bang', and 'Smolin Selection Theory'
references to, 35, 72, 74, 83, 119, 131, 150, 158, 211, 331, 357
MDAP. See 'Moral Discourse and Practice: Some Philosophical Approaches'
meaning of life, 10, 15, 155, 161–64, 202, 204, 254, 293, 301–2, 327, 345. See also 'love' and 'morality'
meaning, theory of, 27–43, 288–89
meditation, 13–14, 203, 206,

memes and memetic evolution, 149–50, 166, 175–76, 181–86, 181–88, 204–5, 258–72, 309–11, 326–27, 364–65

metaethics, 293–302, 313–44

Metaphysical Naturalism, 4, 5, 14, 47, 53, 65–70, 119, 128, 135, 158, 166, 184, 191, 203, 211–12, 221–22, 227–28, 253, 256–57, 258–59, 293, 313–14, 329–31, 336, 342, 345, 381, 411–14

method, 49–61, 67–68, 148–50, 190–91, 203–4, 205–7, 227–52, 282–83, 285–86, 324–25, 411. See also 'historical method', 'first principles', and 'scientific method'

Methodists, 9–10, 18

mind. See 'brain'

miracles, 11, 12, 68, 211–13, 221–24, 227–52, 256, 261, 268, 271, 274–75. See also 'origins of miracle-claims'

modal properties, 128–30

modo hoc fallacy, 130–31, 309, 361–62

Moral Discourse and Practice: Some Philosophical Approaches, 309, 313, 314, 315, 327, 328, 331, 332, 333, 334, 335, 337, 339, 343, 346, 347

moral imperatives, 39–40, 42, 315–18, 324, 331–35, 362–63. See also 'values' and 'morality'

moral relativism, 336–37

morality, 39–40, 115–16, 197, 204–5, 275, 279–80, 293–348, 369–70, 376, 377, 378. See also 'values', 'evil', 'moral imperatives', 'responsibility', 'love', 'happiness', 'ethicology', and 'Golden Rule'

Moreland, J. P., 100–108, 293–314, 317, 320, 324–26, 328–33, 336, 337, 339, 341, 342, 345, 347

Mormonism, 18, 205, 260, 317

multiverse theory, 75–91, 92–93

mysticism, 13–14, 202–8

N

natural law. See 'physical law'

natural selection. See 'evolution'

naturalism. See 'Metaphysical Naturalism'

naturalistic fallacy, 331–33

nature, definition of, 68, 169. See also 'supernatural: definition of'

NDE, 'Near Death Experience', 154–57, 158–59

New Testament, 10–11, 15, 16–17, 236, 242–47, 259–60, 263, 266, 298–99, 300–301, 342

normativity. See 'moral imperatives', 'values', and 'naturalistic fallacy'

Norton, Ed, 128–29, 130
numbers, nature of, 31, 125–26, 181

O

OBE, 'Out of Body Experience'. See 'NDE'
Occam's Razor, 224–26. See also 'simplicity'
Old Testament, 15–16, 249–52, 259, 264–65, 300
opinion, 37–39, 194–97, 198–202, 314, 338, 362–63
origins
of civilization, 175–76
of humankind, 165, 168–73
of life, 165–67
of miracle-claims, 154–56, 212–13, 231–36
of the mind, 173–74, 192
of the universe. See 'universe: origins of'

P

paranormal, the, 154–57, 203–4, 211–13, 221–23, 226–52. See also 'supernatural', 'miracles', and 'atheism'
parenting, 9, 12, 23, 195, 201, 262, 263–64, 278, 280–81, 284, 310, 315–16, 344
parsimony. See 'Occam's Razor'
past lives. See 'reincarnation'
perception, 55–56, 91–95, 127–28, 135–39, 144–48, 149, 153, 156, 173, 177–78, 193–96, 353–60. See also 'sensation' and 'consciousness'
personhood, 103, 107–9, 201–2, 300–301, 308–11, 329–30. See also 'consciousness' and 'human nature'
philosophy, 3–5, 23–26, 33–34, 65, 202–3, 206, 270–71, 306–7, 337, 370, 381, 397, 398, 403
physical law, 68, 76, 77, 79, 80, 86–88, 93, 98–99, 119–20, 122–24, 129–30, 166, 211, 219, 224–26, 236, 274, 278, 325, 357, 406
physicalism, 68, 123, 130, 150–57, 313
Plantinga, Alvin, 43–47, 70, 184–85
Platonism, 31–32, 191, 354. See also 'abstraction and abstract objects'
plausibility, 59, 71–81, 83, 239, 241
Plutarch, 233, 243, 245
political philosophy, 367–406
prediction, 27–29, 36, 40–59, 71, 74–75, 76–77, 78–81, 83, 97, 99, 123, 148, 179, 214–15, 219–20, 288, 340. See also 'prophecy'
Prohibition, 304
prophecy, 247–52
proposition, 49–54
propositions, 27–29, 33, 35, 40–41, 45–46, 57–58, 148,

179, 223, 317, 331–33, 336, 338, 346–47. See also 'moral imperatives'
Proteus Peregrinus, 234
psychopathy, 16, 330, 342–44
punishment, 113, 116, 248, 299, 317. See also 'justice'

Q

qualia, 146–48
Quantum Mechanics, 84, 85, 91, 97, 99, 122, 123, 132, 182

R

reality, 31–32, 52, 114, 137–39, 145, 181–92, 270. See also 'Cartesian Demon' and 'virtual reality'
reason, 53–54, 56, 135, 139, 141, 149, 153, 177–91, 196–97, 339–41, 414. See also 'logic', 'method', 'warrant', 'brain' and 'knowledge'
Red Summer, 305
reductionism, 30–31, 37, 107, 128, 130–34, 143–44, 198, 224–25, 315, 331
reincarnation, 154, 205, 234, 263
Relativity Theory, 85, 91, 98, 123, 182, 214, 216, 219, 225
religion, 3–4, 9–19, 26, 176, 185, 206, 207, 202–8, 228–29, 234–35, 239–40, 257–72, 282, 287, 289, 303, 306, 397, 398, 402, 403, 402–4
resurrection, 158, 227, 234 of Jesus, 224, 242–45
retribution, 17, 116, 317. See also 'justice'
retrofitting, 248–51
rights. See 'human rights'

S

Satan, 18, 172, 261, 266, 288, 343, 391
science, 4, 11, 13, 14, 17, 18, 50, 54–56, 58, 61–68, 76, 79, 80, 83, 91, 128, 129, 130–32, 143, 154, 166, 175, 181–82, 192, 198, 200, 203, 213–26, 231, 232, 261, 262, 296, 309, 324–25, 328, 335, 355, 372, 381, 382–83, 391, 395, 397, 403, 411, 414
scientific method, 54–55, 68, 182–83, 214–26, 324–25, 382–83, 398, 411
Secular Humanism, 14, 206, 272, 293–303, 306–8, 309, 313, 315, 345, 369, 402–3, 405–7, 413–14. See also 'politics' and 'morality'
self. See 'personhood' and 'consciousness'
self-respect, 197, 301, 320–22, 323, 338
sensation, 29–33, 37, 45, 135, 154, 157, 173, 181, 183, 184, 185–88, 190, 193, 280, 342. See also 'perception' and 'synesthesia'
sentences. See 'propositions'
simplicity, 31, 78, 80, 82–83,

84, 92–93, 126, 181, 224–26, 355–56. See also 'reductionism' and 'Occam's Razor'
Smolin Selection Theory, 75, 77–81, 82, 83, 84, 87, 91, 93. See also 'multiverse theory'
Smolin, Lee, 77, 81
socialism. See 'communism'
solar system and solar theory, 87–88, 167, 225–26, 273, 359, 406
solidity, 44, 91, 121, 137–38
soul. See 'brain' and 'life force'
space-time
discussion of, 68, 88, 89–90, 91, 93, 119–21, 123–29, 130, 134. See also 'time, theory of'
references to, 72, 80, 150, 158, 165, 198, 203, 211, 357
speciesism, 330–31, 336–37
spirituality, 3, 18, 202–8, 211, 280, 306, 320, 413, 414
statements. See 'propositions'
substance, 68. See 'matter-energy'
suicide, 163, 206, 279–80, 341–42
supernatural
definition of, 211
references to, 4, 14, 31, 135, 155, 191, 203, 204, 206, 213, 222–23, 227, 229, 232, 247, 271, 293, 298, 326, 411. See also 'causation: supernatural', 'paranormal', and 'god'
superstition, 12, 17, 19, 34–35, 135, 212, 221, 227, 229, 231–36, 262, 266–70, 307, 391
Superstring Theory, 120
Supreme Court of the United States, decisions of, 109–12, 402
surprise, criterion of, 51, 214–15, 220
synesthesia, 156–57

T

Taoism, 11–18, 19, 202–3, 205, 253, 266, 288, 315
taste, aesthetic. See 'beauty', 'opinion', and 'artistic criticism'
taxation, 265, 393, 394–95
Teleological Argument for Atheism, 273–75
See also, 72–73, 74, 78, 172, 222–23, 326
Tertullian, 228, 230
testimony, 56, 58–59, 231–35, 243–44, 382
The Matrix, 32
Thundering Legion, 228, 229
time, theory of, 72, 82, 84–85, 88–95, 119–20. See also 'determinism' and 'space-time'
tolerance, 13, 19, 176, 230, 261, 264–65, 266–67, 287, 306, 308, 320, 402, 414
Torak, 281

Trail of Tears, 305
truth, 41–42, 148–49, 204–7, 331–37. See also 'method', 'knowledge', 'warrant', and 'memes and memetic evolution'
Turing Machine, 141–42

U

U.S. News and World Report, 198–200
universe
origins of, 253–54, 257. See also 'Big Bang'
universe, origins of, 71–95

V

values, 38–40, 145, 194–98, 269–70, 293–302, 313–16, 324–32, 333, 336, 342, 345, 346–47, 362. See also 'morality', 'opinion', and 'moral imperatives'
Vespasian, 233, 236
virtual reality, 52, 136–40, 141, 142–43, 144, 145, 146, 180. See also 'reality'
visions. See 'hallucination'
voting, 100–101, 108, 263, 329, 336, 384–87, 397

W

warrant, 43–47. See also 'reason'
women, treatment of, 16, 205, 263, 294–95, 303–6
words, meaning of, 29–37, 331–32

worldview, concept of, 5, 65

About The Author

Richard Carrier is a philosopher and historian studying ancient science at Columbia University in New York, where he received a Masters degree and a Master of Philosophy in ancient history and is working on his Ph.D. He previously graduated Phi Beta Kappa at UC Berkeley. Mr. Carrier is also a professional writer, teacher, and speaker and translates four languages. His articles have been published in *Biology & Philosophy*, *The History Teacher*, *German Studies Review*, *The Skeptical Inquirer*, and the *Encylopedia of the Ancient World*. He is a veteran of the United States Coast Guard and served as Editor in Chief of the Secular Web for several years, where he has long been one of their most frequently read authors. His popular online essays can be found at *http://www.infidels.org/library/modern/richard_carrier*.

Printed in the United States
66756LVS00004BA/34